Energy, Environment, and Sustainability

This books series publishes cutting edge monographs and professional books focused on all aspects of energy and environmental sustainability, especially as it relates to energy concerns. The Series is published in partnership with the International Society for Energy, Environment, and Sustainability. The books in these series are editor or authored by top researchers and professional across the globe. The series aims at publishing state-of-the-art research and development in areas including, but not limited to:

- Renewable Energy
- Alternative Fuels
- Engines and Locomotives
- Combustion and Propulsion
- Fossil Fuels
- Carbon Capture
- Control and Automation for Energy
- Environmental Pollution
- Waste Management
- Transportation Sustainability

More information about this series at http://www.springer.com/series/15901

Sunita J. Varjani · Edgard Gnansounou
G. Baskar · Deepak Pant
Zainul Akmar Zakaria
Editors

Waste Bioremediation

Springer

Editors
Sunita J. Varjani
Gujarat Pollution Control Board
Gandhinagar, Gujarat
India

Edgard Gnansounou
Swiss Federal Institute of Technology Lausanne (EPFL)
Lausanne
Switzerland

G. Baskar
Department of Biotechnology
St. Joseph's College of Engineering
Chennai, Tamil Nadu
India

Deepak Pant
Department of Environmental Science
School of Earth and Environmental Science, Central University of Himachal Pradesh
Kangra, Himachal Pradesh
India

Zainul Akmar Zakaria
Department of Bioprocess & Polymer Engineering
Institute of Bioproduct Development, University of Technology Malaysia
Johor Bahru, Johor
Malaysia

ISSN 2522-8366 ISSN 2522-8374 (electronic)
Energy, Environment, and Sustainability
ISBN 978-981-10-7412-7 ISBN 978-981-10-7413-4 (eBook)
https://doi.org/10.1007/978-981-10-7413-4

Library of Congress Control Number: 2017958847

Printed on acid-free paper

This Springer imprint is published by Springer Nature
The registered company is Springer Nature Singapore Pte Ltd.
The registered company address is: 152 Beach Road, #21-01/04 Gateway East, Singapore 189721, Singapore

Preface

In recent years, there has been increasing concern over public health threat presented by the introduction of different pollutants in the environment due to anthropogenic activities to a greater extent and natural processes to some extent. Many countries all over the world are currently facing severe problems as these pollutants are haemotoxic, carcinogenic, immunotoxic and teratogenic. There has been an increasing interest in the human health and environmental risk of pollutants present in solid waste or effluent generated by anthropogenic activities. Remediation of polluted sites by applying various physical, chemical, thermal and biological techniques is in practice.

The first international conference on 'Sustainable Energy and Environmental Challenges' (SEEC-2017) was organized under the auspices of 'International Society for Energy and Environmental Sustainability' (ISEES) by the 'Center of Innovative and Applied Bioprocessing' (CIAB), Mohali, held from 26 to 28 February 2017. ISEES was founded at IIT Kanpur in January 2014 with the aim of spreading knowledge in the Fields of Energy, Environment, Sustainability and Combustion. The Society's goal is to contribute to the development of clean, affordable and secure energy resources and a sustainable environment for the society and to spread knowledge in the above-mentioned areas and awareness about the environmental challenges, which the world is facing today. ISEES is involved in various activities such as conducting workshops, seminars and conferences in the domains of its interest. The Society also recognizes the outstanding work done by the young scientists and engineers for their contributions in these fields by conferring them awards under various categories.

This conference provided a platform for discussions between eminent scientists and engineers from various countries including India, USA, South Korea, Norway, Malaysia and Australia. In this conference, eminent speakers from all over the world presented their views related to different aspects of energy, combustion, emissions and an alternative energy resource for sustainable development and cleaner environment. The conference started with four mini-symposiums on very topical themes, which included (i) New Fuels and Advanced Engine Combustion, (ii) Sustainable Energy, (iii) Experimental and Numerical Combustion and

(iv) Environmental Remediation and Rail Road Transport. The conference had 14 technical sessions on topics related to energy and environmental sustainability and a panel discussion on 'Challenges, Opportunities and Directions of Technical Education & Research in the Area of Energy, Environment and Sustainability' to wrap up the three-day technical extravaganza. The conference included 2 plenary talks, 12 keynote talks, 42 invited talks from prominent scientists, 49 contributed talks and 120 posters. A total of 234 participants and speakers attended this three-day conference, which hosted Dr. V. K. Saraswat, Member NITI Ayog, India, as a chief guest for the award ceremony of ISEES. This conference laid out the road map for the technology development, opportunities and challenges in this technology domain. The technical sessions in the conference included Advances in IC Engines and Fuels; Conversion of Biomass to Biofuels; Combustion Processes; Renewable Energy: Prospects and Technologies; Waste to Wealth—Chemicals and Fuels; Energy Conversion Systems; Numerical Simulation of Combustion Processes; Alternate Fuels for IC Engines; Sprays and Heterogeneous Combustion of Coal/Biomass; Biomass Conversion to Fuels and Chemicals—Thermochemical Processes; Utilization of Biofuels; and Environmental Protection and Health. All these topics are very relevant for the country and the world in present context. The Society is grateful to Prof. Ashok Pandey for organizing and hosting this conference, which led to the germination of this series of monographs, which included 16 books related to different aspects of energy, environment and sustainability. This is the first time that such voluminous and high-quality outcome has been achieved by any Society in India from one conference.

The editors would like to express their sincere gratitude to the authors for submitting their work in a timely manner and revising it appropriately at short notice. We would like to express our special thanks to the reviewers for reviewing various chapters of this monograph and providing their valuable suggestions to improve the manuscripts. We acknowledge the support received from various funding agencies and organizations for the successful conduct of the first ISEES conference SEEC-2017, where these monographs germinated. These include Department of Science and Technology, Government of India (special thanks to Dr. Sanjay Bajpai); TSI, India (special thanks to Dr. Deepak Sharma); Tesscorn, India (special thanks to Sh. Satyanarayana); AVL India; Horiba, India; Springer (special thanks to Swati Mehershi); CIAB (special thanks to Dr. Sangwan).

Remediation of polluted sites by various bioremediation technologies is a viable option as conventional physico-chemical methods for remediation seem technically as well as economically challenging. Bioremediation is non-invasive and could be cost-effective in the removal of organic as well as inorganic pollutants present in waste. Green processes by the use of microorganisms or plants are considered as ultimate mechanism for pollutant removal. Hence, bioremediation processes are referred to as eco-friendly, efficient, economically feasible and versatile for pollutant remediation.

This monograph is intended for environmental scientists, microbiologists and biotechnologists. We hope that the book would be of great interest to the professionals, postgraduates and research students involved in research work in the field of waste bioremediation. The main objective of this monograph is to promote a better and more accurate understanding of waste bioremediation besides recent advances and challenges in clean technology development for sustainable development of environment. The book shall include chapters on different aspects of waste bioremediation either solid or liquid both from within India and internationally. Some of the topics covered shall include Organic carbon and pollutants reduction through composting; Polycyclic aromatic hydrocarbons from petroleum oil industry activities: Effect on human health and their biodegradation; E-waste management: Bioremediation; Role of microbial consortia in bioremediation of textile recalcitrant compounds; Bioreduction of hexavalent chromium using moderate thermophilic and thermophilic microorganisms; Bioprocesses for sulphate removal from wastewater; Microbial transformation of heavy metals; Bioremediation of Lithium-ion battery (LIB) waste; Biodigester technology for eco-friendly disposal of night soil; Anaerobic digestion: Factors affecting anaerobic digestion process; Energy recovery with microbial fuel cells: Bioremediation and bioelectricity; Microbial depolymerisation; Mechanism and action of *Aureobasidium pullulan* on biosorption of metals; Bioremediation of leachates; Effectiveness of plant growth promoting rhizobacteria in phytoremediation of chromium-stressed soils and so on.

Gandhinagar, India — Sunita J. Varjani
Lausanne, Switzerland — Edgard Gnansounou
Chennai, India — G. Baskar
Kangra, India — Deepak Pant
Johor Bahru, Malaysia — Zainul Akmar Zakaria

Contents

About the Editors

Dr. Sunita J. Varjani is Scientific Officer at Gujarat Pollution Control Board, Gandhinagar, India. She holds M.Sc. degree in Microbiology (2009) and Ph.D. degree in Biotechnology (2014). Her major areas of research are industrial and environmental microbiology/biotechnology and molecular biology. She has authored 35 publications and has won several awards and honours, including the Young Scientist Award at the AFRO-ASIAN Congress on Microbes for Human and Environmental Health, New Delhi (2014), and Best Paper Awards for oral presentations at national and international conferences in 2008, 2012 and 2013. She is a member of the editorial board of the Journal of Energy and Environmental Sustainability.

Prof. Edgard Gnansounou graduated with M.S. degree in Civil Engineering and Ph.D. degree in Energy Systems at the EPFL. He is a Professor of Modelling and Planning of Energy Systems at the Swiss Federal Institute of Technology Lausanne (EPFL) where he is the Director of the Bioenergy and Energy Planning Research Group. He is currently investigating the techno-economic and environmental assessment of biorefinery schemes based on the conversion of agricultural residues. He is leading research projects in that field in several countries including Brazil, Colombia and South Africa. He has published numerous papers in leading scientific journals. He is a member of the editorial board of Bioresource Technology Journal.

Prof. G. Baskar is currently working as Professor of Biotechnology at St. Joseph's College of Engineering, Chennai, India. He has published more than 100 papers in national and international journals and 5 book chapters. His research expertise is in biofuels, therapeutic proteins, microbial enzymes, nanomedicine and nanocatalysis. He has received the International Bioprocessing Association's Young Scientist Award (2015) and the Indian Society for Technical Education (ISTE)-Syed Sajed Ali National Award (2016).

Dr. Deepak Pant is currently the Dean and Head of Environmental Science, Central University of Himachal Pradesh, Dharamshala. He received the Silver Jubilee Research Fellowship Award in 2003 from Kumaun University, Nainital,

India; the UCOST Young Scientist Award 2009, INSA Fellow (2010), DST-SERC Visiting Fellow (2010) and DST-SERC Young Scientist Award (2011). He has 5 patents in the area of waste management using green techniques and published 10 books, 45 research papers in various national and international journals. He recently received Visitor's Award 2017 from the President of India for his innovations in the area of waste management.

Dr. Zainul Akmar Zakaria is Associate Professor at the Department of Bioprocess and Polymer Engineering, Faculty of Chemical and Energy Engineering, Universiti Teknologi Malaysia (UTM), Malaysia. He is currently a research fellow at the Institute of Bioproduct Development (IBD), UTM, which is a National Centre of Excellence for Plant and Microbial Bioproducts. He has published two research books, four edited books and eight book chapters. He has served as a guest editor for various journals; an evaluator for research grants from the American Chemical Society, Saudi Arabia, Australia and South Africa; as well as an external examiner for Ph.D. candidates from South Africa, India and Pakistan.

Chapter 1
Introduction to Waste Bioremediation

Sunita J. Varjani, Edgard Gnansounou, G. Baskar, Deepak Pant and Zainul Akmar Zakaria

Abstract Incomplete discharge of waste materials into the environment is of concern due to its slow degradability, highly soluble and biomagnification features in animals and plants. Conventional treatment techniques include chemical precipitation, ion-exchange, reverse osmosis and combustion are effective but energy intensive and consumes huge amounts of chemicals which may give rise to secondary problems such as spillage, corrosion and toxicity. The application of biological approach notably from use of microorganisms is an interesting alternative. Microorganisms such as bacteria, yeast and algae are known to survive in waste-containing environments owing to its ability to reduce, accumulate, sequester, absorb and oxidize different types of waste materials into forms, mostly making it less soluble and easily precipitated, that is, less toxic to the environment. This monograph covers biological approaches to remediate waste generated from various industries such as petroleum, electronic, textile, electroplating and landfill site(s). The role of microbes in composting and anaerobic digestion processes is also discussed. Apart from this effectiveness of microbes living in legumes of plant to remediate toxic heavy metals are also reported.

S. J. Varjani (✉)
Paryavaran Bhavan, Gujarat Pollution Control Board,
Gandhinagar 382010, Gujarat, India
e-mail: drsvs18@gmail.com

E. Gnansounou
Bioenergy and Energy Planning Research Group (BPE),
Ecole Polytechnique Fédérale de Lausanne (EPFL), 1015 Lausanne, Switzerland

G. Baskar
Department of Biotechnology, St. Joseph's College of Engineering,
Chennai 600119, India

D. Pant
Department of Environmental Science, Central University
of Himachal Pradesh, Kangra 176206, Himachal Pradesh, India

Z. A. Zakaria
Department of Bioprocess Engineering & Polymer,
Universiti Teknologi Malaysia, 81310 Utm Johor Bahru, Johor, Malaysia

S. J. Varjani et al. (eds.), *Waste Bioremediation*, Energy, Environment, and Sustainability, https://doi.org/10.1007/978-981-10-7413-4_1

Keywords Waste · Bioremediation · Bioprocesses · Biotransformation
Biodegradation · Bioreduction · Biodigester · Phytoremediation
Composting

Bioremediation is the use of microbes or other living things to clean-up contaminated soil and groundwater. This process proceeds from the ability of microbes and plants to carry-out various energy-dependant processes including oxidation, reduction, accumulation and precipitation, of the compounds of interest present in contaminants. Bioremediation occurs both in aerobic or anaerobic conditions where naturally the process requires long completion period. To expedite this condition, scientists and engineers would create a favourable environment for the process to take place where it would normally involve deciding on the best process arrangements (either in situ or ex situ), availability of nutrients (if not, has to be provided), right temperature as well as whether it is best to apply aerobic or anaerobic approaches (would determine whether an air pumping system needs to be installed). One important consideration that must not be taken lightly by all bioremediation operators is the bioavailability of contaminants. This factor is extremely important as it is more important limiting rate in all biodegradation processes. If bioavailability of a contaminant is perceived too low by microbial cells, the induction of catabolic gene system used in biodegradation will not occur; hence, the bioremediation process will not proceed. Some of the factors limiting bioavailability of contaminants for bioremediation includes irreversible sorption of contaminant to soil or sediment, chemical kinetics of desorption, extent of dissolution of contaminants into aqueous phase, diffusion rate through natural organic matters and rate of diffusion within particles. Some contaminated sites may allow in situ bioremediation while some sites may not notably due to a number of reasons such as soil might be too dense to allow amended nutrients to spread evenly underground. This would warrant such contaminants to be excavated and treated ex situ, i.e. above ground, whether in tanks or in a more controlled bioreactor. Depending on whether final major compound(s) arising from bioremediation processes would require a post-treatment step (due to potential hazard), the treated waste may be used as organic compost for agricultural application, pumped back to the ground or simply discharged to surface water or to a municipal wastewater system.

Nevertheless, all these bioremediation ventures will not be possible without early investigation carried out by thousands of scientists, engineers, researchers and students who have sacrificed hours after hours of experimentation in order to justify or proof some concepts which is highly crucial prior to any scaling-up or on-site demonstration attempts.

One example is the study on microbial fuel cells (MFCs). In MFC systems, chemicals are reduced at the cathode, and in some systems it is possible to achieve chemical oxidation at the anode in situations when high concentration of biodegradable organics is present in wastewater. For this to work, however, sufficient electron acceptors should be present at the cathode. For example, if a site is contaminated with petroleum or gasoline, the water can be channelled through consecutive hydraulic chambers similar to that used for zero valent iron walls for treating chlorinated aliphatics in groundwater.

Bacteria have been investigated in various processes such as biological removal of sulphate from water or wastewater. The treatment of sulphate rich wastewater using sulphur reducing bacteria (SRB) is advantageous due to minimal sludge production, ability to perform simultaneous oxidation of organic matter and the reduction of sulphate. Sulphate reduction in bioreactors is affected by parameters such as type of electron donor, COD: sulphate ratio, pH, HRT and reactor configuration. Most of the sulphate-reducing bioreactors are also able to handle fluctuations in COD or sulphate loading rates. The ability of the SRB to overcome feast and famine periods clearly shows the application of this technology for industrial situations.

Another example of laboratory-scale investigation of bioremediation properties using bacteria is the biodegradation of textile dyes either by using single bacterium or bacterial consortium. These systems utilize inducible consortial oxidoreductases for detoxification and mineralization into least/reduced toxic metabolites. Presently, the design of such consortial systems is gaining pervasive importance as they work under adverse conditions. Further, tailoring of process parameters is required for utilizing bacterial consortia for making the process viable and serving necessity of environmental safety and health.

Apart from application of bacterial cells as single or consortium, bacteria can also be immobilized in a suitable carrier such as calcium alginate. Immobilized cells offer robust and continuous application which is an important feature in real life bioremediation application. This approach is definitely advantageous for processes such as reduction of hexavalent chromium to trivalent chromium where the effect of chromium toxicity towards bacterial cells can be significantly reduced. However, further investigation on factors affecting both microbial growth and chromate reduction processes in free and immobilized forms is necessary. Understanding reduction pathways, possible formation of toxic intermediates, and if more stable products are implicated on the process, is also necessary. Another useful application of chromate reduction is from the use of plant growth-promoting bacteria (PGPR) in phytoremediation of chromium polluted soils. Since, the efficiency of specific PGPR strains varies according to the soil type and environmental conditions; an extensive research is required to enhance root colonization and chromium phytoextraction ability of PGP bacteria under soil stressed conditions. In addition, use of consortium instead of monoculture can be undertaken for research in order to enhance the PGPR and chromium phytoextraction process.

Other than that, numerous investigations are currently ongoing on the application of biological processes for the management of electronic and electrical waste (E&E or simply E-waste). Every management strategies concentrated upon organic and inorganic portions of E-waste. Organic part consists of variety of thermosetting plastic(s) with the presence of halogenated material. Microbes are involved in process of dehalogenation in many ways. Microbes can manage leaching of inorganic portion of E-waste which consists of both metallic and non-metallic components.

Plant also offers interesting solution in bioremediation processes. One example is the use of vetiver plant, which is a highly resilient and versatile plant that is used in various application including soil erosion control, water conservation, bioremediation, fragrance oil production, household and energy applications. However, several policy measures are required for successful implementation of this bioremediation system in large scale that needs to overcome several limitations through rigorous research, collaboration from developed and developing countries and large-scale test trials. In addition, growing vetiver for bioremediation and harvesting shoots for energy purpose could generate economic returns and employment.

This monograph highlighted recent studies on the applications of bioremediation for various types of waste including, viz. textile dyes, polyaromatic hydrocarbons, polymer and electric and electronic as well as microbial fuel cells, sulphate removal, hexavalent chromium reduction. Specific topics covered in the manuscript include:

- Energy recovery with microbial fuel cells: bioremediation and bioelectricity
- Bioprocesses for sulphate removal from wastewater
- Microbial depolymerisation
- Bioremediation techniques for E-waste management
- A review on bioremediation potential of vetiver grass
- Organic waste and pollutants reduction through composting
- Role of bacterial consortia in bioremediation of textile recalcitrant compounds
- Polycyclic aromatic hydrocarbons from petroleum oil industry activities: effect on human health and their biodegradation
- Bioreduction of hexavalent chromium using moderate thermophilic and thermophilic microorganisms
- Mechanism and action of *Aureobasidium pullulans* on biosorption of metals
- Bioremediation of leachates
- Microbial transformation of heavy metals
- Bioremediation of lithium ion battery (LIB) waste
- Characterization of leachate and groundwater in and around saduperi municipal solid waste open dump site, Vellore district, Tamil Nadu, India
- Effectiveness of plant-growth promoting rhizobacteria in phytoremediation of chromium stressed soils
- Biomining of natural resources
- Anaerobic digestion: factors affecting anaerobic digestion process
- Biodigester technology for effective and eco-friendly decomposition of night soil.

Author Biographies

Dr. Sunita Varjani is Scientific Officer at Gujarat Pollution Control Board, Gandhinagar, Gujarat, India. Her major areas of research are Industrial and Environmental Microbiology/Biotechnology and Molecular biology. She has delivered talks as invited speaker in many international conferences including the countries India and China. She is recipient of visiting scientist fellowship from EPFL, Lausanne, Switzerland (2017).

Prof. Edgard Gnansounou is a Professor of modelling and planning of energy systems at the Swiss Federal Institute of Technology Lausanne (EPFL) where he is Director of the Bioenergy and Energy Planning Research Group. His current research works comprise techno-economic and environmental assessment of bio-refinery schemes based on conversion of agricultural residues and other renewable energy systems.

Prof. G. Baskar is currently working as Professor of Biotechnology in St. Joseph's College of Engineering, Chennai, India. His current research expertise in biofuels, therapeutic proteins, microbial enzymes nanomedicine and nanocatalysis. The International Bioprocessing Association has conferred Dr. G. Baskar with Young Scientist Award 2015. The Indian Society for Technical Education has conferred him with ISTE-Sayed Sajed Ali National Award 2016 for his research work.

Dr. Deepak Pant is currently the Dean and Head of Environmental Science, Central University of Himachal Pradesh, Dharamshala. He received the Silver Jubilee Research Fellowship Award in 2003 from Kumaun University, Nainital, India; the UCOST Young Scientist Award 2009, INSA Fellow (2010), DST-SERC, Visiting Fellow (2010) and DST-SERC Young Scientist Award (2011), Visitor's Award (2017) from the President of India.

Dr. Zainul Akmar Zakaria is an Associate Professor under Department of Bioprocess and Polymer Engineering, Faculty of Chemical and Energy Engineering, Universiti Teknologi Malaysia (UTM). Currently he is Research Fellow at Institute of Bioproduct Development (IBD), UTM. He has ISI h-index of 11 and has published 2 Research Books, 4 Edited Books and 8 Book Chapters.

Chapter 2
Energy Recovery with Microbial Fuel Cells: Bioremediation and Bioelectricity

Debajyoti Bose, Vaibhaw Kandpal, Himanshi Dhawan, P. Vijay and M. Gopinath

Abstract This chapter presents an overview on microbial fuel cells (MFCs) as a novel electrogenic reactor systems for simultaneous treatment of wastewater and generation of bioelectricity. MFCs work on the principle that organic matter present in wastewater serves as a primary substrate for the bacteria to consume and release electrons, facilitating the treatment of wastewater with simultaneous generation of power. Microbes in the anode chamber generate protons (H^+) and electrons (e^-) through reactions by decomposing the rich organics present in the wastewater and in the process treating the wastewater and producing a value added product which is bioelectricity. When these protons travel through the membrane and the circuit, respectively, power is generated from the system. Given the non-renewable aspect and polluting nature of fossil fuels, MFCs have generated interest among several research communities around the world. Following a historical approach toward this technology, the chapter discusses the various types of microbial fuel cells prevalent and compares the different MFC designs used. The role of proton exchange membrane separating the anodic and cathodic chambers is also explained. It focusses on the principle and working of an MFC and describes the instrumentation and procedure for reporting data. Additionally, the chapter presents benefits, drawbacks, and future scope of research in this field.

Keywords Microbial fuel cell · Energy generation · Nafion 117
Wastewater treatment · Substrate

D. Bose
Department of Electrical Power & Energy Engineering,
College of Engineering Studies, University of Petroleum & Energy Studies,
Bidholi via-Prem Nagar, Dehradun 248007, Uttarakhand, India

V. Kandpal · H. Dhawan · P. Vijay · M. Gopinath (✉)
Department of Chemical Engineering, College of Engineering Studies,
University of Petroleum & Energy Studies, Bidholi via-Prem Nagar,
Dehradun 248007, Uttarakhand, India
e-mail: mgopinath@ddn.upes.ac.in

S. J. Varjani et al. (eds.), *Waste Bioremediation*, Energy, Environment, and Sustainability, https://doi.org/10.1007/978-981-10-7413-4_2

1 The Energy Economy and Energy Needs of Society

1.1 Historical Energy Consumption Trend

During 1700s, primary energy sources originated from water, wind, and firewood or manual labor. Before 1900, energy consumption per capita has not risen so much even with the expansion of coal energy, proposing that the early utilization of coal for the most part balances other fuel utilization. There has been a little increment in energy utilization per capita during World War I, however, between World War II and 1970; there was a tremendous increase in energy utilization per capita. There are reasons like European nations and Japan were reconstructed after World War II, oil industry needed to be created, to give occupations and assessment income to people, and between the periods of 1800–2000 the world population also expanded rapidly.

Around 1900, we started to bore for oil and petroleum gas. By 1950, these "non-renewable energy sources" had displaced the more seasoned energy sources with the exception of water energy sources. Figure 1 gives a brief idea for the same. We are utilizing non-renewable energy sources at a far more prominent rate than they are being made. After 1950, we started to utilize nuclear energy from uranium from its earliest stage. In the course of recent years, utilization of more established sustainable energy sources has expanded and utilization started new sustainable power sources too. Since fossil fuel-based economy will not keep going forever and that their utilization adds to ecological contamination. Sustainable power source which essentially originates from the solar energy gives chances to a boundless, reasonable energy supply with minimum damage to the ecosystem.

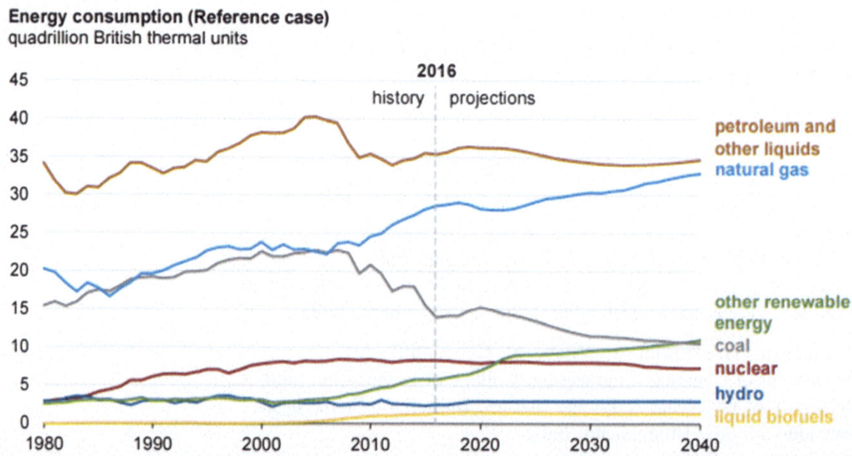

Fig. 1 History and projections of different energy forms (Aelterman et al. 2006)

1.2 The Need for Renewable and Alternate Energy

Renewable and sustainable power sources are those regenerated by natural processes at a rate independent of its utilization pattern. Such as:

Solar Energy: Energy from the Sun is utilized to deliver power, or for heating. In India, current solar energy generation is about 12.3 GW but there are challenges like more land requirement (40–60 MW per 1 km^2), cost of installation which is needed to be overcome as time progresses toward more renewable energy. The program reached an ambitious Solar Alliance during the Paris Agreement of 2015.

Biomass Energy: Biomass as a fuel comprises of natural matter (cellulose, hemicellulose, and lignin) which is mostly agricultural residues; it is non-polluting in nature, and as plants capture CO_2 throughout their lives so such sources are considered carbon neutral. As a fuel, it might incorporate residues from wood, straw, sugarcane, and numerous other by-products of certain agricultural processes.

Biomass itself is a carbon neutral source that is CO_2 from atmosphere which is taken by plants during photosynthesis. This CO_2 is changed over into natural carbon and put away in woody biomass. Trees discharge the stored carbon when they die, rot, or are getting combusted. As the biomass discharges carbon as CO_2, the carbon cycle is maintained. Biomass energy can be extracted by either thermochemical (using heat) or by biochemical (using enzymes, bacteria). India currently generates about 6000 MW of energy from biomass.

Hydropower: Hydropower harbors one of the biggest sustainable power sources. Hydropower plants change over the energy of streaming water into power. This is for the most part done by making dams crosswise over waterways to make expansive repositories or reservoirs and after that discharging water through turbines to deliver power. Hydropower brings about no discharges of harmful gases into the air; however, the way toward making dams can make critical natural issues for water quality and for oceanic and untamed life environment.

Geothermal: Geothermal energy utilizes high temperatures of profound earth outside layer to deliver steam, which then powers turbines and produces power.

There are four major types of geothermal energy resources: hydrothermal, geopressurized brines, hot dry rocks, magma. The estimated potential for geothermal energy in India is about 10,000 MW, but due to the lack of infrastructure, India still not able to produce any amount of energy from earth's crust.

Wind Power: It has been the quickest developing energy source in the course of the most recent decade fundamentally because of exceptionally critical changes in wind energy (or vitality) innovation. Wind power is created by the wind turning streamlined sharp edges of the turbine blades mounted to a center point. The center point is associated with a pole that turns a generator.

Fuel Cells: Fuel cells are electrochemical devices which operate on hydrogen-rich sources as long as it has a supply of the fuel to produce clean energy. This energy can be used to feed power to the grid, automobiles, and even everyday electronics that we use. As of now, delivered power modules or fuel cells join hydrogen and oxygen without burning to create power. The oxygen can be utilized

Table 1 Fuel cell types and their configurations with estimated power output

Fuel cell type	Electrolyte	Operating temperature (°C)	Electrical efficiency	Fuel oxidant	Energy output (single stack)
Alkaline fuel cell	Potassium hydroxide	25–90	60–70%	H_2O_2	300 W–5 kW
Proton exchange membrane fuel cell	Proton exchange membrane	25–80	40–60%	H_2O_2	1 kW
Direct methanol fuel cell	Proton exchange membrane	25–130	20–30%	CH_3OH, O_2, air	1 kW
Phosphoric acid fuel cell	Phosphoric acid	160–200	50–55%	Natural gas, biogas, coal gas, H_2O_2, air	200 kW
Molten carbonate fuel cell	Molten mixture of alkali metal carbonates	620–660	55–65%	Natural gas, biogas, coal gas, H_2O_2	2–100 MW
Solid oxide fuel cell	Ceramic-type membranes	800–1000	60–65 °C	Natural gas, biogas, coal gas, H_2O_2	100 kW

from air itself, while the hydrogen can either be delivered from existing hydrogen infrastructure or renewable energy-based systems. Further, new power devices are being produced that can utilize petroleum products specifically in this context (Table 1).

The advancement of power devices has come about to improve new power device advances, and subsequently, the requirement for vitality and worries over energy security has brought about the improvement of microbial fuel cells (MFCs) advances. Globally, more than one billion people lack drinking water and more than two billion lack adequate sanitation. More than 38 billion of liquid wastewater is generated in urban areas of India; this does not include the industrial, rural, solid wastes, etc. The capacity of waste generated in USA is around 17 GW which is equivalent to energy produced by 17 nuclear plants. Hence, MFCs are developing at much faster rate for renewable energy generation.

1.3 Drive Toward Hydrogen Economy

Our universe comprises a blend of an unlimited cluster of components. Every component has an imperative part in the organization of the world. The most abundant elements in the universe include hydrogen, oxygen, and nitrogen. Hydrogen occupies up to 75% of the universe and has the potential to contribute to environmentally benign energy infrastructure. Apart from assisting other distinctive living species in surviving, hydrogen can be used to generate energy as well. Hydrogen has often been stated as a future fuel in the energy economy trend line. Hydrogen is not found in its pure form on Earth; however, it can be produced from different compounds like biomass, natural gas, alcohols, or water. In all these cases, conversion of hydrogen to its pure form is energy intensive. As a result, hydrogen is used as an energy carrier or storage medium instead of an energy source in itself (Fig. 2).

As shown in Fig. 4, the expression "hydrogen economy" refers to the vision of utilizing hydrogen as a low-carbon energy source, for example, gas as a transport fuel or natural gas as a heating fuel. Hydrogen is appealing in light of the fact that whether it is burned to produce heat or reacted with air in a fuel cell to generate electricity, the by-product is water.

1.3.1 Some Advantages of Hydrogen Economy

1. Abundant energy source and clean form of energy.
2. Non-toxic does not cause any harm or destruction to human health.
3. Efficient energy source to convert a lot of energy for every pound of fuel. This means hydrogen vehicles will have more mileage as compared to an equivalent amount of gasoline.

These are issues that with the hydrogen infrastructure; however, with time, engineers and scientists are expected to resolve them, as with every technology.

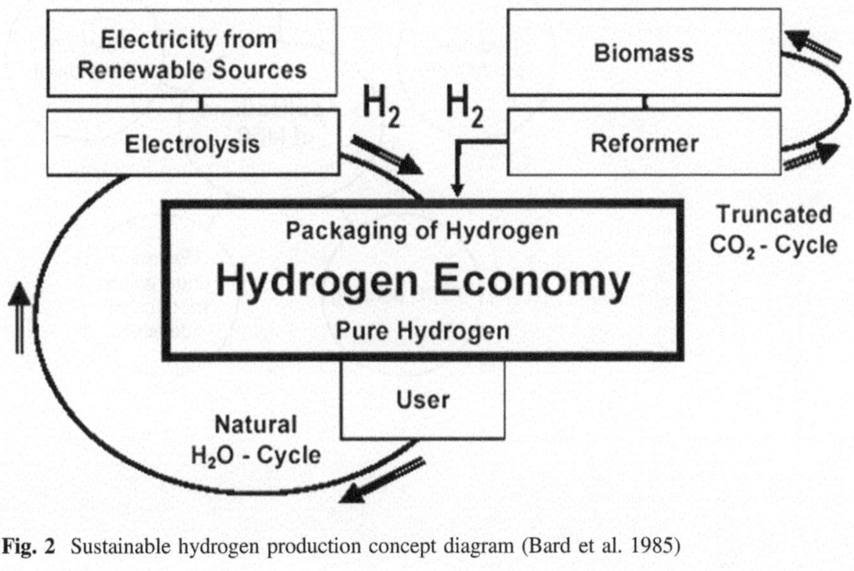

Fig. 2 Sustainable hydrogen production concept diagram (Bard et al. 1985)

Another type of fuel cell is the microbial fuel cell, these are devices where bacteria can grow on one electrode, break down organic matter, while releasing electrons from the organic matter, it is similar to how we get energy, we eat food, we oxidize it, we remove electrons, we send these electrons to the respiratory enzymes and then when we are done with them, we release them to oxygen. We eat and breathe to do this. So when bacteria releases this electron, it creates a potential of about 0.5 V; voltage time current is power, and that is how power is generated from the system. In concept, a real simple system, having two electrodes on either side of the container, bolted together, wastewater is added to the system and you generate electricity.

1.4 The Role of Microbial Fuel Cells as a Technology of Importance

MFCs have been depicted as bioreactors which are the newest approach for generating electricity by oxidation of organic matter present in the wastewater streams (Fig. 3).

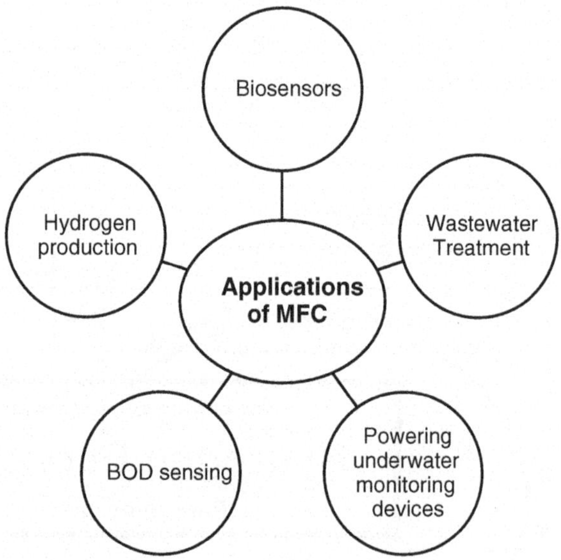

Fig. 3 End use application for MFCs to serve bioremediation and bioelectricity generation

2 Microbial Fuel Cells

Optimizing energy recovery from wastewater is a sustainable approach for wastewater treatment and is also of interests to protect and improve the water environment everywhere. There is a large gap between generation and treatment of wastewater for domestic wastewater in developing nations and to facilitate such energy recovery processes microbial fuel cells can be used. The energy utilization research in microbial fuel cells is growing exponentially, but it suffers from established terminology and methodical framework to analyze system functionality. The sheer diversity in the process employed makes it intricate to compare experimental analysis with each other. Studies of such are interdisciplinary with concepts of microbiology, integrated with electrochemistry to serve as a component of environmental engineering (Table 2).

This chapter reports how the organic matter removal efficiency coupled with simultaneous bioelectricity production has been studied with pure and mixed microbial culture, which are naturally found or inoculated into wastewater streams. And how such alternate forms of energy harvesting by addressing and optimizing key parameters such as temperature, pH, and dissolved solutes can potentially contribute to energy security and economic stability across nations where a dire need of energy conversion crisis exist.

There is a large gap between generation and treatment of wastewater for domestic wastewater in India and to facilitate these renewable energy recovery process; microbial fuel cells can be used. MFCs can be used directly with wastewater or by inoculation with microbial species. Bacteria identified to produce bioelectricity in MFC are metal-reducing bacteria such as *Geobacter metallireducens, Geobacter sulfurreducens, Clostrodium butyricum, and Shewanella putrefaciens.*

In microbial fuel cells, bacteria act as a catalyst and oxidize the organic matter and inorganic matter to produce electricity. These are an older invention than the battery. The electrons produced by the microbes from the substrate are transferred to the anode, which is the negative terminal and onto the cathode, which is the positive terminal. These are linked by conductive materials with a load (resistor). Electrons can be transferred by using electron mediators into the anode usage of

Table 2 List of some industries and the contaminants in its wastewater streams

Sources of industrial wastewater	Characterization of wastewater in the discharge stream
Coal-based thermal power plant	Significant levels of lead, mercury, cadmium, chromium, arsenic, and nitrous compounds
Food industry	High concentration of BOD and suspended solids
Iron and steel industry	Ammonia, cyanide, phenols, benzene, and other organics
Paper and pulp industry	High suspended solids (SS), BOD, chlorinated organic compounds
Petrochemical industry	Mineral oils, phenol, high COD and BOD, chromium

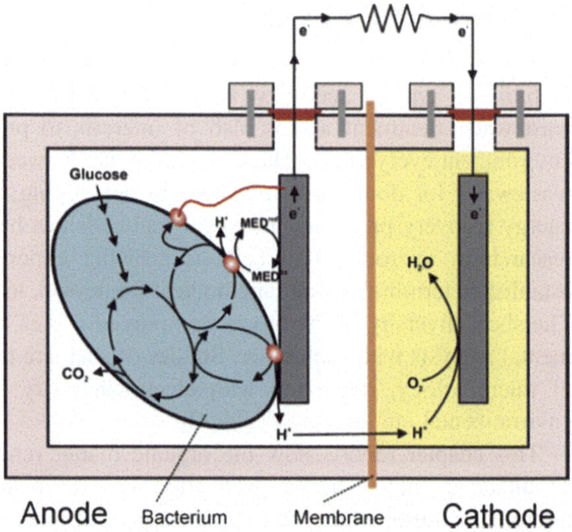

Fig. 4 Transfer of electrons by a bacterium in the anode chamber which then produces a three-phase reaction with electrons, protons, and oxygen at the cathode. Thereby, completing the circuit and producing power. Protons produced in the process migrate to cathode using cation exchange membrane (CEM) (Rabaey et al. 2004)

direct membrane electron transfer has also been shown in studies or by use of nanowires. It can be speculated that further undiscovered means can also facilitate such processes (Fig. 4).

2.1 *Various Designs of Microbial Fuel Cells*

Several configurations are possible with MFC systems, a commonly used and relatively cost-efficient design is the H-shaped two-chambered setup, having two chambers separated by a cation exchange membrane (CEM) such as Nafion (DuPont Co. USA) or a plain salt bridge made from mostly agar and saturated salts (acting as a CEM).

The efficiency in this design lies in selection of the CEM and the reaction chemistry at the cathode in terms of reduction reactions (typically for oxygen). As seen from Figure f, in such setups, the membrane is in the center of the tubes joining the bottle although the system is not restricted to using only tubes, if both the chambers are kept separate. Such systems despite being cost economic are plagued by high internal resistances, thus producing little power. However, such can be used to examine power production using new materials or compound-specific microbial activities.

Some studies have shown high power densities by using ferricyanide as electron acceptor at the cathode. These studies have shown the best fit model for bioelectricity production which is the use ferricyanide as an aqueous catholyte (Catal et al. 2008). However, use of ferricyanide is not sustainable in practice and it must be chemically regenerated. Some geometry of systems includes outer cylindrical reactor along with a concentric inner tube being the cathode as shown in Fig. 5d,

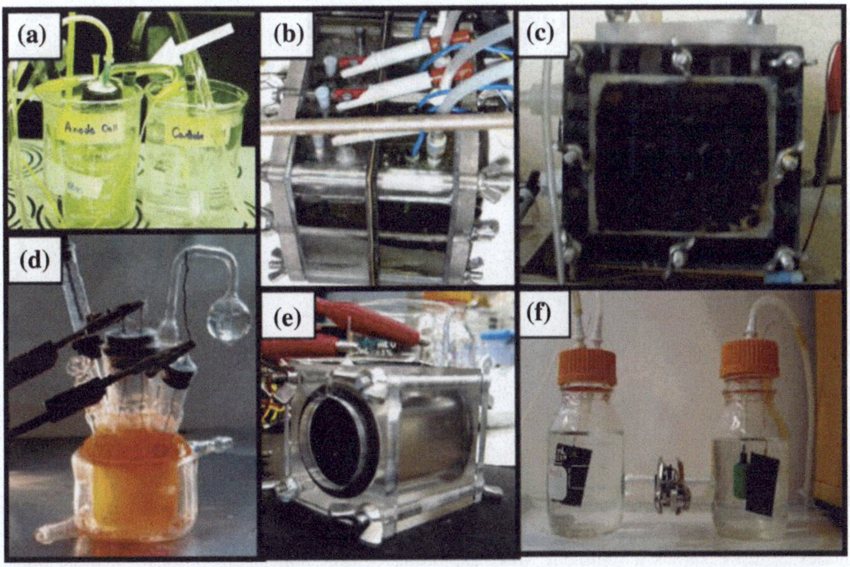

Fig. 5 Different MFC configurations: Setup **a** is a salt bridge MFC using regular glassware (Logan et al. 2006); setup **b** shows batch reactors where membrane separates the anode from the cathode, all bolted together (Rabaey et al. 2005); setup **c** is identical to **b** but has a continuous process as water is recirculated in anode chamber; setup **d** photoheterotrophic type MFC; setup **e** shows a single chambered air-cathode MFC system (Liu and Logan 2004); setup **f** is a two-chambered H-shaped system where gas sparging can be done (Logan et al. 2005)

and having an inner cylindrical reactor (granular media-based anode) with the cathode on the external surface as shown in Fig. 5a.

Another variation is shown in Fig. 5b, which is an up-flow fixed bed biofilm reactor, with continuous fluid flow through the porous anodes toward a membrane that separates the anode and cathode chambers. Some systems as shown in Fig. 5c has been made identical to resemble a hydrogen fuel cell, where the membrane is sandwiched between the cathode and the anode. To increase the current, MFCs can be stacked up and connected in series as shown in Fig. 5e.

Sediment MFCs was initially developed, and it demonstrated that power generation can be sustained by bacteria breaking down the organic matter in the sediment. Since then by developing proper materials, power output has been increased significantly. It is often viewed as an effective bioremediation tool, and it produces bioelectricity at the same time. Some studies used graphite disks platinum mesh electrodes while some have demonstrated its feasibility with modifications at the cathode by increasing the active surface area and using non-corrosive materials (Fig. 6).

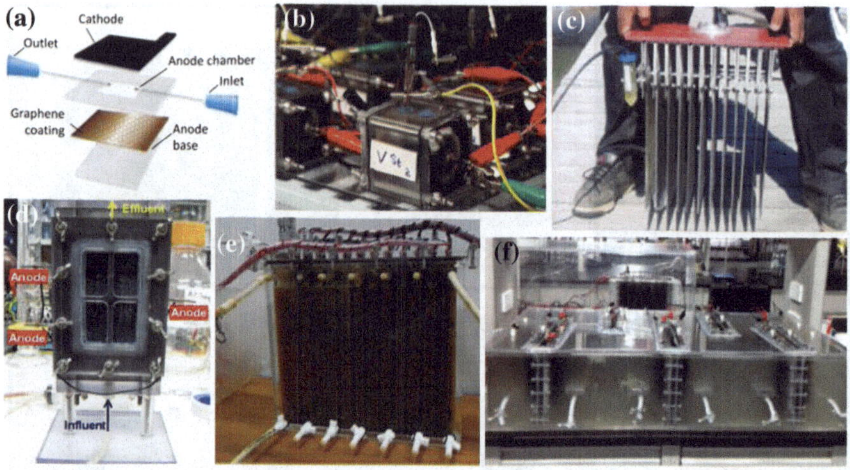

Fig. 6 Design improvement in MFC systems to account limiting factors: Setup **a** shows a micro MFC setup with anode capacity of 25 μL; setup **b** uses an air-cathode system with low-cost brush anodes; setup **c** shows a sediment MFC exploiting benthic soil; setup **d** shows a three brush electrode setup for a continuous MFC reactor; setup **e** shows a 12 assembly stack system; and setup **f** shows a cassette arrangement for baffled MFCs (Logan et al. 2015)

Table 3 Different microbial electrochemical technologies which are either identical or an extension of MFCs

System names based on design	Mechanism	Notable work
Microbial electrolysis cell (MEC)	Production of hydrogen gas at the cathode, and metal reduction	16
Microbial electrosynthesis system (MES)	Produces soluble organics such as acetate	21
Microbial methanogenesis cell (MMC)	Production of methane at the cathode	13
Microbial reverse electrodialysis fuel cell (MRFC)	Power generation using a RED stack, yield is higher than a standard MFC	33
Sediment microbial fuel cell (s-MFC)	Power generation from marine sediments, soil, and mud	15

The above-mentioned systems with varying configurations with bipolar membranes can facilitate effective recovery of both acid and caustic solutions from the cathode and this has commercial value as well. Alternatively, these solutions can be complemented with minerals to allow for a carbon sequestration technology as shown in Table 3.

2.2 Construction Material for Electrodes

Anode: Material for anode construction should be cost-effective, having good electrical conductivity, and stable during operation. One study showed the utility of non-corrosive stainless steel. Copper-based materials are usually avoided, as it leads to toxic reactions with bacteria and leading corrosion current as well. Carbon emerges out to be the most versatile material, available as graphene, graphite rods, and plates or in the form of cloth, paper, and foam. In addition to that, suitable chemical mediators have also been employed, some of which are given in Table 4.

Graphite brushes are relatively inexpensive materials for anode construction, and also graphite felt electrodes as they are much larger surface area. One study found the relation between anodic materials in terms of reactivity in the order of carbon felt > carbon foam > graphite. Materials such as reticulated vitreous carbon (RVC) with different pore sizes have shown improved performance. It must be mentioned that the long-term effect of biofilm growth on anode surfaces has not been adequately examined.

Cathode: The reduction of oxygen is the primary reaction at the cathode and is associated with high cost because of the potential barrier when performed on graphite or carbon-based electrodes, implicating the need for catalysts such as platinum (Pt). In a MFC, activity loss for platinum can occur due to side reactions and other losses.

Some of the alternatives, however, are proposed, but different configuration can be possible. Some of the published literature shows that manganese oxides, iron phthalocyanine (FePc), polypyrrole (Ppy), cobalt tetramethoxyphenylporphyrin (CoTMPP), Fe^{3+} cathode made with ferric sulfate, cathodes based on high active surface area such as graphite-granule cathodes, and recent advancements have shown the favorable use of activated carbon because of its adsorption properties and ease of operation.

Table 4 Chemical mediators added to improve anode performance

Substrate	Microbes	Current density (mA/cm^2)	References
Fe (III) and Mn (IV) linked with neutral red	Mixed culture (anaerobes) from STP	0.175	HaoYu et al. (2007)
Sucrose	Anaerobic sludge from septic tank	0.190	Call et al. (2009)
Food industry wastes	Aerobic sludge	0.025	Rabaey et al. (2004)
Food waste	Anaerobic culture	0.045	Davis and Yarbrough (1962)
Beer brewery wastewater	Anaerobic mixed culture	0.18	Zhen et al. (2015)

Literature studies have shown that the above-employed cathodes have produced less power densities except activated carbon in comparison with platinum. As the above, materials are cost-effective and are easily available compared, this makes them a promising avenue to explore further MFC research.

2.3 Construction Material for Membranes

Membranes are primarily used in double-chambered MFCs, as a process to keep the anode and the cathode fluids separate. These membranes must be permeable in nature to facilitate movement of protons generated at the anode to migrate to the cathode. Membranes also serve as a barrier to avoid transfer of other species in the chamber. For example, these are effective in reducing unwanted substrate flux from anode to the cathode and oxygen from the cathode to the anode, having a positive influence on Coulombic efficiency of the system.

The most commonly used CEM is Nafion (Dupont Co., USA) while, Ultrex CMI-7000 (Membranes International Incorp. NJ, USA) has also been employed (Table 5).

2.4 Use of Wastewater/Xenobiotics for Bioremediation and Simultaneous Bioelectricity Generation

The working and proficiency of MFC to a great extent depend upon the kind of materials utilized for the metabolism of microorganisms. Researchers have revealed use of straightforward and complex natural/inorganic sources which introduced in wastewaters from various sources acting as a substrate for power generation. The utilization of wastewater as a medium for substrate is favorable; to start with, the electrical power is generated, and second, it drives treatment of wastewater. Broad research has demonstrated that these chemical substrates, for example, agro wastes are oxidized by various types of microorganisms and cause power generation. Here, we have talked about some pollutants, which are employed in MFC for power generation (Table 6).

3 Fundamentals of Bioremediation with MFC

3.1 Types of Substrates Used in MFCs

Microbial communities in MFCs can function as a tool for bioremediation. In addition, they also play a significant part in regulating the natural cycles of

Table 5 List of microbial fuel cell membranes, cation exchange membrane (CEM), anion exchange membrane (AEM), bipolar membranes (BPM), and other novel materials

Membrane type	Performance	References
Ultrex^R AMI-7001 (AEM), Nafion^R 117, Ultrex^R CMI-7000 (CEM), Ultrafiltration membranes (UF-0.5 K, UF-1 K, UF-3 K)	• AEM has the best performance with the highest maximum power density (610 mW/m^2) and CE (72%) compared to other membranes • All the membranes have almost the same internal resistance (1230 ± 744 Ω till 1308 ± 718 Ω) except UF-0.5 K (6009 ± 758 Ω) • Nafion has the highest oxygen mass transfer coefficients (k_o = 1.3 × 10^{-4} cm/s)	Fornero et al. (2008)
Ultrex^R AMI 7001, Nafion^R 117, Ultrex^R CM1 7000, Hyflon^R, Zirfon^R, Nylon meshes (NY 11, NY 20, NY 41, NY 6 H), Glass fiber filter (GFAPFF, GFAP40) J-cloth, Celgard^R, SciMat^R	• Ultrex^R AMI 7001, Celgard^R, Nylon meshes (NY 20, NY 41, NY 6 H), GFAP40 and J-cloth have the lowest pH splitting extent • Porous structured GFAP40 has the highest ionic conductivity • Nylon meshes (NY11, NY20, and NY41) and Hyflon^R have the lowest ionic conductivity	Lee et al. (2010)
Fumasep^R FAB (AEM), Nafion^R 117 (CEM), Fumasep^R FBM (BPM), Charge mosaic membrane (CMM)	• AEM has the highest current density among all the membranes. • The BPM with the highest protons/hydroxyl ions transport numbers has the lowest pH increase in cathode, followed by AEM, CMM and CEM	Wang et al. (2008)
Ultrex^R AMI-7001 (AEM) and Ultrex^R CMI-7000 (CEM)	• AEM has higher performance compared to CEM with open circuit voltage and maximum power density at variations of air-cathode pressure	Zhen et al. (2015)

ecosystem including the carbon cycle and water cycle. It is important for interactions in microbial communities so as to develop an effective design of microbial fuel cell. In microbial systems, there is a stronger interaction between the bacteria, due to the coupling of the electron signal with the energetic state of the cells via the electron transport chain. Thus, alternative strategies for analysis and optimization must be undertaken, which take into account the specificity of the interactions between cells in bacterial community. In turn, the relation between biofilm development and reactor operation is of interest to venture into.

In MFCs, substrate is often considered critical to obtain desired power densities. A great variety of substrates can be used in MFCs for power generation depending upon the culture of microbe; it can be pure organics to complex mixtures of organic effluents or introduced substrates for the wastewater. In these studies, the major objective has been the efficient COD removal from the wastewater and its safe

Table 6 Bioremediation studies with natural wastewater sources and power generation using MFCs

Wastes/pollutant	Source of wastewater	Concentration	Power density	References
Hexavalent chromium	Leather industry, metallurgy	39.2 mg/L	2.4 W/m^3	Wang et al. (2008)
Agro wastes (cyanide-rich streams)	Farming, poultry processing, slaughterhouses	1086 COD-mg/L	22.19 W/m^3	Prasertsung et al. (2012)
Cellulose	Wheat straw, rice mill, corn stover	0.5 g COD/l	70 mW/m^3	Clauwaert et al. (2008)
Selenite	Glass manufacturing, electronic industry	25 mg/L	2900 mW/m^2	Cai and Zheng (2013)
Nitrate and sulfite	Fertilizer industry, animal wastes	10.5 mg/L (nitrate) 60 mg/L (sulfite)	31.92 mW/m^2	Catal et al. (2009)

discharge to the environment and bioelectricity generation in the process. Diverse kinds of substrates including single substrate such as glucose, acetate or complex substrates used from the chemical industry, pharmaceutical waste, and from dairy or food wastewater, etc., have been tested as anolyte in MFCs for generation of electricity. With varying reactor configuration, active surface area, inconsistency of methods used, it is intricate to compare the data among each other for system characterization.

The current density is commonly used unit to measure the performance of a MFC. It can be described in terms of surface area of anode (mA/m^2) or volume of the cell (current generate per unit volume, (mA/m^3)).

Some commonly used substrates include the following:

Lignocellulosic Biomass: Lignocellulosic biomass serves as a promising feedstock for cost-effective energy production in MFCs due to renewability and availability. However, cellulosic biomass cannot have direct utilization by the microbes. Conversion to monosaccharides or other compound with low molecular weight is needed.

Acetate: Acetate is a simple substrate commonly used in MFCs due to its inertness toward metabolic conversions (fermentations and methanogenesis) at a temperature. However, the use of acetate rules out the possibility of enriching a diverse microbial community and is not very practical as this contributes to energy costs.

Glucose: A major carbon source that can work as a nutrition source for diverse microbial community.

Dyes: Azo dyes are the most commonly used synthetic dye and are present in the dyes and textile-based industries. These dyes have colors which can lead to severe environmental problem and hence, their removal is of utmost importance. These dyes are not only having high toxicity but can also interrupt the interaction of light

Table 7 Performances of some commonly employed substrates in different MFC systems

Substrate	Concentration	Current density (mA/cm^2)	References
Cellulose	4 g/L	0.02	Nevin et al. (2010)
Acetate	1 g/L	0.8	Zhen et al. (2015)
Azo dye with glucose	300 mg/L	0.09	Han et al. (2017)
Glucose	6.7 mM	0.7	Catal et al. (2008)
Synthetic wastewater	16 g COD/day	0.017	Aldrovandi et al. (2009)

and oxygen with water bodies. Using these dyes as substrate is of interest among some researchers, as this can lead to color removal from the wastewater and simultaneously generate electricity.

Chemical Wastewater: Chemical wastewater can be sued as a substrate, to conduct control experiments in terms of pH, conductivity, and organic loading rate. Several chemical compounds have been used for bacterial growth in wastewater contains reduced sulfur species and redox mediators (e.g., cysteine) that can serve as an abiotic electron donor, and power production can be increased. However, such systems are not practical as energy cost is increasing with the consistent use of these chemicals (Table 7).

3.2 Various Microorganisms Employed in MFC

MFCs are devices which utilize microorganisms and oxidizes organic matter present in substrates biologically, providing electrons to the electrode. A number of substrates having different cultures such as *Pseudomonas, Geobacter, Shewanella, Clostridium,* and *Desulfuromonas* can be used in a MFC for electricity generation and hence for other benefits. The composition of these bacterial communities that can be maintained in an MFC is an area of active research along with factors like their interactions in biofilms and growth kinetics (Table 8).

Aside from the bacterial species helped by mediators to exchange electrons, a few microorganisms can exchange the electrons oxidized from natural matter to anodes and cathodes without a mediator. Among the electrochemically active microscopic organisms are *Aeromonas hydrophila, Shewanella putrefaciens, Rhodoferax,* and *Geobacter.* In MFCs, electrons can be exchanged by means of the surface of bacterial cells; microscopic organisms can exchange electrons through self-delivered mediators; electrons exchange is identified with nanowires created by microorganisms. Without nanowires, a few microorganisms create pili or nanowires on their outer film and hence can exchange their electrons. It is interesting to note

Table 8 MFC studies with pure bacteria cultures inoculated into wastewater streams with appropriate substrates

Microorganisms	Substrate	Power density (mW/m^2)	References
Shewanella putrefaciens	Glucose	355.5	Logan et al. (2006)
Shewanella oneidensis	Lactate	24	Ringeisen et al. (2006)
Escherichia coli	Complex substrate	600	Zhang et al. (2007)
Pseudomonas sp.	Peptone	979	Zhang et al. (2007)
Nocardiopsis sp. *KNU*	Carboxymethyl cellulose (CMC)	162	Ringeisen et al. (2006)

that nowadays so much research is going into nanomaterials while these microbes have been using nanowires for millions of years.

3.3 *Various Wastewater Analysis Parameters and Its Effectiveness in the MFC*

The industrial developments increase in population density and the absence of energy sources are the significant reasons which help in the development and hence improvement of renewable energy technologies in this energy era. Bioelectrochemical systems (BESs) are one of the rising practical advances which is able to generate energy from wastewater. The usage of BESs in the treatment of wastewater additionally helps to control contamination (or pollution) and hence economy of the treatment framework. Not only the electricity generation but also, bioremediation of waste, biosensing and the chemicals (hydrogen, methanol, etc.) generation are likewise significant uses of such BESs (Table 9).

Organic matter is the major component of wastewater, and its amount has been measured as BOD and COD traditionally. Some important parameters related to wastewater are mentioned below.

COD: The COD refers to the chemical oxygen demand, an analysis that measures chemical oxidation of the majority of organic matter present in the wastewater sample. COD measurements are needed for mass balances in wastewater treatment.

BOD: Refers to biochemical oxygen demand. The BOD analysis measures the oxygen used by microbes for oxidation part of the organic matter. The standard BOD analysis takes 5 days (BOD5), but alternatives used are BOD1 and BOD7, respectively. Starting with very high values of BOD and COD's of wastewater the values obtained after putting in a MFC has seen significant reduction even up to more than 60%.

Table 9 Wastewater parameters characteristics, testing these parameters before and after use from MFC gives the effectiveness of bioremediation (Pant et al. 2010)

General parameters		Solvents and alcohols (mg/L)	Volatile fatty acids (mg/L)
pH, TS (kg/m^3), TSS (kg/m^3)	Inorganic compounds	Acetone	Acetate
Conductivity (mS/cm)	Phosphorous (mg/L)	Methanol	Propionate
TCOD (kg/m^3)	Sulfate (mg/L)	Ethanol	Butyrate
SCOD (kg/m^3)	Nitrate (mg/L)	Propanol	–
BOD (kg/m^3)	Nitrogen Ammonia (mg/L)	Butanol	–

TSS: Total soluble solids or **TSS** is the amount of total suspended solids in the sample. In other words, the organic and inorganic materials are in the form of suspension in the wastewater. The TSS is generally not estimated while using a MFC, but it can become a factor to be evaluated as to see effectiveness of MFC's.

FOG: FOG stands for fats, oil, and grease and is used to determine the hydrocarbon components of the wastewater sample.

Bacteriological Analysis: This is done to characterize the bacteria present in the waste sample so as to get the virulent strain which is responsible for power generation. Generally, mixed cultures are more power generating than pure cultures. Such analysis can be done using real-time polymerase chain reaction (RT-PCR).

3.4 Oxidizing and Reducing Agents in the Anode and Cathode

In most of the MFCs, electrons reach the cathode and combine with proton which diffuses from the anode side through a salt bridge or membrane to cathode side and the oxygen is provided from air hence obtained product is water. Chemical oxidizers, for example ferricyanide or Mn (IV), can likewise be utilized despite the fact that these must be replaced or recovered. On account of metal particles, for example, Mn that is diminished from Mn (IV) to Mn (II), microscopic organisms can catalyze the re-oxidation of the metal utilizing dissolved oxygen.

Anode: The simplest materials for anode electrodes are graphite plates or rods, as they are relatively inexpensive, easy to handle and have a defined surface area. To increase the anode performance, different chemical and physical strategies have been followed. One study used fused Mn (IV) and Fe (III) and utilized covalently connected neutral red to intercede the electron exchange to the anode. Electrocatalytic materials, for example polyanilins/Pt composites, have likewise been appeared to enhance the present era through helping the immediate oxidation of microbial metabolites.

Cathode: Due to its good performance, ferricyanide (K_3 [Fe $(CN)_6$]) is very popular as an experimental electron acceptor in microbial fuel cells. The best advantage of ferricyanide is the low over-potential utilizing a plain carbon cathode, bringing about a cathode working potential near its open circuit potential. The best hindrance, in any case, is the inadequate re-oxidation by oxygen, which requires the catholyte to be consistently regenerated. Moreover, the long-term performance of the system can be affected by diffusion of ferricyanide across the cation exchange membrane (CEM) and into the anode chamber.

Oxygen is the most appropriate electron acceptor for a MFC because of its high oxidation potential, accessibility, ease (it is free), supportability, and the absence of a substance waste item (water is shaped as the main finished result).

4 Fundamentals of Bioelectricity Generation with MFC

4.1 Basics of Voltage Generation

Microbial power modules (MFCs) are gadgets that utilize microscopic organisms as the catalysts to oxidize natural and inorganic matters and produce current. Electrons created by the microscopic organisms from these substrates are exchanged to the anode (negative terminal) and stream to the cathode (positive terminal) connected by a conductive material containing a resistor, or worked under a load (i.e., creating power that runs a gadget). By tradition, positive current streams from the positive to the negative terminal, inverse to that of electron stream. Hence, voltage is generated by means of this transfer of electrons and protons.

4.2 Thermodynamics and Electromotive Force

Power is produced in a MFC just if the general response is thermodynamically great or favorable. The response can be assessed as far as Gibbs free energy communicated in units of Joules (J), which is a measure of the maximum work that can be obtained from the response or reaction is calculated as:

$$\Phi(\mathrm{G}_r) = \Phi(\mathrm{G}_r^0) + \mathrm{RT}\ln(Q)$$

where

Φ (G_r)	is the Gibbs free energy for the specific given conditions.
Φ (G_r^0)	is the Gibbs free energy under standard conditions usually defined as 298.15.
K, 1	bar pressure, and 1 M concentration for all species.
R (8.314 J mol^{-1} K^{-1})	is the universal gas constant.

T (K) is the temperature in absolute scale, and
Q (dimensionless) is the reaction quotient calculated as the activities of the products divided by those of the reactants.

The standard reaction Gibbs free energy is generally calculated from the energies of formation of organic compounds in water, available from many sources. For MFC computations, it is more advantageous to assess the response as far as the general cell electromotive force (emf), E_{emf} (V), characterized as the potential contrast between the cathode and anode. This is identified with the work, W (J), created by the cell as:

$$W = E_{emf} Q = -\Phi(G_r)$$

where Q = nF is the charge transferred in the reaction, expressed in Coulomb (C), determined by the number of electrons transferred in the reaction, n is the number of electrons per reaction mol, and F is Faraday's constant (96,485.3 C/mol).

Even the use of Nernst equation is more effective compared to the above ones. The Nernst equation is

$$E_{emf} = E^0_{emf} - (RT/nF)\ln(K)$$

4.3 Standard Electrode Potentials and Efficiencies

Standard Electrode Potentials: The responses happening in a MFC can be broken down as the half-cell responses, or the different responses happening at the anode and the cathode. As per the IUPAC convention, standard conditions (at 298 K, 1 bar, 1 M) are accounted for as a reduction potential, i.e., the response is composed as consuming electrons (Table 10).

For example, $O_2 + 4H^+ + 4e^- \rightarrow 2H_2O$, a reduction reaction hence will take place at cathode

Table 10 Half-cell potentials against SHE for MFCs

Electrode	Reaction	Eo (V)	Conditions	E_{MFC}(V)
Anode	$2HCO^{3-} + 9H^+ + 8e^- \rightarrow CH_3COO^- + 4H_2O$	0.187[a]	HCO^{3-} = 5 mM, CH_3COO^- = 5 mM, pH = 7	−0.296
Cathode	$MnO_2(s) + 4H^+ + 2e^- \rightarrow Mn^{2+} + 2H_2O$	1.23	$[Mn2^+]$ = 5 mM, pH = 7	0.47
	$Fe(CN)_6^{3-} + e^- \rightarrow Fe(CN)_6^{4-}$	0.31	$[Fe(CN)_6^{3-}] = [Fe(CN)_6^{4-}]$	0.361

Therefore,

$$E_{cat} = E^{O}_{cat} - (RT/4F)\ln\left\{1/(p_{O_2})*[H^+]^4\right\}$$

The cell emf is calculated as

$$E_{cell} = E_{cat} - E_{an}$$

where the minus sign is a result of the meaning of the anode potential as reduction response (despite the fact that an oxidation response is happening). The power delivered by a MFC hence relies on upon the decision of the cathode, and this ought to be considered when looking at power densities accomplished by various MFCs. The conditions for the reaction at cathode and anode side should be same; otherwise, different levels of power output will be obtained.

Efficiencies: MFC efficiency depends on the material for electrodes, performance of the membrane, the ohmic and non-ohmic losses, and a favorable thermodynamic reaction. Some important parameters in this context are as follows:

Treatment Efficiency: MFCs have been proposed as a technique to treat wastewater, and along these lines, it is imperative to assess the general performance using biochemical oxygen demand (BOD), chemical oxygen demand (COD), or total organic carbon (TOC) removal. Different elements may likewise be imperative, for example, solvent versus particulate removal, and supplement, or nutrient removal. The choice of treating efficiency is arbitrary but mostly used is COD removal efficiency and can be found out by the ratio between the removed and influent COD. This parameter measures the amount of the accessible "fuel" that has been changed over in the MFC, either into electrical current (by means of the Coulombic efficiency) or biomass (by means of the development yield) or through reaction with electron acceptors (e.g., oxygen, nitrate, and sulfate).

Coulombic Efficiency: The Coulombic efficiency is defined as the ratio of total Coulombs actually transferred to the anode from the substrate to maximum possible Coulombs if all substrate removal produced current. The total Coulombs obtained (Logan et al. 2005) is as:

$$\varepsilon_{Cb} = \frac{M\int_0^{t_b} I\,dt}{Fb_{v_{an}}\Delta COD}$$

where $M = 32$, the molecular weight of oxygen, F is Faraday's constant, $b = 4$ is the number of electrons exchanged per mole of oxygen, v_{an} is the volume of liquid in the anode compartment, and ¢COD is the change in COD over time t_b.

For continuous flow through the system, the Coulombic efficiency is as follows:

$$\varepsilon_{Cb} = \frac{MI}{Fbq\Delta COD}$$

where q is the volumetric influent flow rate, and ¢COD is the difference in the influent and effluent COD. Factors that reduce Coulombic efficiency are competitive processes and bacterial growth. Bacteria unable to utilize the electrode as electron acceptor are likely to use substrate for fermentation and/or methanogenesis.

4.4 Polarization Curves and Open Circuit Voltage

4.4.1 Polarization Curves

A polarization curve describes a relationship between voltage and current. A potentiometer is used to record polarization curve for anode, cathode, and for entire MFC using a potentiometer. Thus, voltage is measured and the current is calculated using the Ohm's law. The polarization curve can be categorically divided into three zones (Fig. 7):

Power Curves: A power curve (or power density curve) is a derivative of the polarization curve, as it is computed from the polarization curve. A power curve depicts power density as a function of current density. Power density for open circuit is zero. Maximum power point (MPP) is position where the power density and current reach the apex value. Beyond the MPP, the power drops due to increase in ohmic resistance and decrease in power production.

Open Circuit Voltage (OCV): Cell voltage that is measured in the absence of current or resistor is known as the open circuit voltage. Ideally, the OCV should approach the cell EMF (Fig. 8).

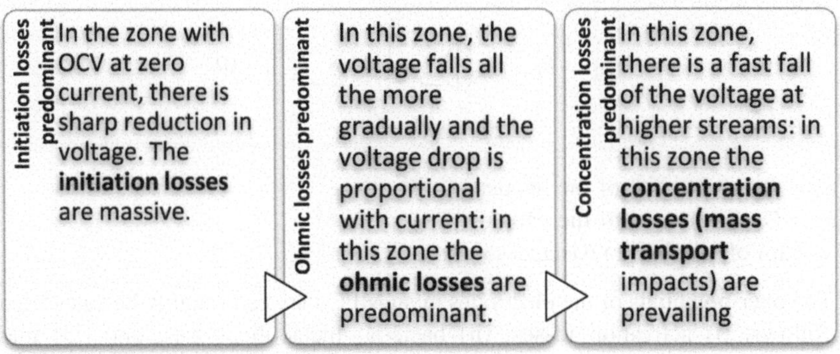

Fig. 7 Overview of losses associated with MFC systems

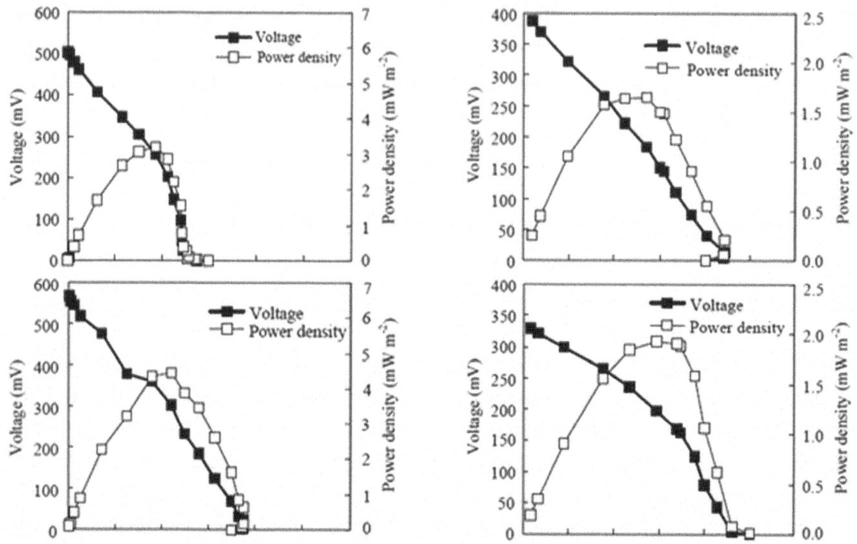

Fig. 8 Polarization and power curves for MFC performance (Lee et al. 2010)

4.5 Types of Losses in MFC

The maximum attainable MFC voltage (emf) is theoretically on the order of 1.1 V (see above). However, the measured MFC voltage is considerably lower due to a number of losses.

The difference between the measured cell voltage and the cell emf is known as overvoltage, and it is the sum of the over-potentials of the anode and the cathode and the ohmic loss of the system.

$$E_{\text{cell}} = E_{\text{emf}} - \left(\left|\sum \eta_{\text{a}}\right| + \left|\sum \eta_{\text{c}}\right|\right) + \text{IR}$$

$\sum \eta_{\text{a}}$ = Over-potential of the anode
$\sum \eta_{\text{c}}$ = Over-potential of the cathode
IR = Sum of all ohmic resistances

The over-potentials of the electrodes in an MFC they can roughly be categorized as follows: (i) activation losses; (ii) bacterial metabolic losses; and (iii) mass transport or concentration losses. Some of the major losses in MFCs are described in the following page.

Ohmic Losses: The ohmic losses (or ohmic polarization) in an MFC occur due to the resistance to electron flow through the membrane electrode assembly (MEA) and the resistance to the flow of ions through the CEM. Ohmic losses can be lessened by limiting the electrode spacing, also by utilizing a membrane that has low resistivity, checking altogether all contacts, and solution conductivity which are at the upper limit tolerated by the bacteria.

Activation Losses: Activation loss (also called activation polarization) occurs whenever a reaction occurs on the electrode surface. This energy loss is due to the activation energy required for an oxidation/reduction reaction to occur. Activation losses increase substantially at lower currents and gradually with increase in the current density. An effective way to lower these losses is to increase the electrode surface area and operating temperature.

Concentration Losses: Loss in concentration (or concentration polarization) happens when the rate of mass transport of a species to or from as far as possible limits current generation. Such happen for the most part at high current densities because of confined mass exchange of synthetic species by diffusion to the cathode surface while at anode, the concentration losses are brought on by constrained supply of diminished species toward the terminal or by restricted release of oxidized species from the cathode surface. This surges the proportion between the oxidized and the lessened species at the terminal surface which can deliver an expansion in the anode potential. At the cathode side, a drop in cathode potential may happen. Mass transport restrictions in the mass liquid can restrain the substrate flux to the biofilm, which is a different kind of concentration loss.

Microbial Losses: In a MFC, microscopic organisms transport electrons from a substrate at a low potential to the last electron acceptor at a higher potential through an outer circuit. In a MFC, the anode is the last electron acceptor and its potential decides the vitality pick up for the microscopic organisms. The more prominent is the distinction between the redox capability of the substrate and the anode potential, the higher is the metabolic vitality pick up for the microbes, yet the lower is the most extreme achievable MFC voltage. To boost the MFC voltage, the capability of the anode ought to be kept as low (negative) as could reasonably be expected. Be that as it may, if the anode potential turns out to be too low, electron transport will be repressed and maturation of the substrate (if conceivable) may give more prominent vitality to the microorganisms.

5 Conclusion

The success of any technology depends upon how it influences the energy market and the perception of the common public since MFCs can produce electricity while removing pollutants and other organic matter from wastewater streams, it can be speculated to offer advantages such as:

1. Low-cost electricity production from everyday waste materials.
2. Bioelectricity will be available all year around, as waste and xenobiotics are readily available.
3. Every household can produce a certain amount of electricity.

4. MFCs can be critical for nations in the African continent where in some places, the energy infrastructure has still not arrived.
5. Working with parallel with bioremediation and generate bioelectricity, making the process sustainable (Fig. 9).

Thus, MFCs represent powerful predictive tools, which will aid the design of systems exploiting bacterial capabilities. In MFC systems, chemicals are reduced at the cathode, and in some systems, it is possible to achieve chemical oxidation at the anode in situations when high concentration of biodegradable organics is present in the wastewater. For this to work, however, sufficient electron acceptors should be present at the cathode. For example, if a site is contaminated with petroleum or gasoline, the water can be channeled through consecutive hydraulic chambers similar to that used for zero-valent iron walls for treating chlorinated aliphatics in groundwater. First section should have the anode, with material of construction such as graphite granules, where the chemical will be oxidized (assuming anaerobic conditions) on the anode providing current to the cathode. The second section would contain a cathode, tube cathodes can be considered, and where oxygen will have the advantage of providing the additional electron acceptor into water to allow for either continued treatment or to increase the concentration of oxygen in the groundwater; this concept was explored with an air-cathode MFC for petroleum-contaminated groundwater. Power generation was 120 mW/m^2 which suggests MFC systems similar to this can be used to enhance bioremediation of petroleum-contaminated groundwater under anaerobic conditions (Fig. 10).

MFCs are being developed utilizing an assortment of materials in a continually expanding quality of designs. These frameworks are worked under a scope of conditions that incorporate contrasts in temperature, electron acceptor, anodic surface area, pH, operation time, and reactor sizing. In some cases, the operating conditions and even electrical components, such as internal resistance, power density, are missing, such has made it difficult to analyze and interpret results among similar systems. However, the list of accomplishments in our understanding of how electricity is produced in an MFC and how effective systems cost with increased power density are impressive. Precious metals, such as platinum, are no

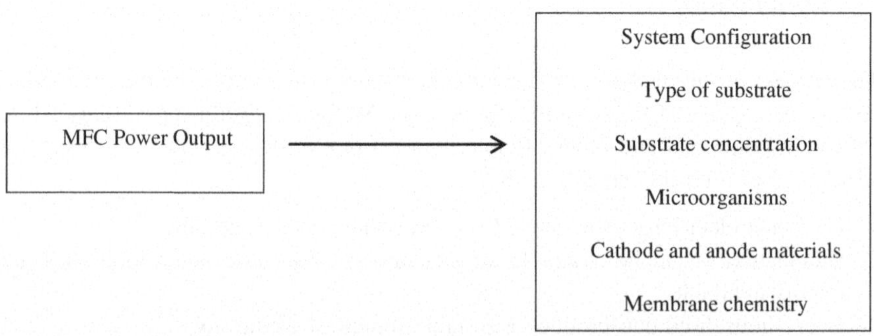

Fig. 9 Factors important for MFC performance and critical for commercial success

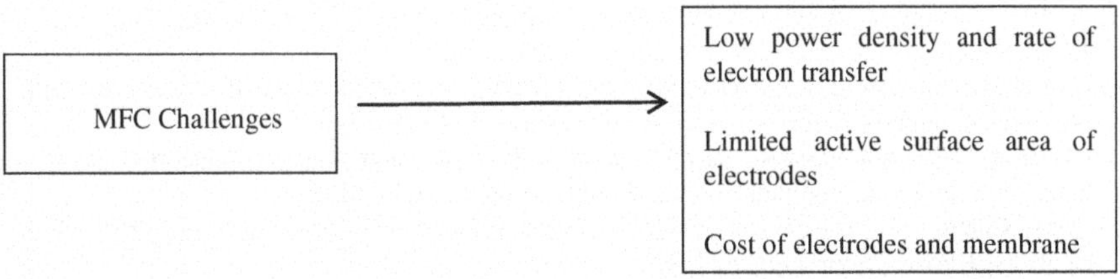

Fig. 10 Limitations associated with MFCs addressing which are a key to commercial success

longer needed on either electrodes, and use of non-precious metals such as cobalt, iron has produced similar power densities when coupled with suitable oxygen reducing agents.

The scope of MFC research is not just restricted to wastewater treatment as modified MFCs called BEAMRs can be used to achieve bio-hydrogen from any biodegradable matter, and such systems have shown potentials to cross the "fermentation barrier" with maximum possible conversion efficiency. For MFCs, the ultimate achievement will be when they can be solely used as a method of renewable energy production; right now, it might face challenges to grow in the shadow of large fossil fuel industries, but advances in power densities, reductions in material costs, and a global need to produce power from non-CO_2 sources will make MFCs practical for electricity production. The progress of such remains in the hands of researchers, who believe that MFC technologies are a part of the bright and promising future made on the foundation of a new generation of electrogenic reactor systems.

References

Aelterman P, Rabaey K, Pham TH, Boon N, Verstraete W (2006) Continuous electricity generation at high voltages and currents using stacked microbial fuel cells. Environ Sci Technol 40:3388–3394

Aldrovandi A et al (2009) Sustainable power production in a membrane-less and mediator-less synthetic wastewater microbial fuel cell. Biores Technol 100:3252–3260

Bard AJ, Parsons R, Jordan J (eds) (1985) Standard potentials in aqueous solution. Marcel Dekker, New York

Cai J, Zheng P (2013) Simultaneous anaerobic sulfide and nitrate removal in microbial fuel cell. Biores Technol 128:760–764

Call D, Merrill MD, Logan BE (2009) High surface area stainless steel brushes as cathodes in microbial electrolysis cells (MECs). Environ Sci Technol 43:2179–2183

Catal T et al (2008) Electricity production from twelve monosaccharides using microbial fuel cells. J Power Sour 175:196–200

Catal T, Bermek H, Liu H (2009) Removal of selenite from wastewater using microbial fuel cells. Biotechnol Lett 31(8):1211–1216

Clauwaert P, van der Ha D, Verstraete W (2008) Energy recovery from energy rich vegetable products with microbial fuel cells. Biotechnol Lett 30(11):1947–1951

Davis JB, Yarbrough HF (1962) Preliminary experiments on a microbial fuel cell. Science 137:615–616
Fornero JJ, Rosenbaum M, Cotta MA, Angenent LT (2008) Microbial fuel cell performance with a pressurized cathode chamber. Environ Sci Technol 42:8578–8584
Han W et al (2017) Simultaneous dark fermentative hydrogen and ethanol production from waste bread in a mixed packed tank reactor. J Clean Prod 141:608–611
HaoYu E, Cheng S, Scott K, Logan B (2007) Microbial fuel cell performance with non-Pt cathode catalysts. J Power Sour 171(2):275–281
Lee et al (2010) Bioelectricity generation and organic removal in microbial fuel cells used for treatment of wastewater from fish-market. J. Environ Eng Manage 20(3):173–180
Liu H, Logan BE (2004) Electricity generation using an air-cathode single chamber microbial fuel cell in the presence and absence of a proton exchange membrane. Environ Sci Technol 38:4040–4046
Logan BE, Murano C, Scott K, Gray ND, Head IM (2005) Electricity generation from cysteine in a microbial fuel cell. Water Res 39:942–952
Logan BE, Rabaey K et al (2006) Microbial fuel cells: methodology and technology. Environ Sci Technol 40(17):2006
Logan BE, Wallack MJ, Kim KY, He W, Feng Y, Saikaly PE (2015) Assessment of microbial fuel cell configurations and power densities. Env Sci Tech Lett 2(8):206–214
Nevin KP, Woodard TL, Franks AE, Summers AM, Lovley DR (2010) Microbial electro synthesis: feeding microbes electricity to convert carbon dioxide and water to multicarbon extracellular organic compounds. mBio 1: e00103-10
Pant D, Van Bogaert G, Diels L, Vanbroekhoven K (2010) A review of the substrates used in microbial fuel cells (MFCs) for sustainable energy production. Bioresour Technol 101:1533–1543
Prasertsung N, Reungsang A, Ratanatamskul C (2012) Alkalinity of cassava wastewater ffed in anodic enhance electricity generation by a single chamber microbial fuel cells. Engr J 16 (5):17–28
Rabaey K, Boon N, Siciliano SD, Verhaege M, Verstraete W (2004) Biofuel cells select for microbial consortia that self-mediate electron transfer. Appl Environ Microbiol 70:5373–5382
Rabaey K, Boon N, Hofte M, Verstraete W (2005) Microbial phenazine production enhances electron transfer in biofuel cells. Environ Sci Technol 39:3401–3408
Ringeisen BR, Henderson E, Wu PK, Pietron J, Ray R, Little B, Biffinger JC, Jones-Meehan JM (2006) High power density from a miniature microbial fuel cell using *Shewanella oneidensis* DSP10. Environ Sci Technol 40(8):2629–2634
Wang G, Huang LP, Zhang YF (2008) Cathodic reduction of hexavalent chromium [Cr(VI)] coupled with electricity generation in microbial fuel cells. Biotechnol Lett 30:1959–1966
Zhang X, Jantama K, Moore JC, Shanmugam KT, Ingram LO (2007) Production of L-alanine by metabolically engineered Escherichia coli. Appl Microbiol Biotechnol 77(2):355–366
Zhen G, Kobayashi T, Lu X, Xu K (2015) Understanding methane bioelectrosynthesis from carbon dioxide in a two-chamber microbial electrolysis cells (MECs) containing a carbon biocathode. Bioresour Technol 186:141–148

Author Biographies

Debajyoti Bose has a bachelor degree in Chemical Engineering and a master degree in Renewable Energy Engineering, his area of interest include biofuels, bioremediation, and other waste to energy practices, presently he is a research fellow at the University of Petroleum & Energy Studies, Dehradun, UK, India.

Vaibhaw Kandpal is currently pursuing his bachelor degree "Chemical Engineering with specialization in Refining & Petrochemicals" from University of Petroleum & Energy Studies, Dehradun, UK, India. He will be graduating on June 2018.

Himanshi Dhawan is currently pursuing her bachelor degree "Chemical Engineering with specialization in Refining & Petrochemicals" from University of Petroleum & Energy Studies, Dehradun, UK, India. She will be graduating on June 2018.

Dr. P. Vijay is currently the Associate Professor and Head, Department of Chemical Engineering at University of Petroleum and Energy Studies. He has a PhD in Chemical Engineering from IIT Madras and has more than twelve years of teaching, research and industrial experience. He is also an active member in IIChE and other various review committee.

Dr. M. Gopinath is presently the Assistant Professor (Sel. Grade) at University of Petroleum & Energy Studies, he has a PhD in Biochemical Engineering and has more than ten years of teaching and research experience. He is an active member of various bodies such as IIChE, BRSI, CAPS and other various review committee.

Chapter 3
Bioprocesses for Sulphate Removal from Wastewater

Luis C. Reyes-Alvarado, Eldon R. Rene, Giovanni Esposito and Piet N. L. Lens

Abstract This chapter highlights and discusses the important research studies that have been carried out previously on sulphate removal from wastewaters under anaerobic conditions. Moreover, the role of electron donor addition on biological sulphate reduction and the beneficial role of sulphate-reducing bacteria (SRB) are reviewed in this chapter. This chapter describes the fundamentals of the anaerobic sulphate reduction process, the factors affecting biological sulphate reduction and the different bioreactor configurations used for sulphate removal from wastewater.

Keywords Biological sulphate reduction · Sulphidogenesis · Bioreactors Electron donors · Sulphate-reducing bacteria · Sulphate-rich wastewater

1 The Sulphate Reduction Process

1.1 Sulphur Cycle

The tenth most abundant element on the surface of the earth is sulphur. Microorganisms require sulphur for the processing of vitamins, amino acids and hormones. Microbes play a major role in the biogeochemical sulphur cycle. The sulphur oxidation states are −2 (sulphide), 0 (elemental sulphur), +4 (sulphite) and +6 (sulphate), among which sulphate is very important for nature. Sulphur biogeochemical pathways are known to interact with those of other elements, especially metals (Pepper et al. 2004).

L. C. Reyes-Alvarado · E. R. Rene (✉) · P. N. L. Lens
UNESCO-IHE Institute for Water Education, P.O. Box 3015,
2601 DA Delft, The Netherlands
e-mail: e.raj@un-ihe.org

G. Esposito
Department of Mechanics, Structures and Environmental Engineering,
University of Cassino, via Di Biasio 43, 03043 Cassino, FR, Italy

S. J. Varjani et al. (eds.), *Waste Bioremediation*, Energy, Environment, and Sustainability, https://doi.org/10.1007/978-981-10-7413-4_3

Wastewaters with high sulphate concentrations are produced through leaching from landfills (Nedwell and Reynolds 1996), and this leaching process can cause an unbalance to the natural biological sulphur cycle by altering the biodegradation pathways and the kinetic rates. A different oxidation state (−2 up to +6) of sulphur can be found in other sources such as wastewater from the textile industry. Cirik et al. (2013) reported that sulphate in textile industries is added to dye baths for regulating the ionic strength. Deep sea venting, volcanic activity, bacterial activities, fossil fuel combustion and industrial emissions are some of the major sources of sulphate in the atmosphere. Sulphate oxidized from sulphur in the atmosphere can be deposited in wet or dry form.

Redox reactions generally characterize the sulphur cycle as shown in Fig. 1. Sulphur can be reduced to sulphide, which in turn can be oxidized to elemental sulphur or sulphate by microbes. However, sulphate can be reduced back to sulphide by sulphate-reducing bacteria (SRB) (Robertson and Kuenen 2006). The release of sulphur aerobically or anaerobically from its organic form is known as sulphur mineralization.

Chemoautotrophic and heterotrophic microorganisms, including bacteria and fungi under aerobic conditions oxidize sulphur to sulphate or thiosulphate. Phototropic or chemolithotrophic bacteria fix carbon dioxide (CO_2) by utilizing light energy in order to oxidize sulphide to sulphur or sulphate. When there is an imbalance in the reductive or oxidative pathways, an accumulation of intermediates such as elemental sulphur, iron sulphide or hydrogen sulphide occurs. The process of sulphur disproportionation is an energy-generating process carried out by SRB, wherein elemental sulphur or thiosulphate acts as both the electron donor and the electron acceptor and results in the formation of sulphate and sulphide, respectively (Tang et al. 2009).

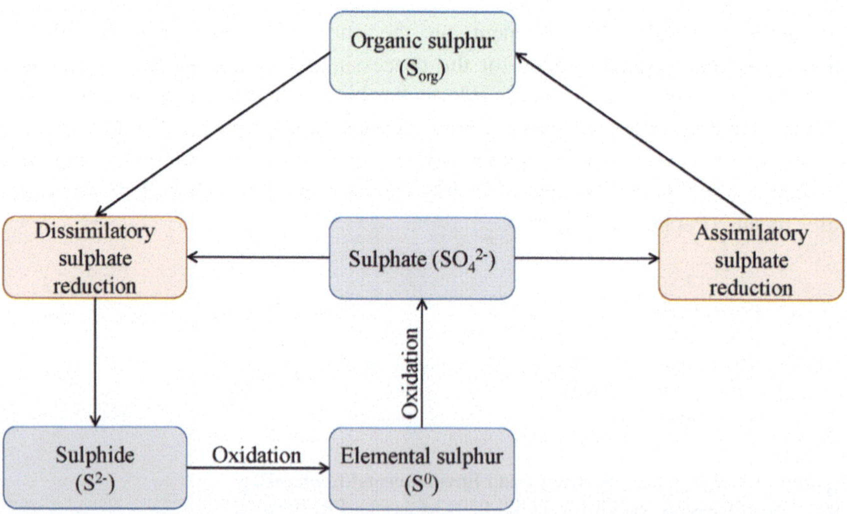

Fig. 1 Biological sulphur cycle

1.2 Biological Sulphate Reduction

Two anaerobic microbial degradation pathways (Fig. 2) are well documented in the literature. The sulphate removal can be assimilatory or dissimilatory (Fig. 3). In the assimilatory pathway, sulphate is reduced to sulphide, in small quantities, and later the sulphide is converted to cysteine. This amino acid is the source of other biological sulphur-containing molecules (Leustek et al. 2000). The dissimilatory pathway is confined to two archaeal and five bacterial genera, wherein the terminal electron acceptor (sulphate) produces large quantities of sulphide and the process is also known as sulphidogenesis (Grein et al. 2013). The two pathways have a similar starting point: the activation of sulphate by reaction with adenosine-5′-Triphosphate (ATP) forming adenosine-5′-phosphosulphate (APS). Sulphate adenylyl transferase (SAT) acts as a catalyst in this step, also referred to as ATP sulphurylase (Taguchi et al. 2004).

Sulphate is reduced using an electron donor to produce sulphide and SRB are responsible for this process (Hao et al. 1996). Table 1 summarizes sulphate reduction (Eqs. 1–14) using electron donors such as lactate, propionate, acetate and hydrogen.

In the first step of sulphate reduction, the exogenous sulphate is transferred through the bacterial cell membrane into the cell. In the next step, the ATP sulphurylase facilitates the sulphidogenesis process in the cell membrane (Fig. 3). In the presence of sulphate, the highly activated molecule APS, and pyrophosphate

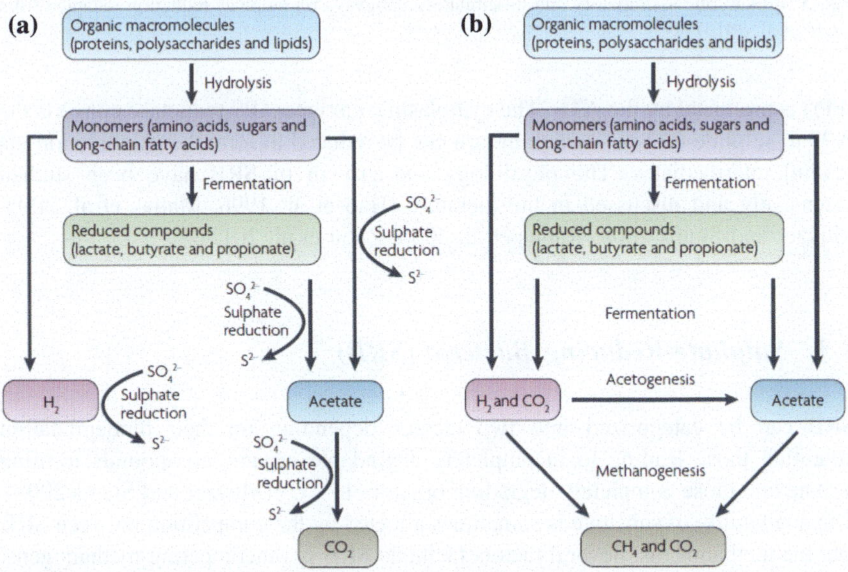

Fig. 2 Pathway for anaerobic degradation of organic substrates: **a** sulphidogenesis and **b** methanogenesis (adapted from Muyzer and Stams 2008)

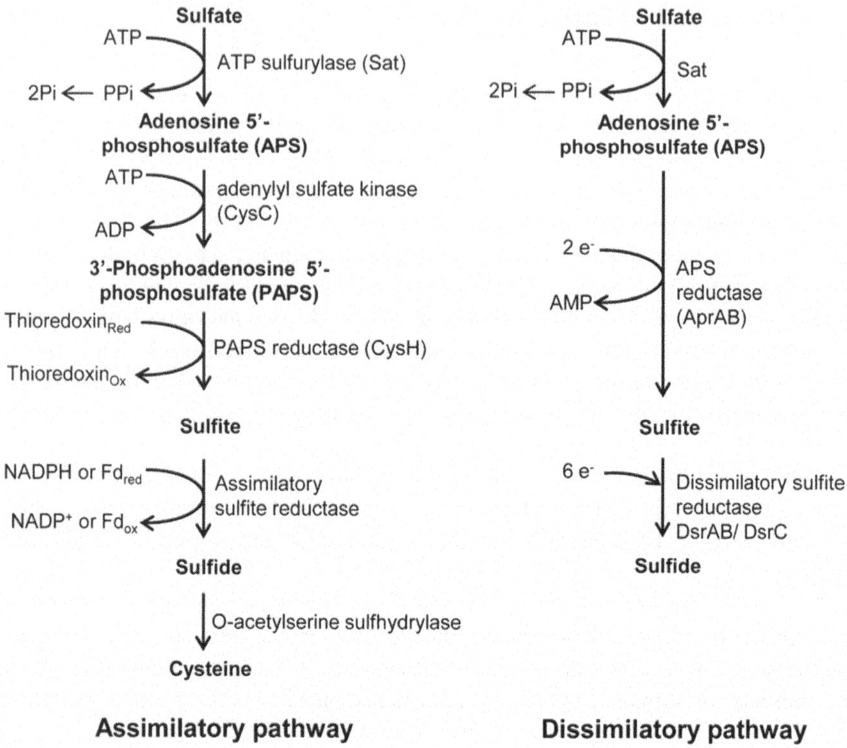

Fig. 3 Prokaryotic assimilatory and dissimilatory pathways for sulphate reduction (adapted from Grein et al. 2013)

(PPi) is produced by the ATP. The cytoplasmic enzyme APS reductase converts the APS to sulphite (SO_3^{2-}), which in turn can be reduced to form the sulphide ion via several intermediates. The physiology and growth of SRB have been studied extensively and discussed in the literature (Hao et al. 1996; Matias et al. 2005; Muyzer and Stams 2008; Rabus et al. 2006; Zhou et al. 2011).

1.3 Sulphate-Reducing Bacteria (SRB)

SRB can be categorized into two classes depending on their biodegradation potential: those leading to incompletely degradable organic compounds forming acetate and those completely degrading organics to CO_2 (Muyzer and Stams 2008). The availability of substrate is sometimes affected by the competition between SRB and the methanogens. Several factors facilitate SRB in outcompeting methanogens. These factors include anaerobic respiration in the presence of sulphate as the final electron acceptor leading to more energy for the growth of SRB and at conditions

Table 1 Stoichiometric reactions involved in sulphidogenesis

Sulphidogenesis	$\Delta G^{0\prime}$ (kJ. reaction^{-1})	
Glucose + SO_4^{2-} $\rightarrow$ 2 CH_3COO^- + HS^- + 2 HCO_3^- + 3 H^+	−358.2	(1)
2 $CH_3CHOHCOO^-$ + SO_4^{2-} $\rightarrow$ 2 CH_3COO^- + HS^- + 2 HCO_3^- + H^+	−160.1	(2)
2 $CH_3CHOHCOO^-$ + 3 SO_4^{2-} $\rightarrow$ 6 HCO_3^- + HS^- + H^+	−255.3	(3)
$CH_3(CH_2)_2COO^-$ + 0.5 SO_4^{2-} $\rightarrow$ 2 CH_3COO^- + 0.5 HS^- + 0.5 H^+	−27.8	(4)
$CH_3(CH_2)_2COO^-$ + 3 SO_4^{2-} + 2 H_2 $\rightarrow$ CH_3COO^- + HS^- + HCO_3^- + 2 H_2O	−198.4	(5)
$CH_3CH_2COO^-$ + 0.75 SO_4^{2-} $\rightarrow$ CH_3COO^- + HCO_3^- + 0.75 HS^- + 0.25 H^+	−37.7	(6)
$CH_3CH_2COO^-$ + 1.75 SO_4^{2-} $\rightarrow$ 3 HCO_3^- + 1.75 HS^- + 0.25 H^+	−85.4	(7)
$CH_3CH_2COO^-$ + SO_4^{2-} + H_2 $\rightarrow$ CH_3COO^- + HS^- + HCO_3^- + 2 H_2O	−75.8	(8)
2 CH_3CH_2OH + SO_4^{2-} $\rightarrow$ 2 CH_3COO^- + HS^- + H^+ +H_2O	−22	(9)
2 CH_3OH + SO_4^{2-} $\rightarrow$ 2 $HCOO^-$ + HS^- + H^+ + 2 H_2O	−108.3	(10)
CH_3COO^- + SO_4^{2-} $\rightarrow$ 2 HCO_3^- + HS^-	−48	(11)
$HCOO^-$ + SO_4^{2-} + H^{-+} $\rightarrow$ HS^- + 4 HCO_3^-	−144	(12)
4 CO + SO_4^{2-} + 4 H_2O $\rightarrow$ HS^- + 4 HCO_3^- + 3 H^+	−212	(13)
4 H_2 + SO_4^{2-} + H^+ $\rightarrow$ HS^- + 4 H_2O	−151.9	(14)

difficult for methanogens to survive. In addition, SRB are able to consume substrates to very low concentrations because of their high affinity for hydrogen and acetate (Rabus et al. 2006). It is noteworthy to mention that SRB have a higher specific growth rate compared to methanogens (Moestedt et al. 2013). *Desulfovibrio* species has a high affinity for hydrogen and this is contemplated to be the rational for outcompeting hydrogenotrophic methanogens under sulphidogenic conditions (Widdel 2006). *Desulfobacter*, *Desulfobulbus*, *Desulfococcus*, *Desulfocarcina*, *Desulfomaculum*, *Desulfonema* and *Desulfovibro* use sulphate as the terminal electron acceptor, using acetate, lactate and methanol as the electron donor (Pepper et al. 2004). Polymeric compounds such as protein, starch and cellulose are not utilized directly by SRB as the substrates, but they depend on other microorganisms to ferment these compounds to products which can be used as substrates by the SRB.

An analysis of 16S ribosomal ribonucleic acid (rRNA) by Muyzer and Stams (2008) grouped SRB into seven different lineages, two of which were archaea and five were bacteria (Fig. 4). From an application viewpoint, the SRB are classified by their oxygen sensitivity and their affinity towards using sulphate as an electron acceptor. Besides, different species within the same genus can also show great differences in their selectivity towards a particular type of electron donor. For example, the genus *Desulfotomaculum* can use acetone, aniline, acetate, catechol, ethanol, indole and other carbon sources for its growth. The genus *Desulfovibrio* uses lactate, pyruvate and hydrogen as electron donors for its growth. Thus, depending on the species, different organic substrates can be used by the SRB, leading to acetate or CO_2 as the end product.

The ecology, bioenergetics and physiology of SRB have been discussed in a number of review articles (Barton and Fauque 2009; Gibson 1990). SRB are known to exist in different environments such as: anoxic estuarine sediment, acid mine water, saline water, freshwater and in several soil types. The temperature range at which they grow is also diverse, and thermophilic SRB have been isolated at temperatures >60 °C in deep aquifers (Hao et al. 1996). According to Mizuno et al. (1998), hydrogen-consuming and lactate-consuming SRB can be enumerated using

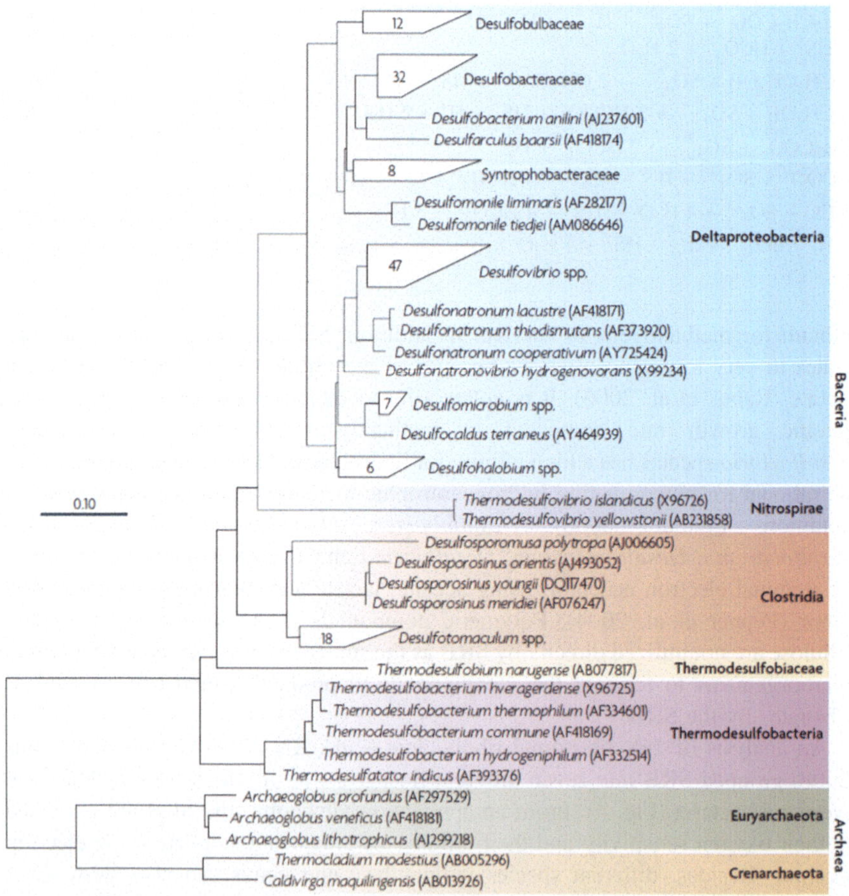

Fig. 4 Phylogenetic tree on 16S ribosomal RNA (rRNA) sequence of SRB species (adapted from Muyzer and Stams 2008)

the most probable number (MPN) counting technique, while qualitatively, the presence of SRB can be confirmed by the presence of black FeS precipitates.

2 Electron Donors for SRB

Al-Zuhair et al. (2008) determined the appropriate sulphate concentration required for pure cultures of SRB and their sulphate reduction capacities. As shown in Table 2, several electron donors have been studied as energy and carbon sources for the growth of SRB. The commonly used electron donors for biological sulphate removal are products from fermentation, monomers or cell components from other sources. As shown in several studies, hydrogen is one of the most important substrates for SRB. According to Widdel (2006), *Desulfovibrio* species has a high affinity for hydrogen, and this property facilitates them to outcompete hydrogenotrophic methanogens in sulphate-rich environments.

Furthermore, SRB require nitrogen, phosphorus and iron. Complex organic carbon sources can also serve as energy sources for SRB. Sewage sludge was one of the first carbon sources considered because of its complexity (Butlin et al. 1956). van Houten et al. (1996) studied the use of synthesis gas (H_2, CO and CO_2 mixture) as energy source, and a stimulation of biological sulphate reduction was observed in the presence of SRB. A wide variety of commonly available organic waste matrices such as mushroom, leaf mulch, wood chips, sewage sludge, sawdust, compost, animal manure, whey, vegetable compost and other agricultural waste has also been used as carbon sources and electron donors for biological sulphate reduction (Liamleam and Annachhatre 2007).

Table 2 Compounds used as energy substrates by SRB (adapted from Hansen 1993)

Compound	Substrates
Inorganic	Hydrogen, carbon monoxide
Monocarboxylic acids	Lactate, acetate, butyrate, formate, propionate, isobutyrate, 2- and 3-methyl butyrate, higher fatty acids up to C18, pyruvate
Dicarboxylic acids	Succinate, fumarate, malate, oxalate, maleinate, glutarate, pimelate
Alcohols	Methanol, ethanol, propanol, butanol, ethylene glycol, 1,2- and 1,3-propanediol, glycerol
Amino acids	Glycine, serine, cysteine, threonine, valine, leucine, isoleucine, aspartate, glutamate, phenylalanine
Miscellaneous	Choline, furfural, oxamate, fructose, benzoate, 2-, 3- and 4-OH-benzoate, cyclohexanecarboxylate, hippurate, nicotinic acid, indole, anthranilate, quinoline, phenol, p-cresol, catechol, resorcinol, hydroquinone, protocatechuate, phloroglucinol, pyrogallol, 4-OH-phenyl acetate, 3-phenylpropionate, 2-aminobenzoate, dihydroxyacetone

2.1 Organic Polymeric Compounds

2.1.1 Starch

Potato is a staple food in Europe and other parts of the world. It is produced in large quantities in the Netherlands, and according to a recent report, the estimated potato production in the year 2011 was ~73 million metric tons (FAOSTAT 2011). Table 3 shows the composition of fresh potato. They contain ~70–80% water and starch counts for 16–24% of the total weight. Potato is considered to be the fourth most important tuberous food crop in the world after wheat, rice and maize. Kang and Weiland (1993) ascertained the biodegrading characteristics of potato using batch tests and reported that ~90% potato could be degraded at 35 °C, at a substrate to inoculum ratio of 0.8.

2.1.2 Cellulose

Cellulose exists in abundance on the earth under different forms. The anaerobic degradation of cellulose begins with depolymerisation and is followed by solubilisation. Besides, the degradation products of cellulose (i.e. cellobiose) can be converted to CH_4 and CO_2 through acidogenesis, acetogenesis and methanogenesis. When anaerobic microorganisms excrete cellulosomes from their cell wall, they are attached to the cellulose particles. Several studies have argued whether the bacterial hydrolysis or methanogenesis phase is the rate-limiting step for the biotransformation of polymeric compounds rich in cellulose content (Jeihanipour et al. 2011). Recent studies have demonstrated the anaerobic digestion of cellulose under mesophilic and thermophilic conditions (O'Sullivan et al. 2008; Xia et al. 2012). According to Yang et al. (2004), in batch experiments, thermophilic cellulose

Table 3 Proximate composition of different potatoes (adapted from Hoover 2001)

Starch source	Starch yield (%)	Size (μm) and shape	Amylose content (%)	Total lipid (%)	Phosphorus (% dsb)		Nitrogen (%)
					Org	Inorg	
Solanum tuberosum (potato)	32	15–110 oval, spherical	25.4	0.19	0.089	0.001	0.1
Ipomea babatas (sweet potato)	30	2–42 round, oval and polygonal	19.1	0.06–0.6	0.012	–	0

Note dsb-double-strand break

digestion is less effective compared to mesophilic conditions. In that study, 16% volatile solids were removed under thermophilic conditions, while 52% volatile solids were removed under mesophilic conditions in 30 d.

2.1.3 Proteins

The hydrolysis products of proteins under anaerobic conditions include peptides and amino acids, which are further degraded to ammonium, carbon dioxide, short-chain fatty acids and hydrogen. According to Örlygsson et al. (1994), the hydrolysis of protein is frequently affected by the electron acceptor availability. The hydrolysis rate under anaerobic conditions is lower than that observed under aerobic conditions. Deamination is the initial step of the degradation of protein, which is favoured under aerobic rather than anaerobic conditions (Shao et al. 2013). Only a few members of the *Desulfobacterium*, *Desulfotomaculum* and *Desulfovibrio* genera utilize amino acids. Baena et al. (1998) reported that the addition of thiosulphate to the growth media enabled to optimize the degradation of amino acids by an SRB (*Desulfovibrio aminophilus sp. nov.* DSM 12254), indicating the role played by sulphate and thiosulphate in the degradation of proteinaceous compounds.

2.1.4 Chitin

Chitin is a polymer, occurring naturally as a white, hard, inelastic and nitrogenous polysaccharide. It can be found in the exoskeletons of crabs and shrimps in the alpha-chitin form (Rinaudo 2006). The beta-chitin form has a higher affinity for solvents than the alpha form because it has weak hydrogen bonds (Pillai et al. 2009). One of the derivatives of chitin is chitosan which is obtained by the alkaline deacetylation of chitin in a strong alkaline solution.

2.2 *Selection of Electron Donors for Biological Sulphate Reduction*

According to van Houten et al. (1996), the following three criteria's can be used for selecting a suitable electron donor for biological sulphate reduction: (i) high sulphate efficiency complemented by a low COD effluent concentration, (ii) electron donor availability and (iii) cost of sulphate converted to sulphide. Thermodynamic parameters such as physiological free energies $\Delta G^{0\prime}$ (kJ mol^{-1}) of the sulphidogenesis (Table 1) are also important for the sulphate reduction efficiency.

2.2.1 Efficiency of Sulphate Removal

The information presented in Table 4 indicates that acetate (28.5 g L^{-1} d^{-1}), ethanol (21 g L^{-1} d^{-1}) and hydrogen (51 mmol L^{-1} d^{-1}) support the highest sulphate redection rates. On the other hand, lactate (0.41 g L^{-1} d^{-1}) and molasses (<6.5 g L^{-1} d^{-1}) support lower sulphate removal rates. The electron donors summarized in Table 4 indicate different affinities of SRB for the carbon source (Stams et al. 2005). Additionally, the bioreactor configuration also plays an important role in determining the sulphate reduction efficiency (Kijjanapanich et al. 2014). The key factor affecting the sulphidogenesis during long-term bioreactor operation is the capability of retaining the active SRB in the bioreactor.

2.2.2 Availability and Cost of Electron Donor

Lactate and molasses are cost-effective, but they are not completely oxidized by SRB. They also generate high COD concentrations in the effluent (Liamleam and Annachhatre 2007). Hydrogen and ethanol are used as electron donors when the sulphate load exceeds 200 kg SO_4^{2-} h^{-1}. Besides, hydrogen and ethanol are not cost-effective; however, due to safety reasons, ethanol is preferred over hydrogen.

Table 4 Sulphate reduction at different operational conditions and different bioreactors (adapted from Liamleam and Annachhatre 2007)

Electron donors	Temperature (°C)	Reactor type	SO_4^{2-} removal rate (g. L^{-1} d^{-1})
Acetate	35	Packed bed	15–20
Acetate	33–35	EGSB	28.5
CO	35	Packed bed	2.4
Ethanol	33	EGSB	21
Glucose/acetate	35	Anaerobic digester	1.92
H_2/CO_2	30	Gas-lift	30
H_2/CO	35	Packed bed	1.2
Lactate	Room temp	Plug flow	0.41
Molasses	30	UASB	4.3
Molasses	27	CSTR	0.84
Molasses	35	Anaerobic RBC	0.35
Molasses	31	Packed bed	6.5
Synthesis gas	30	Gas-lift	12–14

Note CO carbon monoxide; *EGSB* expanded granular sludge bed; *UASB* upflow anaerobic sludge blanket; *RBC* rotating biological contactor; *CSTR* continuous stirred tank reactor

2.3 Environmental Parameters Affecting Sulphate Reduction

2.3.1 Temperature

A very important factor in the anaerobic digestion process is temperature, and the performance of bioreactors usually varies depending on the operating temperatures and the adaptability of the microbes to different temperature ranges (Table 5). SRB comprise both mesophilic and thermophilic strains which are affected by temperature. Weijma et al. (2000) showed a significant increase in sulphate reduction when temperature was increased from 20 to 35 °C, but bacterial activity decreased at 40 °C. According to Tsukamoto et al. (2004), the efficiency of acid mine drainage (AMD) treatment was not affected by temperature due to the acclimatization of SRB to low-temperature conditions over prolonged time. Thermophilic processes lead to H_2S stripping, thereby reducing its concentration (S^{2-}) in the liquid phase. Therefore, the treatment of sulphate-rich wastewater can be more efficient at temperatures ranging between 55 and 70 °C (Sarti and Zaiat 2011).

2.3.2 pH and S^{2-} Concentration

SRB show high specific activities in the pH range of 5.0–8.0. Beyond this range, their metabolic activity reduces and inhibition effects set in (Dvorak et al. 1992). Sheoran et al. (2010) reported that some SRB are able to remove 38.3% of sulphate from the influent at a pH of 3.3, but their removal performance dropped when the pH was reduced to ~3.0. The hydrogen sulphide and bicarbonate present in the system buffers the solution pH; however, the buffering capacity depends on the type

Table 5 Temperature range for the growth of a number of SRB (adapted from Tang et al. 2009)

SRB	Temperature (°C)	
	Range	Optimum
Desulfobacter	28–32	
Desulfobulbus	28–39	
Desulfomonas	–	30
Desulfosarcina	33–38	
Desulfovibrio	25–35	
Thermodesulforhabdus norvegicus	44–74	60
Desulfotomaculum luciae	50–70	
Desulfotomaculum solfataricum	48–65	60
Desulfotomaculum thermobenzoicum	45–62	55
Desulfotomaculum thermocisternum	41–75	62
Desulfotomaculum thermosapovorans	35–60	50
Desulfacinum infernum	64	–

and quantity of organic end products (Dvorak et al. 1992). This can cause inhibition of anaerobic digestion (AD) and could lead to a complete failure of the process. The effect of sulphide is believed to be caused by non-ionized H_2S, because neutral molecules can permeate through the cell membranes (Sarti and Zaiat 2011).

2.3.3 Hydraulic Retention Time (HRT)

The HRT determines the time allowable for the SRB to adapt to the environment, initiate growth and metabolic activity, thereby increasing the amount of sulphate or COD reduced. In bioreactors for sulphate reduction, a long HRT may lead to high sulphate reduction efficiencies and a complete oxidation of the electron donor used with minimal residual acetate concentrations (Kaksonen et al. 2006). However, according to Sheoran et al. (2010), a short HRT may reduce the time available for the SRB to metabolize the substrate and could lead to complete washout of biomass from the reactor.

3 Bioreactors for Sulphate Reduction

Several branches of biotechnology use bioreactors, such as the production of biofuels (Ozmihci and Kargi 2008), food products (Genari et al. 2003), pharmaceutical compounds (John et al. 2007) and the treatment of environmental pollutants (Show et al. 2011). Anaerobic wastewater treatment systems use mixed microbial consortia, which is somewhat different compared to other biotechnological processes where isolation or/and sterilization is required (Goršek and Tramšek 2008). Setting different steps of a process in one stage can make the process more attractive in terms of process intensification. Therefore, the use of flocs, granules and biofilms is of great interest in the field of environmental biotechnology. This is possible by facilitating solid–liquid–gas separation, and these coupled to the reactor configuration make the separation of the three phases and downstream processing feasible.

Laboratory-scale experiments have shown promising results for the treatment of wastewater rich in sulphate by using different bioreactor configurations. A variety of bioreactors (Fig. 5) such as expanded granular sludge bed reactors (EGSB), fluidized bed reactors (FBR), gas-lift bioreactors (GLB), inverse fluidized bed reactors (IFB), membrane bioreactors (MBR), sequencing batch reactors (SBR) and upflow anaerobic sludge blanket (UASB) reactors have been applied for sulphate reduction in wastewaters (Sheoran et al. 2010). Numerous two-stage processes combining anaerobic biological sulphate reduction with an aerobic step have also been used at the laboratory scale (du Preez et al. 1992; Maree and Hill 1989). Upflow packed bed reactors (Fontes Lima and Zaiat 2012; Peixoto et al. 2011),

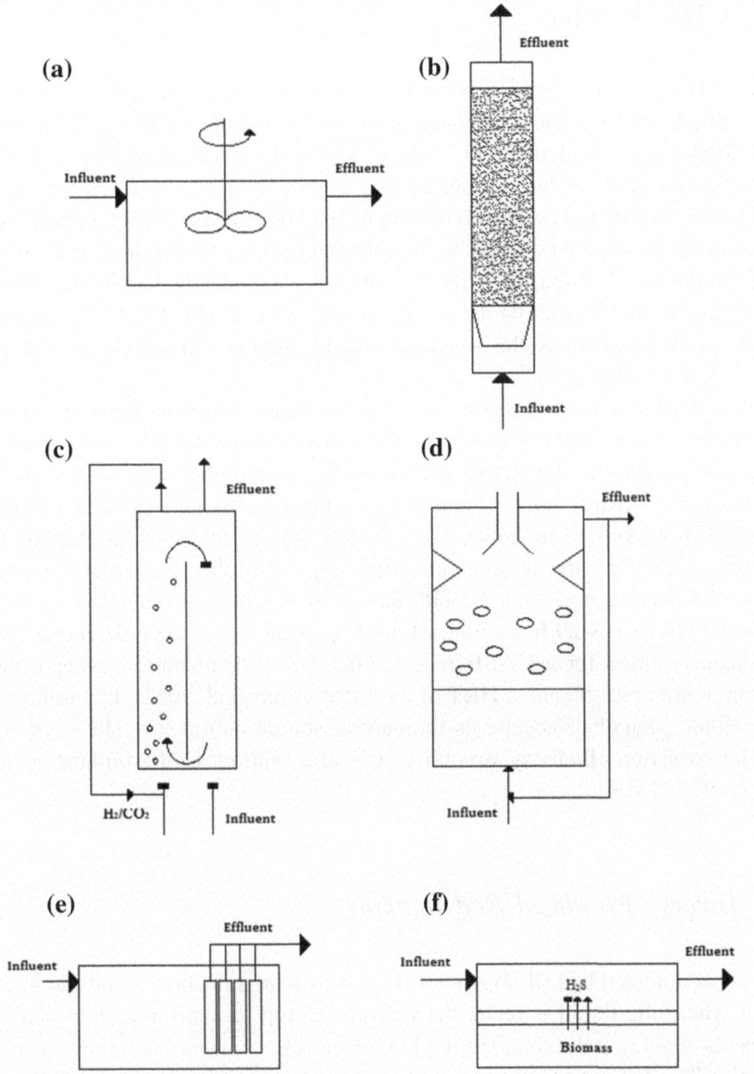

Fig. 5 Schematic representation of a continuous stirred tank reactor (**a**), packed bed reactor (**b**), gas-liquid bioreactor (**c**), upflow anaerobic sludge blanket reactor (**d**), immersed membrane bioreactor (**e**) and extractive membrane bioreactor (**f**) (adapted from Papirio et al. 2013a, b)

stirred tank reactors (Kieu et al. 2011), sulphate-reducing columns (Baskaran and Nemati 2006) and biofilm reactors (D'Acunto et al. 2011) have also been studied. These bioreactors have shown efficient sulphate reduction efficiencies alongside the selective removal of heavy metals from effluents by sulphide precipitation (Sampaio et al. 2010; Villa-Gomez et al. 2011).

3.1 UASB Reactor

The UASB reactor is a mature technology for wastewater treatment (van Lier et al. 2015). The UASB reactor is operated at upward velocities <2 m h^{-1} (Hulshoff Pol et al. 2004; van Haandel et al. 2006). A UASB is considered as a high-rate bioreactor because of its capability to deconvolute the solid retention time (SRT) from the HRT. The deconvolution of the HRT from the SRT is possible due to granule formation that ensures high biomass concentrations (van Lier et al. 2015; van Haandel et al. 2006). UASB reactors are designed to handle a volumetric loading rate of 4–15 kg COD. m^3 d^{-1} (van Lier et al. 2015). The UASB bioreactor (Fig. 5d) is intensified by the gas–liquid–solid separator placed at the top with a shape of an inverted funnel.

The UASB reactor has been used for sulphate reduction from sulphate-rich wastewaters. A long HRT is beneficial for acetate consumption by SRB. Increasing the mixing capacity of the UASB reactor by applying higher recirculation rates and increasing the upward velocity can increase the performance of SRB (Alphenaar et al. 1993). According to Lopes et al. (2008), using sucrose as an electron donor and at pH <7.0, the sulphate reduction efficiency was higher in a UASB (>50%) at a HRT ~23 h compared to a CSTR that showed sulphate reduction efficiencies <38% at a HRT of ~20 h. In contrast, high sulphate reduction efficiencies (>80%) have been reported for a UASB reactor after 500 d of operation using lactate as electron donor at 25 °C and a HRT of 24 h (Bertolino et al. 2012). In another study, using either ethanol or acetate as the carbon source and at low HRT (>6 h), the sulphate reduction efficiency was only 30% at a limiting COD/sulphate ratio of 1 (Jing et al. 2013).

3.2 Inverse Fluidized Bed Reactor

An inverse fluidized bed (IFB) reactor (Fig. 6) is a modification of the fluidized bed reactor where the liquid is recirculated from the top (downwards recirculation) of the reactor making it different from a UASB bioreactor, where liquid is recirculated upwards. The IFB reactor uses a carrier material lighter than water onto which the sulphate-reducing biofilm attaches. In an IFB reactor, the growing biofilm on the carrier material is advantageous as it results in a higher surface area for biomass growth, leading to high biomass concentrations and low space requirements for the reactors. However, biomass growth in the liquid suspension is difficult to avoid due to shear forces and abrasion between the carrier materials (Davey and O'toole 2000; Escudié et al. 2011).

Industrial and municipal wastewater treatments have been using biofilm-based reactors since the last decades. Several studies have been done to establish the factors that affect biofilm formation and growth. In an IFB, the volumetric

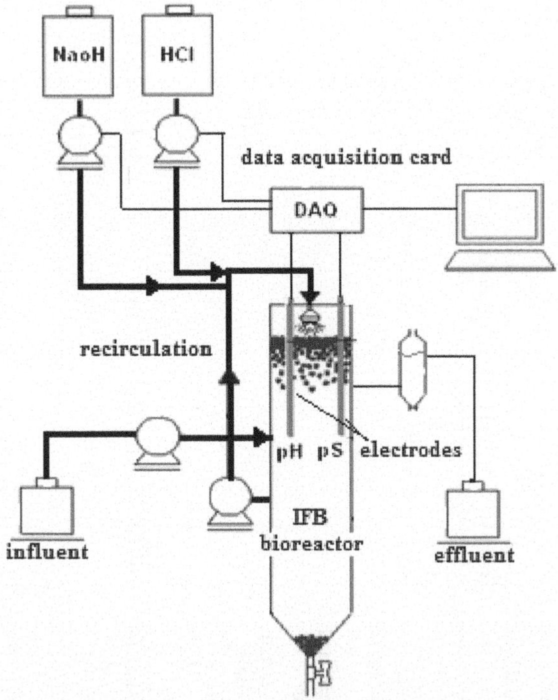

Fig. 6 Schematic of an inverse fluidized bed reactor (adapted from Villa-Gomez et al. 2014)

conversion rate of the pollutant depends on the liquid velocity and the substrate concentration. According to the results obtained by Eldyasti et al. (2012), the substrate concentration has a greater effect on the diffusion rates than the liquid velocities, whereas Diez Blanco et al. (1995) showed contradictory results indicating a reduction in the external mass transfer velocity when the liquid velocity was increased. From a hydrodynamics point of view, Andalib et al. (2012) found that the diffusion rates were affected more by the liquid flow rate, i.e. under turbulent flow conditions, rather than under laminar flow.

Table 6 summarizes the recent studies on sulphate reduction using IFB bioreactors. Recent studies on sulphate reduction use anaerobic sludge from methanogenic bioreactors as inoculum and low-density polyethylene is the most preferred carrier material for IFB bioreactors (Table 6). Till to date, sulphate reduction has been studied at the lowest HRT of 0.37 d. In a study by Villa-Gomez et al. (2011), the sulphate reduction efficiencies were 74 and 38% at a COD/sulphate ratio of 5 and 1, respectively. The influence of different electron donors on the sulphate reduction has been studied in the IFB bioreactor where the most commonly used electron donors were acetate, propionate, butyrate, ethanol and lactate. These electron donors have been studied at different COD/sulphate ratios, however, sulphate reduction is hampered at ratios <1.0 and optimal at ratios >1.0 (Papirio et al. 2013a; Villa-Gomez et al. 2011).

Table 6 Sulphate reduction efficiency reported in inverse fluidized bed bioreactors (adapted from Reyes-Alvarado et al. 2017)

Source of the inoculum	Carrier material	HRT (d)	Electron donor	COD removal efficiency (%)	Sulphate reduction efficiency (%)	COD/sulphate ratio	References
Granular sludge from a UASB reactor treating malting process effluent (Central de Malta, Grajales, Puebla, Mexico)	Low-density (267 $kg.m^{-3}$) polyethylene (0.4 mm diameter)	1–0.7	Mixture of VFA: acetate or lactate, propionate and butyrate	90	73	1.67–0.67	Celis-García et al. (2007)
Granular sludge from a UASB reactor treating paper mill wastewater (Industriewater Eerbeek B.V., Eerbeek, the Netherlands)	Low-density (400 $kg.m^{-3}$) polyethylene (500–1000 μm)	2	Ethanol-lactate: 2:1–1:0 ratio	80	28	0.6	Celis et al. (2009)
Sulphate-reducing biofilm	Low-density (400 $kg.m^{-3}$) polyethylene (500 μm diameter)	2–1	Ethanol-lactate: 2:1–1:0 ratio	50–54	30–41	0.8	Gallegos-Garcia et al. (2009)
Anaerobic sludge from a digester treating activated sludge from a domestic wastewater treatment plant (De Nieuwe Waterweg in Hoek van Holland, the Netherlands)	Low-density polyethylene (3 mm diameter)	1–0.37	Lactate	*R*1 = 14–34 and *R*2 = 35–68	*R*1 = 56–88 and *R*2 = 17–68	*R*1 = 5 and *R*2 = 1	Villa-Gomez et al. (2011)
Methanogenic granular sludge from a full-scale anaerobic digester fed with buffalo manure and dairy wastewater	Polypropylene pellets (3–5 mm diameter)	1	Lactate	*R*1 = 35–64 and *R*2 = 6–61	*R*1 = 18–30 and *R*2 = 1–63	*R*1 = 0.67 and *R*2 = 0.67–4.0	Papirio et al. (2013a)

(continued)

Table 6 (continued)

Anaerobic granular sludge from Biothane Systems International (Delft, the Netherlands)	Low-density polyethylene beads (3 mm diameter)	0.64	Ethanol	Not reported	75	1.88	Kijjanapanich et al. (2014)
Anaerobic sludge from Biothane Systems International (Delft, the Netherlands)	Low-density polyethylene beads (3 mm diameter)	1	Ethanol	75 (pH 7.0) and 58 (pH 5.0)	74 (pH: 7.0) and 50 (pH: 5.0)	1	Janyasuthiwong et al. (2016)
Anaerobic sludge from a reactor digesting waste activated sludge at Harnaschpolder (the Netherlands)	Low-density polyethylene beads (4 mm diameter)	1–0.5	Lactate	72–86	16–74	0.71–1.82	Reyes-Alvarado et al. (2017)

Kijjanapanich et al. (2014) showed that sulphate reduction is possible with an efficiency of 75–85% regardless of the bioreactor configuration, e.g. IFB, UASB or gas-lift anaerobic membrane bioreactor, and at a HRT of 0.64 d. In that study, the IFB bioreactor reached steady state after 20 d of operation compared to the UASB bioreactor that required 35 d (Kijjanapanich et al. 2014). Sulphate reduction efficiencies of ~50% are also possible at low pH (5.0) in an IFB bioreactor and at a COD/sulphate ratio of 1 (Janyasuthiwong et al. 2016). The sulphidogenesis is robust to transient feeding conditions using lactate as an electron donor, at an HRT of 0.5 d (Reyes-Alvarado et al. 2017). Reyes-Alvarado et al. (2017) showed that the sulphate reduction is more affected by the COD/sulphate ratio (i.e. at values <1) than to ten successive (10×) transient feeding condition applied to the IFB. For instance, the average sulphate reduction efficiency was 67 (±15)% during the feast periods and this performance was comparable to that of the IFB bioreactor operation under continuous operating conditions (61 ± 15%) (Reyes-Alvarado et al. 2017).

3.3 Factors Affecting Bioreactor Performance

3.3.1 Characteristics of the Organic Substrate

Different organic substrates can be used as carbon source and electron donor for sulphate reduction (Table 2). The characteristics of the substrate are important for the bioreactor performance, mainly because the anaerobic biodegradability and the composition (VS, COD and TS) of organic matter are interconnected. Furthermore, the concentration of the substrate introduced into the reactor can also affect the metabolic activity of the microbes (Raposo et al. 2011). The VS content of organic substrates is not essentially the same, because of different proteins, lipids and carbohydrates content, which represent the soluble and the easily biodegradable part. The lignin composition represents the almost non-biodegradable part of the VS. Therefore, the biodegradation and the solubility of the electron donors depend on their cellulose and lignin content which means that the hydrolysis–fermentation rate is also affected (Houbron et al. 2008). The cellulose and chitin crystallinity or degree of polymerization show different rates of degradation, and this depends on the content or pretreatment done prior to their use in the methanogenesis or a sulphidogenic process. The crystallinity or degree of polymerization refers to the order of the molecules in the polymer such as the α, β and γ-cellulose (Foston 2014). Similarly, even though the COD content of electron donors with heterogeneous characteristics is different, it is also an important factor because it helps in controlling the growth rate of the SRB.

3.3.2 Particle Size of Electron Donors

The particle size is normally considered as an important design factor because a reduction in the particle size will increase the surface area which in turn improves the performance of the biological process (De la Rubia et al. 2011; Mshandete et al. 2006). Sometimes, hydrolytic-fermentative bacteria find it difficult to biodegrade the organic solid waste because of its size, and therefore, it is suggested to cut or break them to allow more surface area for these microorganisms to metabolize. Since the initial hydrolysis process may take time, it is important to provide an adequate/favourable environment for the SRB to increase their metabolic activity. Failure to do so might pose a delay in the start-up or even complete failure of the bioprocess.

3.3.3 Source of the Inoculum

The adaptation of the inoculum to the bioreactor depends a lot on its origin (Behera et al. 2007). Sources from thermophilic, mesophilic and halophilic conditions adapt differently when introduced into the bioreactors. For the sulphate reduction process, if the origin of the inoculum is from sulphate-rich environments, it will be easier for the microorganisms to adapt themselves to the bioreactor conditions because of the sulphate content. Whereas for sulphate deprived environments, it will take a while for the microorganisms to adapt to their new environment. But the time of adaptation depends on the syntrophic network established in the inoculum (Alon et al. 1999; Barkai and Leibler 1997). An inoculum from wastewater treatment plants can vary in its characteristics due to different operating conditions and daily variations, but it is mostly preferred because they all share common characteristics (Table 6). The effects of the inoculum origin, concentration, activity and storage have been reported in the literature (Raposo et al. 2011). In general, start-up of sulphate-reducing bioreactors could be enhanced by the introduction of inoculum from sulphate-rich origins.

3.3.4 Physical and Chemical Conditions in a Bioreactor

Physical and chemical operational conditions of the reactors affect the sulphate reduction process. Physical conditions, such as volume, temperature and stirring speed have significant effects on the biodegradation rate. Zagury et al. (2006) studied the effect of chemical conditions such as headspace gas concentrations, pH and alkalinity adjustments on the biodegradation of substrates. The majority of the bioreactor experiments are performed under mesophilic conditions (20–45 °C) and a few under thermophilic conditions (45–60 °C). Thermophilic conditions are sometimes avoided due to cost implications. Although the effect of stirring is contested, for organic solid electron donors, the stirring process will favour its contact with the SRB, increasing the microbial activity and facilitating better sulphate reduction.

3.3.5 Biomass Morphology

Different morphological features of biomass can develop during bioreactor operation. The factors that might affect the cell performance and behaviour are the ones increasing stress, which at the same time can also affect the syntrophic structure.

(a) *Flocs*: Flocs are a conglomeration of cells and microcolonies enmeshed in exopolymers, related to the hydrodynamics, wastewater composition and dissolved oxygen concentrations (Dangcong et al. 1999). One advantage is the fast diffusional transport compared with those in granules or biofilms (Morgan-Sagastume et al. 2008). The filamentous bacteria play an important role in these structures. Mielczarek et al. (2012) reported that during warm periods, activated sludge flocs preserve an open structure. A high concentration of filamentous bacteria might cause settling problems.
(b) *Granules*: The anaerobic granular sludge develops spontaneously due to auto-immobilization. This aggregation occurs in the absence of any support material, in contrast to biofilms. As a single bacterium is not able to degrade organic matter to methane, different bacteria, as present in granules, develop a complex and unique microbial ecosystem.
(c) *Biofilms*: Planktonic cells are found in media as free-floating microorganisms, but their attachment to surfaces enhances their survival in diverse environments (Rivas et al. 2007). A biofilm can be defined as a complex coherent structure of cells and cellular products, like extracellular polymers (exopolysaccharide), which form and grow spontaneously attached on a static suspended solid surface (Davey and O'toole 2000).

The process of biofilm formation is a multistage development, which includes attachment of microbes to the surface, cell to cell adhesion and proliferation, maturation and detachment. In the bioreactor, osmolarity, pH, oxygen and temperature are other environmental variables that can also influence the initial biomass attachment, apart from the nature of the support material used (Ishii et al. 2008). For instance, the stratification of microbial communities in biofilms can be influenced by the electron acceptors and donors (Satoh et al. 2009), as well as the degradability of the carbon source (Shen et al. 2013).

4 Conclusions

The variations in concentrations of sulphate and carbon source, pH and the presence of competing ions affect the biochemical activities in wastewater treatment systems. The use of sulphate-reducing bacteria (SRB) technology for sulphate-rich wastewater treatment is advantageous due to minimal sludge production, ability to perform simultaneous oxidation of organic matter and the reduction of sulphate. Sulphate reduction in bioreactors is affected by parameters such as the type of electron donor, COD/sulphate ratio, pH, HRT and reactor configuration. Sulphate reduction efficiencies >90% are achieved when COD is not a limiting factor in a bioreactor, and a COD/sulphate ratio >2.0 is recommended in such cases. Most of the sulphate-reducing bioreactors are also able to handle fluctuations in COD or sulphate-loading rates. The ability of the SRB to overcome feast and famine periods clearly shows the application of this technology for industrial wastewaters.

Acknowledgements This research was funded by the European Union through the Erasmus Mundus Joint Doctorate Programme, Environmental Technologies for Contaminated Solids, Soils and Sediments [ETeCoS^3, grant agreement FPA no. 2010-0009].

References

Alon U, Surette MG, Barkai N, Leibler S (1999) Robustness in bacterial chemotaxis. Nature 397:168–171. https://doi.org/10.1038/16483

Al-Zuhair S, El-Naas MH, Al-Hassani H (2008) Sulfate inhibition effect on sulfate reducing bacteria. J Biochem Technol 1:39–44

Alphenaar AP, Visser A, Lettinga G (1993) The effect of liquid upward velocity and hydraulic retention time on granulation in UASB reactors treating wastewater with a high sulphate content. Bioresour Technol 43:249–258. https://doi.org/10.1016/0960-8524(93)90038-D

Andalib M, Zhu J, Nakhla G (2012) A new definition of bed expansion index and voidage for fluidized biofilm-coated particles. Chem Eng J 189–190:244–249. https://doi.org/10.1016/j.cej.2012.02.065

Baena S, Fardeau ML, Labat M, Ollivier B, Garcia JL, Patel BK (1998) *Desulfovibrio aminophilus sp. nov.*, a novel amino acid degrading and sulfate reducing bacterium from an anaerobic dairy wastewater lagoon. Syst Appl Microbiol 21:498–504. https://doi.org/10.1016/S0723-2020(98)80061-1

Barkai N, Leibler S (1997) Robustness in simple biochemical networks. Nature 387:913–917. https://doi.org/10.1038/43199

Barton LL, Fauque GD (2009) Biochemistry, physiology and biotechnology of sulfate-reducing bacteria. Adv Appl Microbiol. pp. 41–98. doi:https://doi.org/10.1016/S0065-2164(09)01202-7

Baskaran V, Nemati M (2006) Anaerobic reduction of sulfate in immobilized cell bioreactors, using a microbial culture originated from an oil reservoir. Biochem Eng J 31:148–159. https://doi.org/10.1016/j.bej.2006.07.007

Behera SK, Rene ER, Murthy DVS (2007) Performance of upflow anoxic bioreactor for wastewater treatment. Int J Environ Sci Technol 4:247–252. https://doi.org/10.1007/BF03326281

Bertolino SM, Rodrigues ICB, Guerra-Sá R, Aquino SF, Leão VA (2012) Implications of volatile fatty acid profile on the metabolic pathway during continuous sulfate reduction. J Environ Manage 103:15–23. https://doi.org/10.1016/j.jenvman.2012.02.022
Butlin KR, Selwyn SC, Wakerley DS (1956) Sulphide production from sulphate-enriched sewage sludges. J Appl Bacteriol 19:3–15. https://doi.org/10.1111/j.1365-2672.1956.tb00036.x
Celis LB, Villa-Gómez D, Alpuche-Solís AG, Ortega-Morales BO, Razo-Flores E (2009) Characterization of sulfate-reducing bacteria dominated surface communities during start-up of a down-flow fluidized bed reactor. J Ind Microbiol Biotechnol 36:111–121. https://doi.org/10.1007/s10295-008-0478-7
Celis-García LB, Razo-Flores E, Monroy O (2007) Performance of a down-flow fluidized bed reactor under sulfate reduction conditions using volatile fatty acids as electron donors. Biotechnol Bioeng 97:771–779. https://doi.org/10.1002/bit.21288
Cirik K, Dursun N, Sahinkaya E, Çinar Ö (2013) Effect of electron donor source on the treatment of Cr(VI)-containing textile wastewater using sulfate-reducing fluidized bed reactors (FBRs). Bioresour Technol 133:414–420. https://doi.org/10.1016/j.biortech.2013.01.064
D'Acunto, B., Esposito, G., Frunzo, L., Pirozzi, F., 2011. Dynamic modeling of sulfate reducing biofilms. Comput Math Appl 62:2601–2608. doi:https://doi.org/10.1016/j.camwa.2011.07.064
Dangcong P, Bernet N, Delgenes J, Moletta R (1999) Aerobic granular sludge-a case report. Water Res 33:890–893. https://doi.org/10.1016/S0043-1354(98)00443-6
Davey, M.E., O'toole, G.A., 2000. Microbial biofilms: from ecology to molecular genetics. Microbiol Mol Biol Rev 64:847–867. doi:https://doi.org/10.1128/MMBR.64.4.847-867.2000
De la Rubia MA, Fernández-Cegrí V, Raposo F, Borja R (2011) Influence of particle size and chemical composition on the performance and kinetics of anaerobic digestion process of sunflower oil cake in batch mode. Biochem Eng J 58–59:162–167. https://doi.org/10.1016/j.bej.2011.09.010
Diez Blanco V, García Encina PA, Fdz-Polanco F (1995) Effects of biofilm growth, gas and liquid velocities on the expansion of an anaerobic fluidized bed reactor (AFBR). Water Res 29:1649–1654. https://doi.org/10.1016/0043-1354(95)00001-2
du Preez LA, Odendaal JP, Maree JP, Ponsonby M (1992) Biological removal of sulphate from industrial effluents using producer gas as energy source. Environ Technol 13:875–882. https://doi.org/10.1080/09593339209385222
Dvorak DH, Hedin RS, Edenborn HM, McIntire PE (1992) Treatment of metal contaminated water using bacterial sulfate reduction: results from pilot-scale reactors. Biotechnol Bioeng 40:609–616
Eldyasti A, Nakhla G, Zhu J (2012) Influence of particles properties on biofilm structure and energy consumption in denitrifying fluidized bed bioreactors (DFBBRs). Bioresour Technol 126:162–171. https://doi.org/10.1016/j.biortech.2012.07.113
Escudié R, Cresson R, Delgenès J-P, Bernet N (2011) Control of start-up and operation of anaerobic biofilm reactors: an overview of 15 years of research. Water Res 45:1–10. https://doi.org/10.1016/j.watres.2010.07.081
Fontes Lima DM, Zaiat M (2012) The influence of the degree of back-mixing on hydrogen production in an anaerobic fixed-bed reactor. Int J Hydrogen Energy 37:9630–9635. https://doi.org/10.1016/j.ijhydene.2012.03.097
Foston M (2014) Advances in solid-state NMR of cellulose. Curr Opin Biotechnol 27:176–184. https://doi.org/10.1016/j.copbio.2014.02.002
Gallegos-Garcia M, Celis LB, Rangel-Méndez R, Razo-Flores E (2009) Precipitation and recovery of metal sulfides from metal containing acidic wastewater in a sulfidogenic down-flow fluidized bed reactor. Biotechnol Bioeng 102:91–99. https://doi.org/10.1002/bit.22049
Genari AN, Passos FV, Passos FML (2003) Configuration of a bioreactor for milk lactose hydrolysis. J Dairy Sci 86:2783–2789. https://doi.org/10.3168/jds.S0022-0302(03)73875-2
Gibson GR (1990) Physiology and ecology of the sulphate-reducing bacteria. J Appl Bacteriol 69:769–797. https://doi.org/10.1111/j.1365-2672.1990.tb01575.x

Goršek A, Tramšek M (2008) Kefir grains production-an approach for volume optimization of two-stage bioreactor system. Biochem Eng J 42:153–158. https://doi.org/10.1016/j.bej.2008.06.009

Grein F, Ramos AR, Venceslau SS, Pereira IAC (2013) Unifying concepts in anaerobic respiration: insights from dissimilatory sulfur metabolism. Biochem Biophys Acta-Bioenerg 1827:145–160. https://doi.org/10.1016/j.bbabio.2012.09.001

Hansen TA (1993) Carbon metabolism of sulfate-reducing bacteria. In: Odom JM, Singleton JR (eds) The sulfate-reducing bacteria: contemporary perspectives. Springer, New York, NY, pp 21–40. doi:https://doi.org/10.1007/978-1-4613-9263-7_2

Hao OJ, Chen JM, Huang L, Buglass RL (1996) Sulfate-reducing bacteria. Crit Rev Environ Sci Technol 26:155–187. https://doi.org/10.1080/10643389609388489

Hoover R (2001) Composition, molecular structure, and physicochemical properties of tuber and root starches: a review. Carbohydr Polym 45:253–267. https://doi.org/10.1016/S0144-8617(00)00260-5

Houbron E, González-López GI, Cano-Lozano V, Rustrían E (2008) Hydraulic retention time impact of treated recirculated leachate on the hydrolytic kinetic rate of coffee pulp in an acidogenic reactor. Water Sci Technol 58:1415–1421. https://doi.org/10.2166/wst.2008.492

Hulshoff Pol LW, De Castro Lopes SI, Lettinga G, Lens PNL (2004) Anaerobic sludge granulation. Water Res 38:1376–1389. https://doi.org/10.1016/j.watres.2003.12.002

Ishii S, Shimoyama T, Hotta Y, Watanabe K (2008) Characterization of a filamentous biofilm community established in a cellulose-fed microbial fuel cell. BMC Microbiol 8:1–12. https://doi.org/10.1186/1471-2180-8-6

Janyasuthiwong S, Rene ER, Esposito G, Lens PNL (2016) Effect of pH on the performance of sulfate and thiosulfate-fed sulfate reducing inverse fluidized bed reactors. J Environ Eng 142:1–11. https://doi.org/10.1061/(ASCE)EE.1943-7870.0001004

Jeihanipour A, Niklasson C, Taherzadeh MJ (2011) Enhancement of solubilization rate of cellulose in anaerobic digestion and its drawbacks. Process Biochem 46:1509–1514. https://doi.org/10.1016/j.procbio.2011.04.003

Jing Z, Hu Y, Niu Q, Liu Y, Li Y-Y, Wang XC (2013) UASB performance and electron competition between methane-producing archaea and sulfate-reducing bacteria in treating sulfate-rich wastewater containing ethanol and acetate. Bioresour Technol 137:349–357. https://doi.org/10.1016/j.biortech.2013.03.137

John RP, Nampoothiri KM, Pandey A (2007) Fermentative production of lactic acid from biomass: an overview on process developments and future perspectives. Appl Microbiol Biotechnol 74:524–534. https://doi.org/10.1007/s00253-006-0779-6

Kaksonen AH, Plumb JJ, Robertson WJ, Riekkola-Vanhanen M, Franzmann PD, Puhakka JA (2006) The performance, kinetics and microbiology of sulfidogenic fluidized-bed treatment of acidic metal- and sulfate-containing wastewater. Hydrometallurgy 83:204–213. https://doi.org/10.1016/j.hydromet.2006.03.025

Kang H, Weiland P (1993) Ultimate anaerobic biodegradability of some agro-industrial residues. Bioresour Technol 43:107–111. https://doi.org/10.1016/0960-8524(93)90168-B

Kieu HTQ, Müller E, Horn H (2011) Heavy metal removal in anaerobic semi-continuous stirred tank reactors by a consortium of sulfate-reducing bacteria. Water Res 45:3863–3870. https://doi.org/10.1016/j.watres.2011.04.043

Kijjanapanich P, Do AT, Annachhatre AP, Esposito G, Yeh DH, Lens PNL (2014) Biological sulfate removal from construction and demolition debris leachate: effect of bioreactor configuration. J Hazard Mater 269:38–44. https://doi.org/10.1016/j.jhazmat.2013.10.015

Leustek T, Martin MN, Bick J-A, Davies JP (2000) Pathways and regulation of sulphur metabolism revealed through molecular and genetic studies. Annu Rev Plant Physiol Plant Mol Biol 51:141–165. https://doi.org/10.1146/annurev.arplant.51.1.141

Liamleam W, Annachhatre AP (2007) Electron donors for biological sulfate reduction. Biotechnol Adv 25:452–463. https://doi.org/10.1016/j.biotechadv.2007.05.002

Lopes SIC, Dreissen C, Capela MI, Lens PNL (2008) Comparison of CSTR and UASB reactor configuration for the treatment of sulfate rich wastewaters under acidifying conditions. Enzyme Microb Technol 43:471–479. https://doi.org/10.1016/j.enzmictec.2008.08.001

Maree JP, Hill E (1989) Biological removal of sulphate from industrial effluents and concomitant production of sulphur. Water Sci Technol 21:265–276

Matias PM, Pereira IAC, Soares CM, Carrondo MA (2005) Sulphate respiration from hydrogen in bacteria: a structural biology overview. Prog Biophys Mol Biol 89:292–329. https://doi.org/10.1016/j.pbiomolbio.2004.11.003

Mielczarek AT, Kragelund C, Eriksen PS, Nielsen PH (2012) Population dynamics of filamentous bacteria in Danish wastewater treatment plants with nutrient removal. Water Res 46:3781–3795. https://doi.org/10.1016/j.watres.2012.04.009

Mizuno O, Li YY, Noike T (1998) The behavior of sulfate-reducing bacteria in acidogenic phase of anaerobic digestion. Water Res 32:1626–1634. https://doi.org/10.1016/S0043-1354(97)00372-2

Moestedt J, Nilsson Påledal S, Schnürer A (2013) The effect of substrate and operational parameters on the abundance of sulphate-reducing bacteria in industrial anaerobic biogas digesters. Bioresour Technol 132:327–332. https://doi.org/10.1016/j.biortech.2013.01.043

Morgan-Sagastume F, Larsen P, Nielsen JL, Nielsen PH (2008) Characterization of the loosely attached fraction of activated sludge bacteria. Water Res 42:843–854. https://doi.org/10.1016/j.watres.2007.08.026

Mshandete A, Björnsson L, Kivaisi AK, Rubindamayugi MST, Mattiasson B (2006) Effect of particle size on biogas yield from sisal fibre waste. Renew Energy 31:2385–2392. https://doi.org/10.1016/j.renene.2005.10.015

Muyzer G, Stams AJM (2008) The ecology and biotechnology of sulphate-reducing bacteria. Nat Rev Microbiol 6:441–454. https://doi.org/10.1038/nrmicro1892

Nedwell DB, Reynolds PJ (1996) Treatment of landfill leachate by methanogenic and sulphate-reducing digestion. Water Res 30:21–28. https://doi.org/10.1016/0043-1354(95)00128-8

O'Sullivan C, Burrell PC, Clarke WP, Blackall LL (2008) The effect of biomass density on cellulose solubilisation rates. Bioresour Technol 99:4723–4731. https://doi.org/10.1016/j.biortech.2007.09.070

Örlygsson J, Houwen FP, Svensson BH (1994) Influence of hydrogenothrophic methane formation on the thermophilic anaerobic degradation of protein and amino acids. FEMS Microbiol Ecol 13:327–334. https://doi.org/10.1111/j.1574-6941.1994.tb00079.x

Ozmihci S, Kargi F (2008) Ethanol production from cheese whey powder solution in a packed column bioreactor at different hydraulic residence times. Biochem Eng J 42:180–185. https://doi.org/10.1016/j.bej.2008.06.017

Papirio S, Esposito G, Pirozzi F (2013a) Biological inverse fluidized-bed reactors for the treatment of low pH- and sulphate-containing wastewaters under different COD conditions. Environ Technol 34:1141–1149. https://doi.org/10.1080/09593330.2012.737864

Papirio S, Villa-Gomez DK, Esposito G, Pirozzi F, Lens PNL (2013b) Acid mine drainage treatment in fluidized-bed bioreactors by sulfate-reducing bacteria: a critical review. Crit Rev Environ Sci Technol 43:2545–2580. https://doi.org/10.1080/10643389.2012.694328

Peixoto G, Saavedra NK, Varesche MBA, Zaiat M (2011) Hydrogen production from soft-drink wastewater in an upflow anaerobic packed-bed reactor. Int J Hydrogen Energy 36:8953–8966. https://doi.org/10.1016/j.ijhydene.2011.05.014

Pepper IL, Rensing C, Gerba CP (2004) Environmental microbial properties and processes, In: Environmental monitoring and characterization. Elsevier, USA, pp 263–280. doi:https://doi.org/10.1016/B978-012064477-3/50016-3

Pillai CKS, Paul W, Sharma CP (2009) Chitin and chitosan polymers: chemistry, solubility and fiber formation. Prog Polym Sci 34:641–678. https://doi.org/10.1016/j.progpolymsci.2009.04.001

Rabus R, Hansen TA, Widdel F (2006) Dissimilatory sulfate- and sulfur-reducing prokaryotes. In: The prokaryotes. Springer New York, NY, pp 659–768. doi:https://doi.org/10.1007/0-387-30742-7_22

Raposo F, Fernández-Cegrí V, de la Rubia MA, Borja R, Béline F, Cavinato C, Demirer G, Fernández B, Fernández-Polanco M, Frigon JC, Ganesh R, Kaparaju P, Koubova J, Méndez R, Menin G, Peene A, Scherer P, Torrijos M, Uellendahl H, Wierinck I, de Wilde V (2011) Biochemical methane potential (BMP) of solid organic substrates: evaluation of anaerobic biodegradability using data from an international interlaboratory study. J Chem Technol Biotechnol 86:1088–1098. https://doi.org/10.1002/jctb.2622

Reyes-Alvarado LC, Okpalanze NN, Kankanala D, Rene ER, Esposito G, Lens PNL (2017) Forecasting the effect of feast and famine conditions on biological sulphate reduction in an anaerobic inverse fluidized bed reactor using artificial neural networks. Proc Biochem 55:146–161. https://doi.org/10.1016/j.procbio.2017.01.021

Rinaudo M (2006) Chitin and chitosan: properties and applications. Prog Polym Sci 31:603–632. https://doi.org/10.1016/j.progpolymsci.2006.06.001

Rivas L, Dykes GA, Fegan N (2007) A comparative study of biofilm formation by Shiga toxigenic *Escherichia coli* using epifluorescence microscopy on stainless steel and a microtitre plate method. J Microbiol Methods 69:44–51. https://doi.org/10.1016/j.mimet.2006.11.014

Robertson, L.A., Kuenen, J.G., 2006. The colorless sulfur bacteria. In: The prokaryotes. Springer New York, NY, USA, pp 985–1011. doi:https://doi.org/10.1007/0-387-30742-7_31

Sampaio RMM, Timmers RA, Kocks N, André V, Duarte MT, van Hullebusch ED, Farges F, Lens PNL (2010) Zn-Ni sulfide selective precipitation: the role of supersaturation. Sep Purif Technol 74:108–118. https://doi.org/10.1016/j.seppur.2010.05.013

Sarti A, Zaiat M (2011) Anaerobic treatment of sulfate-rich wastewater in an anaerobic sequential batch reactor (AnSBR) using butanol as the carbon source. J Environ Manage 92:1537–1541. https://doi.org/10.1016/j.jenvman.2011.01.009

Satoh H, Odagiri M, Ito T, Okabe S (2009) Microbial community structures and in situ sulfate-reducing and sulfur-oxidizing activities in biofilms developed on mortar specimens in a corroded sewer system. Water Res 43:4729–4739. https://doi.org/10.1016/j.watres.2009.07.035

Shao L, Wang T, Li T, Lü F, He P (2013) Comparison of sludge digestion under aerobic and anaerobic conditions with a focus on the degradation of proteins at mesophilic temperature. Bioresour Technol 140:131–137. https://doi.org/10.1016/j.biortech.2013.04.081

Shen Z, Zhou Y, Wang J (2013) Comparison of denitrification performance and microbial diversity using starch/polylactic acid blends and ethanol as electron donor for nitrate removal. Bioresour Technol 131:33–39. https://doi.org/10.1016/j.biortech.2012.12.169

Sheoran AS, Sheoran V, Choudhary RP (2010) Bioremediation of acid-rock drainage by sulphate-reducing prokaryotes: a review. Miner Eng 23:1073–1100. https://doi.org/10.1016/j.mineng.2010.07.001

Show KY, Lee DJ, Chang JS (2011) Bioreactor and process design for biohydrogen production. Bioresour Technol 102:8524–8533. https://doi.org/10.1016/j.biortech.2011.04.055

Stams AJM, Plugge CM, de Bok FAM, van Houten BHGW, Lens P, Dijkman H, Weijma J (2005) Metabolic interactions in methanogenic and sulfate-reducing bioreactors. Water Sci Technol 52:13–20

Taguchi Y, Sugishima M, Fukuyama K (2004) Crystal structure of a novel zinc-binding ATP Sulfurylase from *Thermus thermophilus* HB8. Biochemistry 43:4111–4118. https://doi.org/10.1021/bi036052t

Tang K, Baskaran V, Nemati M (2009) Bacteria of the sulphur cycle: an overview of microbiology, biokinetics and their role in petroleum and mining industries. Biochem Eng J 44:73–94. https://doi.org/10.1016/j.bej.2008.12.011

Tsukamoto TK, Killion HA, Miller GC (2004) Column experiments for microbiological treatment of acid mine drainage: low-temperature, low-pH and matrix investigations. Water Res 38:1405–1418. https://doi.org/10.1016/j.watres.2003.12.012

van Haandel A, Kato MT, Cavalcanti PFF, Florencio L (2006) Anaerobic reactor design concepts for the treatment of domestic wastewater. Rev Environ Sci Bio/Technol 5:21–38. https://doi.org/10.1007/s11157-005-4888-y

van Houten RT, van der Spoel H, van Aelst AC, Hulshoff Pol LW, Lettinga G (1996) Biological sulfate reduction using synthesis gas as energy and carbon source. Biotechnol Bioeng 50:136–144

van Lier JB, Zee FP, Frijters CTMJ, Ersahin ME (2015) Celebrating 40 years anaerobic sludge bed reactors for industrial wastewater treatment. Rev Environ Sci Bio/Technology 14:681–702. https://doi.org/10.1007/s11157-015-9375-5

Villa-Gomez D, Ababneh H, Papirio S, Rousseau DPL, Lens PNL (2011) Effect of sulfide concentration on the location of the metal precipitates in inversed fluidized bed reactors. J Hazard Mater 192:200–207. https://doi.org/10.1016/j.jhazmat.2011.05.002

Villa-Gomez DK, Cassidy J, Keesman KJ, Sampaio R, Lens PNL (2014) Sulfide response analysis for sulfide control using a pS electrode in sulfate reducing bioreactors. Water Res 50:48–58. https://doi.org/10.1016/j.watres.2013.10.006

Weijma J, Stams AJM, Hulshoff Pol LW, Lettinga G (2000) Thermophilic sulfate reduction and methanogenesis with methanol in a high rate anaerobic reactor. Biotechnol Bioeng 67:354–363

Widdel F (2006) The genus *Desulfotomaculum*. In: The prokaryotes. Springer, New York, NY, pp 787–794. doi:https://doi.org/10.1007/0-387-30744-3_25

Xia Y, Cai L, Zhang T, Fang HHP (2012) Effects of substrate loading and co-substrates on thermophilic anaerobic conversion of microcrystalline cellulose and microbial communities revealed using high-throughput sequencing. Int J Hydrogen Energy 37:13652–13659. https://doi.org/10.1016/j.ijhydene.2012.02.079

Yang Y, Tsukahara K, Yagishita T, Sawayama S (2004) Performance of a fixed-bed reactor packed with carbon felt during anaerobic digestion of cellulose. Bioresour Technol 94:197–201. https://doi.org/10.1016/j.biortech.2003.11.025

Zagury GJ, Kulnieks VI, Neculita CM (2006) Characterization and reactivity assessment of organic substrates for sulphate-reducing bacteria in acid mine drainage treatment. Chemosphere 64:944–954. https://doi.org/10.1016/j.chemosphere.2006.01.001

Zhou J, He Q, Hemme CL, Mukhopadhyay A, Hillesland K, Zhou A, He Z, Van Nostrand JD, Hazen TC, Stahl DA, Wall JD, Arkin AP (2011) How sulphate-reducing microorganisms cope with stress: lessons from systems biology. Nat Rev Microbiol 9:452–466. https://doi.org/10.1038/nrmicro2575

Author Biographies

Luis C. Reyes-Alvardo completed his PhD under the Erasmus Mundus Joint Doctorate Programme, Environmental Technologies for Contaminated Solids, Soils and Sediments ($ETeCoS^3$).

Eldon R. Rene is Senior Lecturer in Resource Recovery Technology at UNESCO-IHE Institute for Water Education, Delft, The Netherlands.

Giovanni Esposito is Associate Professor of Sanitary and Environmental Engineering at University of Cassino, Italy.

Piet N. L. Lens is Professor of Environmental Biotechnology at UNESCO-IHE Institute for Water Education, Delft, The Netherlands.

Chapter 4
Microbial Depolymerization

Anvita Sheel and Deepak Pant

Abstract Depolymerization is a tertiary recycling technique which represents the transformation of polymer chain into monomer units along with oligomers. Depolymerization of polymers with heterofunctional groups can be targeted by microorganisms, especially by fungi. Polyethylene terephthalate (PET), polyurethane (PU), polyester are among such polymers. Hydrolysis is the main cause behind microbial degradation, and hydrophobic surface binding of involving enzyme is responsible for microscopic description. Microbial participation techniques can be modified by using a chemical-assisted hybrid technique, and even halogenated polymer is dechlorinated by this technique.

Keywords PET · PU · PVC, depolymerization · Hydrophobic surface binding Chemical–biological

1 Introduction

Microbial depolymerization of polymers includes various techniques that involve microbes for transformation of polymer chain back into its monomer units along with oligomers. Polymer degradation alters the polymer properties because of the bond scissions and successive transformations via chemical, physical, or biological reactions.

Depolymerization of plastics can be categorized as:

(i) Photo-oxidative
(ii) Thermal
(iii) Mechanochemical
(iv) Ozone-induced

A. Sheel · D. Pant (✉)
Department of Environmental Sciences, Central University of Himachal Pradesh, Dharamshala, Himachal Pradesh 176215, India
e-mail: dpant2003@gmail.com; dpant2003@yahoo.com

S. J. Varjani et al. (eds.), *Waste Bioremediation*, Energy, Environment, and Sustainability, https://doi.org/10.1007/978-981-10-7413-4_4

(v) Catalytic
(vi) Biodegradation

Photo-oxidative degradation results in the decomposition of polymer by the action of light both UV and visible. The most damaging UV wavelength for a specific plastic depends on the bonds present, and therefore, the maximum degradation occurs at different wavelengths for different types of plastics, e.g., polyethylene (PE) degrades at around 300 nm, whereas polypropylene (PP) requires the wavelength of 370 nm for its degradation. The reaction initiates by the absorption of light energy by the suitable groups in the polymer chain; this results in chain scission leading to the formation of monomers and oligomers. Photodegradable polymers need to have a photo-responsive group (either already present in chain or by addition) (Kyrikou and Briassoulis 2007). Polymer degradation happens mainly in the ether parts of the soft segments, where photo-irradiation generates aldehyde, ester, propyl, formate, and end groups (Nagai et al. 2005). Both photochemical and thermal degradations are classified as **oxidative** degradation. The points of differences between these are (a) the sequence of initiation steps leading to auto-oxidation cycle and (b) thermal reactions occur throughout the bulk of the polymer sample, whereas photochemical reactions occur only on the surface (Singh and Sharma 2008). Many additional polymers depolymerize at high temperature, e.g., polymethylmethacrylate (PMMA) and PE have been converted almost quantitatively back to the monomer Thermal degradation (above 200 °C) leads to chain scission and mainly depends on impurities like head-to-head units, unsaturation sites (Ramis et al. 2004).

In **mechanochemical** degradation, breakdown of molecular chains under shear or mechanical force is often followed by a suitable chemical reaction. The mechanodegradation of polymers in melts occurs via free radical pathways. When polymers are subjected to shear, macroalkyl radicals are formed which accelerate oxidation. This may occur either in polymer melt during processing phase or under mechanical stress conditions (e.g., cross-linking rubber at lower temperatures). Under such conditions, chain-breaking electron acceptor (CB-A) antioxidants have noted to be relatively more effective than thermal oxidation (Scott 1984). When polymethylmethacrylate (PMMA) is degraded mechanochemically using nitroxides as chain termination agents, macroradicals are produced which are used in polymerization reactions (Schmidt-Naake et al. 2002). PVC when dechlorinated mechanochemically, using oxide powders such as Fe_2O_3, CaO, SiO_2 and Al_2O_3, in air, has shown to decrease the molecular weight of PVC (Inoue et al. 2004). Atmospheric ozone, even in every modest concentration, speeds up the aging of polymeric materials (Kefeli et al. 1971). When polymers are exposed to ozone, many carbonyl and unsaturated carbonyl products, based on ketones, aliphatic esters, and lactones as well as aromatic carbonyl associated with the styrene phase, are formed rapidly, followed by gradual formation of ether, hydroxyl, and terminal vinyl groups (Allen et al. 2003). Ozone mainly interacts with polymer chain aromatic rings, C=C bonds, or saturated hydrocarbon links. The degradation reaction proceeds via formation of unstable intermediates like peroxy radicals that can isomerize or degrade, thereby causing decomposition of macromolecules (Ghosh and Ray 2004). Ozone attacks polystyrene rather slowly compared to unsaturated polymers. Reaction products (i.e., 18% ketone, 35% peroxide, and 47% acid) are similar to those expected after free radical chain oxidation processes (Singh and Sharma 2008). The addition of a catalyst lowers the temperature of decomposition, improves the quality of products obtained from pyrolysis of plastic wastes, and also enables a given selectivity to a certain product to be achieved. Solid acid catalysts like **zeolites** favor hydrogen transfer reactions due to the presence of multiple active sites. The access of molecules to catalyst's reactive sites is limited to the pore size as well as the growth of end products inside the pores. Zeolite catalysts, therefore, may produce molecular sieving and shape selectivity (Marcilla et al. 2003). Polyolefins are thermally or catalytically degraded into gases and oils. Zeolite, non-zeolite catalyst (Lin and Yen 2005), transition metal catalysts, i.e., Cr, Mo, Ni, Fe, Co supported over Al_2O_3 and SiO_2 (Williams and Bargi 2004) and Pt–Co, Pt–Mo supported over SiO_2 (Gimouhopoulos et al. 2000), are major catalytic combinations employed for degradation. Biodegradation involves biochemical transformation and mineralization of polymers using microorganisms. It is nowadays described as the sustainable technique due to its easy and environmental-friendly nature.

2 Microbial Participation

Biodegradation of a polymeric material is a degradation brought about by the enzymatic action of naturally occurring microorganisms like bacteria; fungi yeast etc. into metabolites such as H_2O, CH_4, CO_2, biomass etc. (Lenz 1993; David et al. 1994; Chandra et al. 1998; Mohanty et al. 2000). In biodegradation of polymers, almost complete transformation of material happens, whereas in ***biodeterioration*** or ***biocorrosion,*** the change is observed only on polymer structure or in its composition (Gu et al. 2003). In both cases, structural integrity is lost due to reduction in molecular weight.

Microorganisms gain energy by catalyzing energy-producing chemical reactions that transport electrons away from the pollutant polymer. The organic part is oxidized, and the chemical that gains the electrons is reduced. The energy achieved from these electron transfers is then utilized along with some electrons and carbon from the contaminant, for the production of more cells. Several factors including accessibility to water, oxygen usage, minerals, temperature, pH, redox potential, carbon, and energy source influence the growth of microorganisms (Sand 2003) (Table 1).

Table 1 Conditions for microorganism survival (Mohan 2011)

S. no.	Factor	Condition range
1.	Temperature	−5 to +116 °C
2.	pH	0–13
3.	Redox potential	(−)55 to (+)850 mV
4.	Saltness	Ultrapure H_2O to saturated H_2O
5.	Radiation	Biofilms on UV lamps, irradiation units, and nuclear power plants
6.	Nutrient conc.	From 10 μg/L in drinking/purified water up to any carbon sources

2.1 *Role of Enzymes*

Enzymes are present in every living cell and therefore in all the microorganisms. Microorganisms are highly adaptive to environment and secrete both ***endo-enzymes*** and ***exo-enzymes*** that attack the substrate and cleave the molecular chains into segments (Lugauskas et al. 2003; Huang et al. 1990). Amounts of various enzymes produced by different microorganisms vary with species and even within the strains of the same species. The enzymes have proteins with complicated chemical structure, high molecular weights, and hydrophilic groups such as –OH, –COOH, and $-NH_2$ (Potts 1978) which can attack and eventually destroy almost anything.

In presence of enzymes, accelerated reaction rates (of 10^6–10^{20}) without creating production of undesirable products can be observed (Lenz 1993). Generally, over 2000 types of enzymes are present in a biological system, each one performing one specific chemical function. The following enzymatic factors greatly influence the susceptibility of polymers toward microbial attack:

(i) enzyme availability;
(ii) sites in the contaminant/polymer for enzyme attack;
(iii) enzyme specificity for that contaminant;
(iv) presence of co-enzyme if needed.

Microorganisms are incapable of directly transporting the lengthy and hydrophobic polymer chains into their cells via outer cell membranes. To utilize polymer molecules as carbon source, microbes excrete extracellular enzymes, which depolymerize the polymers outside the cells. Both extracellular and intra-cellular *depolymerases* enzymes are actively involved in biological degradation of polymers. A bacterium could constantly produce all of the enzymes required for degradation or else could start enzyme synthesis as required to metabolize when needed (Albertsson et al. 1987).

2.2 *Degradation Mechanisms*

During degradation, exo-enzymes from microorganisms break down complex polymers into short chains or smaller hydrophilic molecules such as dimers, oligomers, and monomers, able to pass the outer semi-permeable bacterial membranes and to be utilized as carbon and energy sources (Gu 2003). The following are the steps toward any polymeric degradation and can work individually or jointly (Fig. 1).

2.2.1 Solubilization

Solubilization is dependent on the hydrophilicity of the polymer. Hydration of polymer chains takes place when structure stabilized by van der Waals forces and hydrogen bonds disrupt. After hydration, the polymer chains may become water soluble and/or the polymer backbone may be cleaved by chemical- or enzyme-catalyzed hydrolysis to result in the loss of polymer strength (Ishigaki et al. 1999). In case of cross-linked polymers, the polymer strength may be reduced by cleavage of either of the polymer backbone, cross-linker, or pendent chains. In non-swellable (non-water absorbing) polymer systems, the decrease in molecular weight may lead to the loss of coherence between polymer chains (Park et al. 2011). In a study exploring biodegradation of poly aromatic hydrocarbons (PAHs), enhancement of solubility and biodegradability of hydrophobic hydrocarbons was achieved using biosurfactants (Barkay et al. 1999)

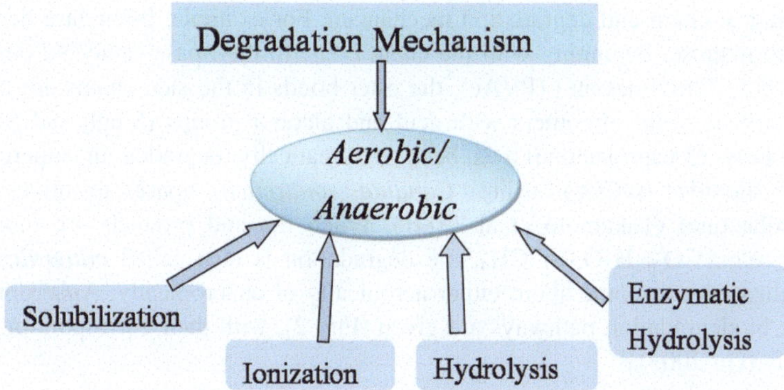

Fig. 1 Degradation mechanism

2.2.2 Ionization

Some water-insoluble polymers can be solubilized by ionization or by protonation of a pendent group. Poly acids become soluble at high pH (Chambliss 1983), and poly (vinyl acetate phthalate) (PVAP) and hydroxyl propyl methyl cellulose phthalate are ionized at a lower pH (cellulose acetate phthalate becomes water soluble at a pH > 6).

2.2.3 Hydrolysis

Water-insoluble polymers with pendent anhydride and/or ester groups may be solubilized if anhydride or esters hydrolyze to form ionized acids on the polymer chain. For example, poly (methacrylate) and poly (methyl methacrylate) are water insoluble but become water soluble upon hydrolysis of the pendent esters and subsequent ionization of the carboxylic group (Murthy et al. 1986). For hydrolysis to occur, the polymer has to contain hydrolytically unstable bonds, reasonably hydrophilic for contact with water. Polyesters are degraded mainly by simple hydrolysis via non-enzymatic, random hydrolytic ester cleavage, and its duration is determined by the initial molecular weight and chemical structure of the polymer.

2.2.4 Enzyme-Catalyzed Hydrolysis

Enzymes act as catalysts for particular reactions or a series of reactions, such as reduction, oxidation, esterification, hydrolysis, synthesis, and molecular interconversions. The enzymatic degradation of some natural polymers occurs by the unzipping or chain-end degradation mechanism. For example, β-amylase degrades starch to maltose, beginning with the chain ends. In the lipase-catalyzed degradation of poly (vinyl acetate) (PVAc), the ester bonds in the side chains are broken particularly to yield oligomers with acid and alcohol groups (Singh and Sharma 2008). Poly (3-caprolactone) has been enzymatically degraded in supercritical carbon dioxide ($scCO_2$) using *Candida antarctica* lipase to give oligo (3-caprolactone) (Takamoto et al. 2001). When the end products are inorganic species, e.g., CO_2, H_2O, or CH_4, the degradation is also called ***mineralization.*** Mineralization can come about either aerobically or anaerobically. Anaerobic and aerobic biodegradation pathways are given (Fig. 2), with their corresponding final products (Gu 2003).

2.3 *Aerobic Respiration*

In aerobic respiration, microbes use O_2 to oxidize part of the carbon in the polymer to CO_2, and rest of the carbon is used to produce new cell mass. In the process, O_2

Fig. 2 Aerobic and anaerobic degradation pathways

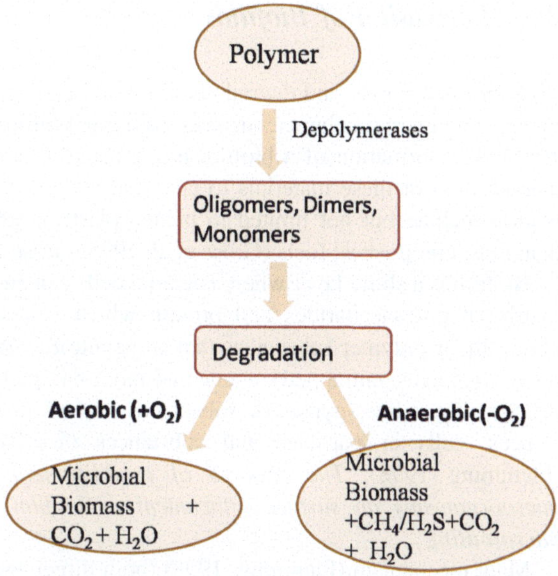

gets reduced, producing H_2O. Thus, the major by-products of aerobic respiration are carbon dioxide, water, and an increased population of microorganisms. The by-products produced during conversion which enter into the microbial cell can be utilized as the energy source (Shimao 2001).

2.4 Anaerobic Respiration

In anaerobic respiration, nitrate ($NO3^-$), sulfate (SO_4^{2-}) metals such as iron (Fe^{3+}) and manganese (Mn^{4+}), or even CO_2 can accept electrons from the degraded contaminant (National Research Council 1993). Hence, anaerobic respiration uses inorganic chemicals as electron acceptors. By-products of anaerobic respiration are new cell matter, H_2S, N_2 gas, $CH_{4,}$ and reduced forms of metals.

Given that thermodynamically, O_2 is a more efficient electron acceptor than both $SO_2^{4-}$and CO_2; **aerobic processes yield much more energy** and are capable of supporting a greater population of microorganisms than anaerobic processes. It is important to note that biodeterioration and degradation of polymer substrate can rarely arrive at 100% because a small fraction of the polymer will always be incorporated into microbial biomass, humus, and other natural products (Alexander 1977; Narayan 1993).

2.5 Formation of Biofilms

Microbes deteriorate and degrade both natural and synthetic polymers by adhering to and colonizing polymer surfaces forming biofilms (Gu et al. 2000; Mitchell 1996). The formation of a biofilm is a precondition for substantial corrosion and deterioration of these materials to take place. Polymers susceptible to biofilm formation include but not limited to paints, plastics, adhesives, sealants, lubricating materials, composites, fuels (Gross et al. 1995; Gross 1993; Lugauskas et al. 2003).

Biofilm is a slimy layer where bacterial cells can encase themselves in a hydrated matrix of polysaccharides and protein which is composed of water (80–95%), extracellular polymer substances that constitute 85–98% of the organic matter, the microorganisms, entrapped organic and inorganic particles (e.g., humic substances, debris, clay, silica, gypsum), substances sorbed to extracellular polymeric substances, cells or particles and substances dissolved in the interstitial water (Flemming 1998). *The process of establishment of complex community of microorganisms on surface attachment as biofilm is known as* ***biofouling*** *or* ***microfouling***

Main mechanism (Flemming 1998) (both direct and indirect) through which the structure and function of synthetic polymeric materials can be damaged by biofilms includes as follows:

(i) coating the surface, masking surface properties, and contaminating adjacent media such as H_2O;
(ii) increasing the leaching additives and monomers out of the polymer matrix;
(iii) attack by enzymes or radicals of biological origin to polymer and additives, leading to both embrittlement and loss of mechanical stability;
(iv) accumulating water and penetrating the polymer matrix with microbial filaments, causing swelling and increased conductivity;
(v) excretion of lipophilic microbial pigments that lead to unwanted colors in the polymer.

2.6 Properties Deterioration

Mechanical properties of polymers are principally dictated by length of the polymer chains and are weakened by chain scissions. A single endo-cleavage in a polymer chain can reduce the molar mass up to 50% and hence cause significant alterations in the mechanical properties such as embrittlement, increased modulus, change in weight, dimensions. When plasticizers are removed from the plastic materials by microorganisms, embrittlement can also occur, whereas deterioration in electric properties such as dielectric constant, insulation resistance power factor, and dielectric strength are observed because of the surface growth and pH variations resulted by excreted metabolic products.

3 Biodegradation of Plastics Using Microorganisms

The potential of microbes for plastic degradation first came to light in 1961 when Fuhs reported that several microorganisms were capable of consuming paraffin as a carbon source. Plastic-degrading microbes were isolated from a multiple sources such as rhizosphere soil of mangroves, marine water, polythene buried in the soil, and soil at the dumping sites. Comparative study of polyethylene and paraffin degradation revealed that bacteria had utilized polyethylene as carbon source (Jen-hou and Schwartz 1961). The mechanism put forward by Gautam et al. (2007) reported that paraffin molecules first get converted to an alcohol by a *monooxygenase* enzyme; then alcohol is oxidized into an aldehyde by alcohol *dehydrogenase* enzyme. Further an aldehyde *dehydrogenase* coverts aldehyde to fatty acid, which then undergoes β-oxidation inside cells (Scheme 1).

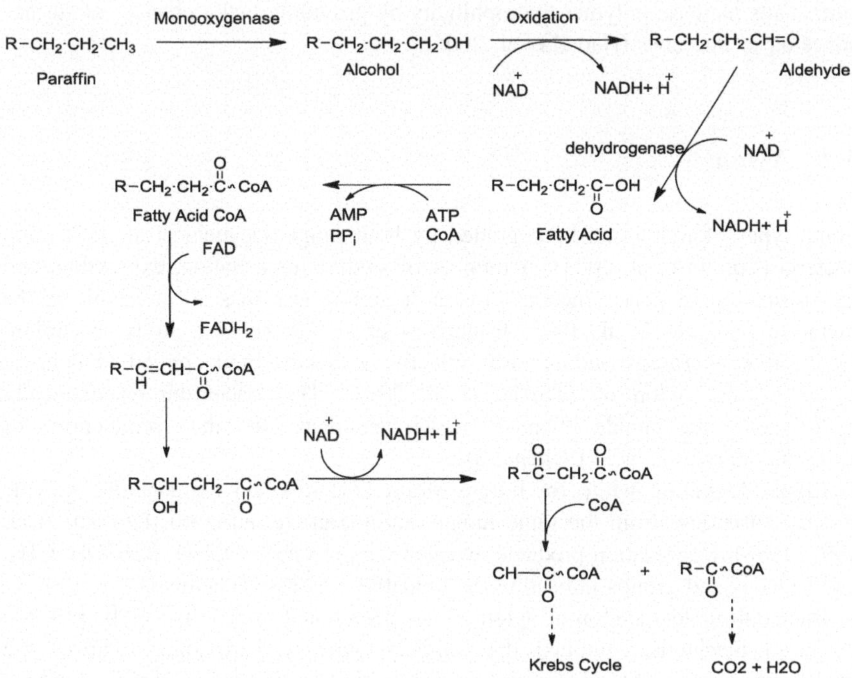

Scheme 1 Paraffin biodegradation pathway (Gautam et al. 2007)

3.1 Polyethylene (PE)

Hydrophobic nature of polyethylene (PE) restricts its bioavailability (Mahalakshmi et al. 2012). It was found that lower molecular hydrocarbon (less than 620) oligomers favor the growth of microorganisms while this is not possible in high molecular weight PE. To increase the effectiveness of the process, it is often coupled with some form of physical and chemical treatments like photo-oxidation, nitric acid oxidation, UV irradiation, and thermal treatment which favor its biodegradation (Hadad et al. 2005). PE molecules when treated with UV radiations form radical. Oxygen absorbed from atmosphere results in formation of hyperoxides which ultimately bring in C=O group in polymer (Scheme 2).

Various modifications affect PE biodegradation (Table 2). Chemical pretreatment of polyethylene with 0.5 molar nitric acid and sodium hydroxide speeds up the subsequent biodegradation using *Pseudomonas* sp. (Nwachkwu et al. 2010). Genera associated with PE biodegradation are *Gordonia* and *Nocardia, Bacillus, Lysinibacillus, pseudomonas, staphylococcus, streptococcus, micrococcus, streptomyces, rhodococcus, proteus, listeria, vibrio, bravibacillus, serratia, nocardia, diplococcus, moraxella, penicillium, arthrobacter, aspergillus, phanerochaete, chaetomium,* and *Gliocladium* (Bonhommea et al. 2003; Koutny et al. 2006a, b; Arutchelvi et al. 2008; Restrepo- Flórez et al. 2014; Grover et al. 2015). Non-ionic surfactants increase polymer hydrophilicity by providing better chances of surface adhesion to microbes (Hadad et al. 2005).

3.2 Nylon

Some types of nylon can be degraded by both fungi (Deguchi et al. 1997) and bacteria (Tomita et al. 2003). A number of studies for isolation, expression, and transformation of genes encoding nylon-degrading enzymes are available in the literature (Kakudo et al. 1993; Kanagawa et al. 1989). In a study, bacterium *Geobacillus thermocatenulatus* was effectively isolated and applied to degrade nylon 12 and nylon 66 (Tomita et al. 2003). This bacterium enzymatically hydrolyzed the amide bonds, which resulted in the production of 12-aminododecanoic acid (Scheme 3).

Lignin-degrading white rot fungus strain IZU-154 and extracellular enzyme fungus peroxidase from the same fungal strain degrade nylon 66 (Deguchi et al. 1997, 1998). Degradation products revealed the presence of CHO, NHCOH, CH_3, and $CONH_2$ end groups indicating the oxidative process (Scheme 4).

Bacterial biodegradation of nylon 66 has been noted to be a hydrolytic process. Proposed mechanism implied that initially, enzyme peroxidase oxidizes the methylene group next to N atom in polymer backbone forming –CH radical, which then undergoes oxidation (Nomura et al. 2001). The enzymes associated with nylon

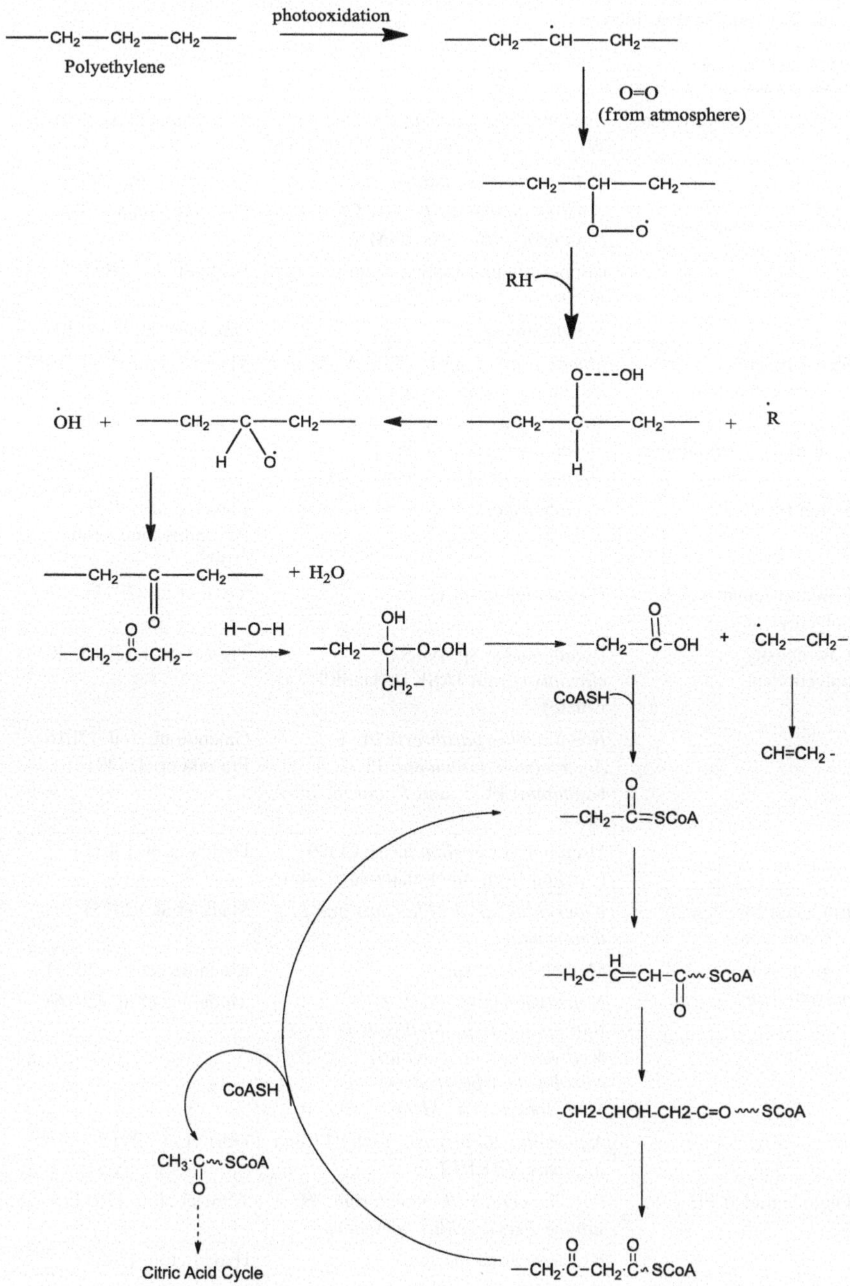

Scheme 2 Biodegradation of PE (Gautam et al. 2007)

Table 2 PE-degrading microbes

Polyethylene type/ modification	Microbes	References
Unmodified polyethylene	*Lysinibacillus fusiformis* VASB14/ WL *Bacillus cereus* strain VASB1/TS	Shahnawaz et al. (2016)
	P. fluorescens, S. aureus, and *A. niger*	Thomas et al. (2015)
	Bacillus amyloliquefaciens BSM-1, *B. amyloliquefaciens* BSM-2	Das and Kumar (2015)
	Copper-binding laccase from *R. Ruber*	Santo et al. (2013)
	S. marcescens	Odusanya et al. (2013)
Starch-based polyethylene	*Mucor rouxii* NRRL 1835, *A. flavus* and strains of *Streptomyces*	Mahalakshmi et al. (2012)
Lignocelluloses Heat treated degradable	*Streptomyces badius* 252 and *Streptomyces setonii* 75Vi2 *Streptomyces viridosporus* T7A	Pometto et al. (1992)
Starch blended	*P. chrysosporium, Streptomyces* sp.	Flavel et al. (2006), Arvanitoyannis et al. (1998)
Low molecular weight polyethylene	*Pseudomonas* sp. E4	Yoon et al. (2012)
Low-density polyethylene	*Pseudomonas* sp. AKS2 *clavatus* (strain JASK1) (landfill isolate)	Tribedi et al. (2015, 2014)
	Brevibacillus parabrevis PL-1, *Acinetobacterbaumannii* PL-2, *A. baumannii* PL-3, and *P. citronellolis* PL-4	Gajendiran et al. (2016), Pramila et al. (2012)
	Streptomyces coelicoflavus 15399T (isolated from oil-contaminated site)	Duddu et al. (2015)
Irradiated low-density polyethylene	*Aspergillus* sp., *P. Lilacinus* and *L. theobromae*	Sheik et al. (2015)
High-density polyethylene	*Methylobacter* sp.	Muenmee et al. (2015)
	Nitrosomonas sp. AL212, *Burkholderia* sp., *Methylobactor* sp., *Methylococcus capsulatus, Nitrobacter winogradkyi, Methylocystic* sp., *Methylocella* sp.	Muenmee et al. (2016)
	Aspergillus tubingensis VRKPT1 and *A. flavus* VRKPT2	Devi et al. (2015)
Photodegraded PE	*Bacillus cereus, B. megaterium, B. subtilis Brevibacillus borstelensis*	Abrusci et al. (2011)
	*B. borstelensis*707	Hadad et al. (2005)
Modified Low-density PE [blend of titania (TiO_2) and starch]	*P. aeruginosa*CA9, *Burkholderia seminalis* CB12, and *Stenotrophomonas pavanii* CC18	Mehmood et al. (2016)

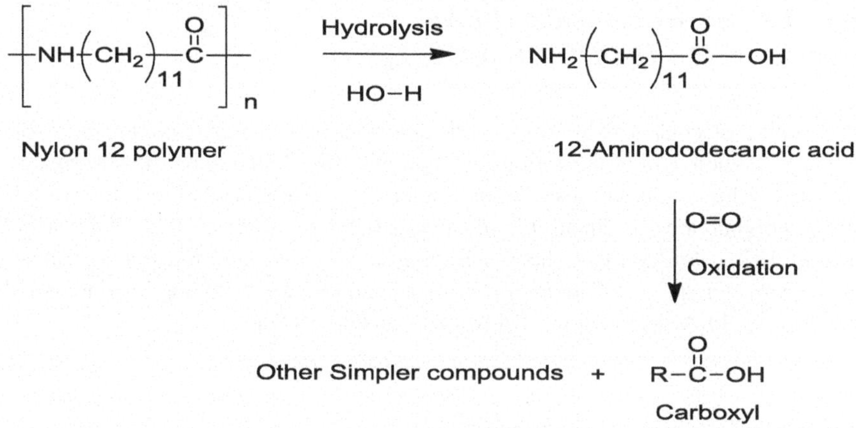

Scheme 3 Nylon 12 degradation

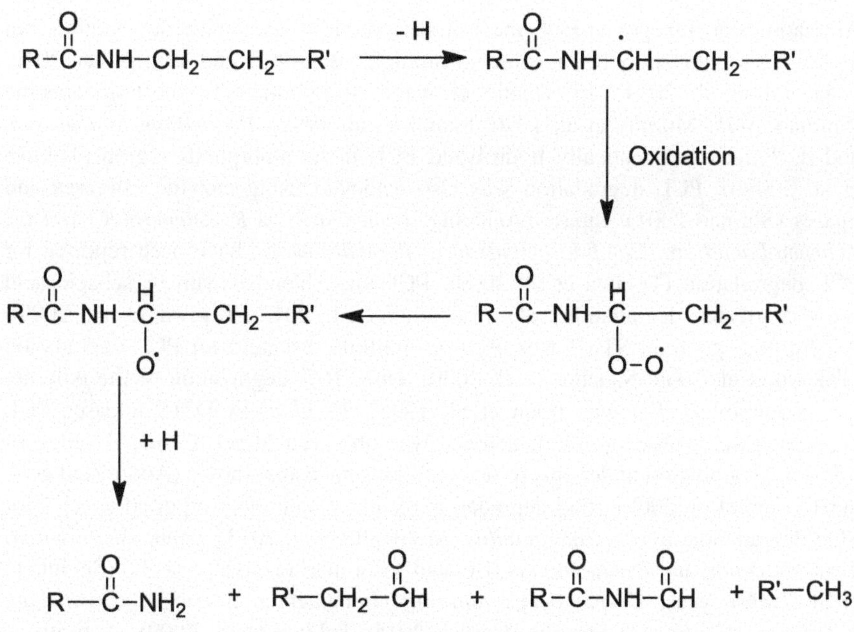

Scheme 4 Oxidative process for Nylon degradation

degradation are 6-aminohexanoate cyclic dimer hydrolase (F-EI), 6-aminohexanoate-dimer hydrolase (F-EII), and 6-aminohexanoate oligomer hydrolase (F-EIII) (extracted from Flavobacterium sp. KI72; Kakudo et al. 1993; Kinoshita et al. 1977, 1981; Negoro 1992).

3.3 *Polyethylene Adipate (PEA) ($[-OCH_2CH_2OOC(CH_2)_4CO-]_n$)*

PEA with molecular weight 3000 can be degraded by using *penicillium sp.* as sole source of carbon. Out of the various other strains, 14-3 found to be showed greatest activities. This strain can also degrade aliphatic polyesters like polyethersulfone (PES), polybutylene succinate (PBS), and polybutylene adipate (PBA) (Tokiwa and Suzuki 1974). Furthermore, lipases from *Rhizopus delemar*, *Rhizopus arrhizus*, *Achromobacter* sp., and *Candida cylindracea* and esterase from hog liver are noted for showing PEA degradation (Tokiwa and Suzuki 1977a, b).

3.4 *Polycaprolactone (PCL) ($[-OCH_2CH_2CH_2CH_2CH_2CO-]_n$)*

Aliphatic ester linkages in PCL make it susceptible to enzymatic degradation, but hydrolysis of this polymer is time-consuming (Woodruff and Hutmacher 2010; Gajanand et al. 2014). Enzymatic cleavage of polymers is substrate specific (Shimao 2001; Murphy et al. 1996; Schink et al. 1992). *Penicillium funiculosum* and *A. Flavus* enzymatically hydrolyzed PCL in its amorphous region (Tokiwa et al. 2009b). PCL degradation was also achieved using enzyme esterases and lipases (Shimao 2001). Lipase-producing species such as *R. delemar, R. Arrizus, Achromobacter* sp., *Candida cylindracea, Penicillium* sp., have been reported for PCL degradation (Tokiwa et al. 2009). PCl when blended with % sebacic acid showed greater degradation rate (Salgado et al. 2012; Tokiwa et al. 2009). Aspergillus sp. strain ST-01 was used on multiple instances for PCL degradation (Tokiwa et al. 2009; Sanchez et al. 2000). Only 10% degradation of the polymer was noted by *P. lilacinus* (Oda et al. 1995). *P. lilacinus* D218 releases PCL depolymerase, whose optimum activity was observed at 30 °C in pH range of 3.5–4.5. Degradation under anaerobic conditions have also shown (Abou-Zeid et al. 2001; Yagi et al. 2009) good degradation results. Some other reported species for PCL degradation are *Bacillus pumilus (*Motiwalla et al. 2013); genus *Alcanivorax, Tenacibaculum* and *Pseudomonas* (isolated from marine environments) (Sekiguchi et al. 2009); yeast *Cryptococcus laurentii* (Benedict et al. 1983); *Clostridium botulinum, Alcaligenes faecalis* (Caruso 2015; Tokiwa et al. 2009). Activity of phytopathogenic fungi against PLC is suspected to be attributed to the presence of cutinases.

3.5 *Polybutylene Succinate (PBS) ($[-O(CH_2)_4OOC(CH2)_2CO-]_n$) and Polyethylene Succinate (PES) ($[-O(CH_2)_2OOC(CH_2)_2CO-]_n$)*

Strain *Amycolatopsis* sp. HT-6 has found to degrade PBS, PHB, PCL (Pranamuda et al. 1995). Actinomycetes (thermophilic) that degrade PBS and PES are *Microbispora rosea, Excellospora japonica, Excellospora viridilutea* (Jarerat and Tokiwa 2001; Duddu et al. 2015; Seretoudi et al. 2002; Hoang et al. 2007). Other microbes and genus reported are *Pseudomonas* sp. AKS2 strain (Qiu et al. 2003; Liu et al. 2012; Tribedi and Sil 2013), *Saccharomonospora, Streptomyces, Microbispora, Thermoactinomyces,* and *Actinomadura* (Tseng et al. 2007), *Streptomyces* sp. (Calabia and Tokiwa 2004) *Bacillus* sp. TT96 (Tokiwaet al. 2009), Mesophilic *Bacillus, Paenibacillus, Bacillus pumilus, B. subtilis,* and *Paenibacillus amylolyticus* (Tezuka et al. 2004), *Rhizopus delemar* (Seretoudi et al. 2002). Biostimulation helps in degradation by forming a new microbial community (Tribedi and Sil 2013). Hydrocarbon-degrading microbes from oil-contaminated marine environment have also been identified (Hazen et al. 2010). Microbial serine proteases have also been reported for PES biodegradation (Lim et al. 2005).

3.6 *Poly (β-Propiolactone) (PPL) ($[-OCH_2CH_2CO-]_n$)*

Microbes belonging to Bacillus sp. have been employed for degrading this polymer (Nishida et al. 1998). Some other strains from thermophilic *Streptomyces* identified for PPL degradation are *Variovorax paradoxus, Acidovorax* sp, *Sphingomonas paucimobilis, Rhizopus delemar* (Tokiwa and Suzuki 1977a, b), poly (3-hydroxybutyrate) depolymerase (Tokiwa and Calabia 2006).

3.7 *Aliphatic–Aromatic Copolyesters (AACs)*

Aliphatic–aromatic copolyesters generally consisting of **polyethylene terephthalate (PET), PCL, polyethylene isophthalate (PEIP), and polybutylene terephthalate** were hydrolyzed by *Rhizopus delemar* lipase (Tokiwa and Suzuki 1981), although the susceptibility decreased with increased aromatic content. Kleeberg et al. (1998) reported *Thermobifida fusca* and its enzyme hydrolase for AAC degradation. This enzyme is highly similar to triacylglycerol lipase from *Streptomyces albus* G and triacylglycerol acylhydrolase from *Streptomyces sp.* M11 (Kleeberg et al. 2005). PET films were degraded when treated with (a) cutinase from *Humilica insolens* and *Fusarium solani pisi* (Donelli et al. 2010) observed 97% weight loss; (b) Pseudomonas mendocina with 5% weight loss (Heumann et al. 2006) and cracks on fiber surface (Zhang et al. 2004). Enzymes from

Aspergillus sp., *Beauveria sp*., *Thermobifida fusca*, and commercial enzymes (TEXAZYME PES sp5 and Lipase PS) increased the hydrophilicity of PET fibers (Heumann et al. 2006; Ronkvist et al. 2009; Naharjan et al. 2006). Besides, the microbial capability of forming biofilm over PET film pointed out the possibility of degradation under natural conditions (Webb et al. 2009). Furthermore, to increase biodegradability of PET, studies have been done to modify the polymer in order to decrease the intermolecular cohesion. Nowak et al. (2011) modified PET films with Bionolle polyester in the presence of filamentous fungi *Penicillium funiculosum.* After 84-day incubation period, considerable reduction in quantity of aromatic rings was observed. In another study, series of copolyesters were prepared using waste PET beverage bottles and L-lactic acid. Samples were incubated in thermophilic sludge at 55 °C in alkaline environment for a period of 394 days. This modified copolyester was more inclined to hydrolytic biodegradation than PET waste. Copolyester mineralization gave methane-rich biogas in 69% and 34% of theoretical amount for two samples, respectively. The rate of anaerobic biodegradation decreased with higher starting content of aromatic sequences (Hermanová et al. 2015). The degradation of transparent PET sheets by microbes of *Nocardia* species and esterase was studied. Esterase was found to be involved in the biodegradation. Although the biodegradation was slow and weak, microbes and esterase could act on the polyethylene terephthalate (Sharon and Sharon 2017).

3.8 Poly (3-Hydroxybutyrate) (PHB) ($[-O(CH_3)CHCH_2CO-]_n$)

First reports of PHB-degrading microbes were from Bacillus, Streptomyces, and Pseudomonas species (Chowdhury 1963). Mostly, microbes that utilize PHB do so at ambient temperatures and they have been isolated from soil *Pseudomonas lemoigne*, *Comamonas sp*., *Acidovorax faecalis*, *Aspergillus fumigatus*, *Variovorax paradoxus,* sludge *Pseudomonas, Ilyobacter delafieldi*, *Alcaligenes faecalis,* and sea and lake water *Pseudomonas stutzeri*, *Comamonas testosterone* (Lee 1996). A thermophilic *Streptomyces* sp. shows higher PHB-degrading activity than thermotolerant and thermophilic strains from same species (Calabia and Tokiwa 2004). *Aspergillus* sp. degraded 90% of PHB film in five (Sanchez et al. 2000). Actinomycetes associated with PHB, PCL, and PES depolymerization are members of the genera *Streptomyces, Microbispora, Actinomadura, Saccharomonospora, and Thermoactinomyces* (Tseng et al. 2007). Another study done to assess the biodegradation pattern of blends of polylactic acid (PLA) and polyhydroxybutyrate (PHB) at different weight ratios showed biodegradability of blends increased as ratio of PHB increases (Zhang and Thomas 2011). Biocomposites prepared form PHB and potato peel waste fermentation residue also showed complete degradation in eight months (Wei et al. 2015).

3.9 *Polylactic Acid (PLA) ($[-O(CH_3)CHCO-]_n$)*

Polylactic acid, a biodegradable plastic is used freely in medicine industry. T16-1 strain of *Actinomadura keratinilytica* NBRC 104111 has been illustrated for producing enzyme that degrades PLA. 60% degradation of PLA films have been observed by *Amycolatopsis* sp. over a period of fourteen days (Babu et al. 2013; Sukhumaporn et al. 2012; Garlotta 2002; Pranamudaet al. 1997). *Bacillus amyloliquefaciens, a mesophilic bacteria,* are also known to degrade PLA (Prema and Uma 2013). Masaki et al. (2005) used lipase enzyme purified from strain *Cryptococcus* sp. S-2 for PLA degradation. When dye having PLA films as a constituent was inspected, enzymes from *Aneurinibacillus migulanus* showed the same activity as proteinaseK for PLA degradation (Chaisu et al. 2015). PLA depolymerase enzyme is also reportedly produced by *Pseudomonas tamsuii* TKU015 (Liang et al. 2016).

3.10 *Polyurethanes (PUR)*

Polyurethanes are both polyether PUR and polyester PUR (Shimao 2001). Enzyme proteases, esterases, ureases, and lipases have been noted to bring about PUR hydrolysis. *Aspergillus terreus, Trichoderma* sp., and *Chaetomium globosum* have also been reported for the producing ureases and esterase that hydrolyze PUR (Bhardwaj et al. 2013; Howard 2012). A total of 22 fungal strains were screened, and up to 95% urease activity was found in most of them (Loredo-Trevino et al. 2012). Microbial species identified as polyurethanes degraders are *Candida ethanolica, Fusarium solani* (Schink et al. 1992; Zafar et al. 2013), *Aspergillus fumigatus, Emericella, Arthrographis kalrae, Lichthemia, Fusarium solanii, Plectosphaerella, Phoma, Nectria, A. niger, Corynebacterium* sp., *Neonectria, Thermomyces, P. aeruginosa, Alternaria, Bacillus* and *Comamonas; C. acidovorans* TB-35, *Pestalotiopsis microspora* (Bhardwaj et al. 2013; Howard 2012; Akutsu et al. 1998; Zafar 2013; Flavel et al. 2006; Shimao 2001). Proteolytic enzymes—urease and papain—also degrade polyurethane used in medical devices. Ma and Wong (2013) investigated the esterase activity of A. Flavus. Different experiments have illustrated the important role of Genes*pue*A and *pue*B in PUR degradation (Howard 2012). *pue*A helps in enhancing PUR degradation by increasing the cellular density as shown by gene silencing experiments (Howard et al. 2007).

The various plastic-degrading microbes are summarized in Table 3.

Table 3 Plastic-degrading microbes

S. no.	Plastic	Microorganisms	References
1	Polyethylene adipate (PEA)	*Penicillium* sp., *Rhizopus delemar, Rhizopus arrhizus, Achromobacter* sp., and *Candida cylindracea*	Tokiwa and Suzuki (1977a, b)
2	Polycaprolactone (PCL)	*Penicillium funiculosum, A. flavus, R. delemar, R. arrizus, Achromobacter* sp., *Candida cylindracea, Penicillium* sp., *Aspergillus* sp., *Cryptococcus laurentii, Bacillus pumilus, Alcanivorax, Tenacibaculum, Pseudomonas, Clostridium botulinum, Alcaligenes faecalis, P. lilacinus*	Tokiwa et al. (2009), Benedict et al. (1983), Motiwalla et al. (2013), Caruso (2015), Tokiwa et al. (2009), Oda et al. (1995)
3	Polybutylene succinate (PBS) Polyethylene succinate (PES)	*Amycolatopsis* sp., *Microbispora rosea, Excellospora japonica, Excellospora viridilutea, Pseudomonas* sp., *Saccharomonospora, Streptomyces, Microbispora, Thermoactinomyces Actinomadura, Bacillus* sp., *Bacillus pumilus, B. subtilis, Paenibacillus amylolyticus, Rhizopus delemar*	Pranamuda et al. (1995), Qiu et al. (2003), Liu et al. (2012), Tribedi and Sil (2013), Tseng et al. (2007), Tokiwa et al. (2009), Tezuka et al. (2004), Hazen et al. (2010), Seretoudi et al. (2002)
4	Poly (βpropiolactone) (PPL)	*Bacillus* sp., *Acidovorax* sp., *Variovorax paradoxus, Sphingomonas paucimobilis, Rhizopus delemar, Streptomyces* sp., *Thermobifida fusca, Streptomyces sp., Humicola Insolens, Aspergillus sp., Beauveria sp. Fusarium solani, Pseudomonas mendocina, Penicillium funiculosum, Nocardia sp.*	Nishida et al. (1998), Tokiwa and Suzuki (1977a, b), Tokiwa and Calabia (2006), Tokiwa and Suzuki (1981), Kleeberg et al. (2005), Donelli et al. (2010), Heumann et al. (2006), Ronkvist et al. (2009), Zhang et al. (2004), Nowak et al. (2011), Sharon and Sharon (2017)
5	Poly (3-hydroxybutyrate) (PHB)	*Bacillus, Pseudomonas, Streptomyces* species, *Comamonas sp., Pseudomonas lemoigne, Aspergillus fumigatus, Acidovorax faecalis, Variovorax paradoxus, Alcaligenes faecalis, Pseudomonas, Ilyobacter delafieldi, Comamonas testosterone, Pseudomonas*	Chowdhury (1963), Lee (1996), Calabia and Tokiwa (2004), Sanchez et al. (2000), Tseng et al. (2007)

(continued)

Table 3 (continued)

S. no.	Plastic	Microorganisms	References
		stutzeri, Actinomadura, Microbispora, Streptomyces, Thermoactinomyces, Saccharomonospora	
7	Polylactic Acid (PLA)	*Actinomadura keratinilytica, Bacillus amyloliliquefaciens, Amycolatopsis, Cryptococcus* sp, *Aneurinibacillus migulanus, Pseudomonas tamsuii*	Pranamudaet al. (1997), Garlotta (2001), Sukhumaporn et al. (2012), Babu et al. 2013), Prema and Uma (2013), Masaki et al. (2005), Chaisu et al. (2015), Liang et al. (2016)
8	Polyurethanes (PUR)	*Trichoderma* sp., *Aspergillus terreus, Chaetomium globosu, Fusarium solani, Candida ethanolic, C. acidovorans, A. niger, P. aeruginosa, Emericella, Nectria, Lichthemia, Fusarium solanii, Corynebacterium* sp., *Neonectria, Arthrographis kalrae, Plectosphaerella, Phoma, A. flavus, Pestalotiopsis microspore, Comamonas acidovorans* TB-35, *Aspergillus fumigatus, E. coli, P. chlororaphis, Seudomonas chlororaphis, Thermomyces, Alternaria*	Bhardwaj et al. (2013), Howard (2012), Schink et al. (1992,b), Zafar et al. (2013), Howard (2012), Shimao (2001), Flavel et al. (2006), Zafar (2013), Akutsu et al. (1998), Ma and Wong (2013), Howard (2012), Shigeno-Akutsu et al. (1999)
9	Polyethylene (PE)	*Pseudomonas* sp., *Gordonia Nocardia, Staphylococcus, Serratia, Phanerochaete, Penicillium, Arthrobacter, Aspergillus, Streptomyces, Lysinibacillus, Nocardia, Gliocladium, Listeria, Vibrio, Streptococcus, Bacillus, Rhodococcus, Micrococcus, Proteus, Diplococcus, Chaetomium, Bravibacillus, Moraxella, Mucorrouxii, A. flavus, Streptomyces badius* 252 and *Streptomyces setonii*75Vi2, *Streptomyces viridosporus* T7A, *R.ruber, P. chrysosporium, Streptomyces* sp., *S. marcescens, Bacillus*	Nwachkwu et al. (2010), Restrepo- Flórez et al. (2014), Bonhommea et al. (2003), Grover et al. (2015), Koutny et al. (2006a, b), Arutchelvi et al. (2008) Mahalakshmi et al. (2012), Pometto et al. (1992), Basnett et al. (2012), Yoon et al. (2012), Odusanya et al. (2013), Das and Kumar (2015), Gajendiran et al. (2016), Thomas et al. (2015) Pramila et al. (2012), Flavel et al. (2006), Muenmee et al. (2015, 2016), Duddu et al. (2015); Abrusci et al. (2011) Devi et al. (2015), Hadad et al. (2005)

(continued)

Table 3 (continued)

S. no.	Plastic	Microorganisms	References
		amyloliquefaciens BSM-1, *B. amyloliquefaciens* BSM-2, *A. clavatus.*, *P. fluorescens, S. aureusA. niger, Brevibacillus parabrevis* PL-1, *Acinetobacter baumannii* PL-2, *A. baumannii* PL-3 and *P. citronellolis* PL-4, *Methylobacter*sp. or *Methylocella*sp., *Nitrosomonas*sp AL212, *Nitrobacter winogradkyi, Burkholderia*sp., *Methylobactor*sp., *Methylococcus capsulatus, Methylocystic*sp. *Streptomyces coelicoflavas, Bacilluscereus, B.megaterium, B. subtilis, Brevibacillus borstelensis, B. megaterium, Aspergillus tubingensis*VRKPT1, *B. borstelensis*707, *P. aeruginosa* CA9, *Burkholderia seminalis* CB12, *Stenotrophomonas pavanii* CC18; *P. lilacinus, L. theobromae, Lysinibacillus fusiformis* VASB14/WL and *Bacillus cereus* VASB1/TS.	Mehmood et al. (2016), Shahnawaz et al. (2016)
10	Polypropylene	*Pseudomonas chlororaphis, Pseudomonas stuzeri, Vibrio species*	Alariqi et al. (2006)
11	Polyvinyl Chloride	*As. Fumigatus, Phanerochaete chrysosporium, Lentinus tigrinus, As. Niger, Aspergillus sydowii*	Kirbas et al. (1999), Ali et al. (2014)
12	Polystyrene	*Azotobacter beijerinckii* HM121	Nakamiya et al. (1997)
13	Nylon	*Geobacillus thermocatenulatus, Flavo bacterium* sp. KI72, *Pseudomonas* sp. strain NK87	Tomita et al. (2003), Kakudo et al. (1993), Kanagawa et al. (1993), Kinoshita et al. (1981), (1977), Negoro et al. (1992)
14	PUR Foam & PS Foam	*Pseudomonas aeruginosa*, sp. *Cladosporium resina, Aspergillus niger, Aspergillus versicolor, Penicillium funiculosum, Chaetomium globosum, Aspergillus flavus*	Edmonds and Cooney (1968), Expandable Polystyrene Molders Association (2004)

(continued)

Table 3 (continued)

S. no.	Plastic	Microorganisms	References
15	PET and other Aliphatic–aromatic copolyesters	*Rhizopus delemar* lipase, *Thermobifida fusca, Humilica insolens* *Aspergillus sp., Beauveria sp., Thermobifida fusca,* commercial enzymes (TEXAZYME PES sp5 and Lipase PS), *Fusarium solani pisi Pseudomonas mendocina* and *Fusarium solani, Penicillium funiculosum* *Nocardia* sp., esterase	Tokiwa and Suzuki (1981), Kleeberg et al. (1998), Donelli et al. (2010), Heumann et al. (2006), Ronkvist et al. (2009), Heumann et al. (2006), Nowak et al. (2011), Sharon and Sharon (2017)

4 Physiochemical Characteristics Influencing Depolymerization

Polymer biodegradation is a heterogeneous process since polymers are multicomponent systems. These components can provide nutrients for some microbes (Kyrikou and Briassoulis 2007). All polymers are biodegradable to some degree because of the organic fraction of their composition like resins, hardener. Polymeric material structures and compositions govern their biodegradation. Single polymer can contain different structural elements, e.g., random or alternating copolymers. Also, varying structures like branched network can directly impact the polymer chain cleavage. Substantial amount of qualitative and semi-quantitative information has been gathered documenting the factors that impact the rate of degradation of synthetic polymers in biological environments (Fig. 3).

4.1 Abiotic Factors

Environmental conditions such as temperature, humidity, pH, salinity, sunlight, water, presence/absence of oxygen, culture conditions, and stress significantly impact the degradation process and also control microbial population and enzyme activity (Gu 200). Increased temperature and moisture content increase the rate of hydrolysis and microbial activity, which in turn amplify the chain scissions subsequently increasing the available polymer sites for attack. This results in faster degradation (Ho et al. 1999). Chain scissions from photodegradation reduce the molecular weight of polymer and improve the approachability of microbes to the polymer (Stevens 2003).

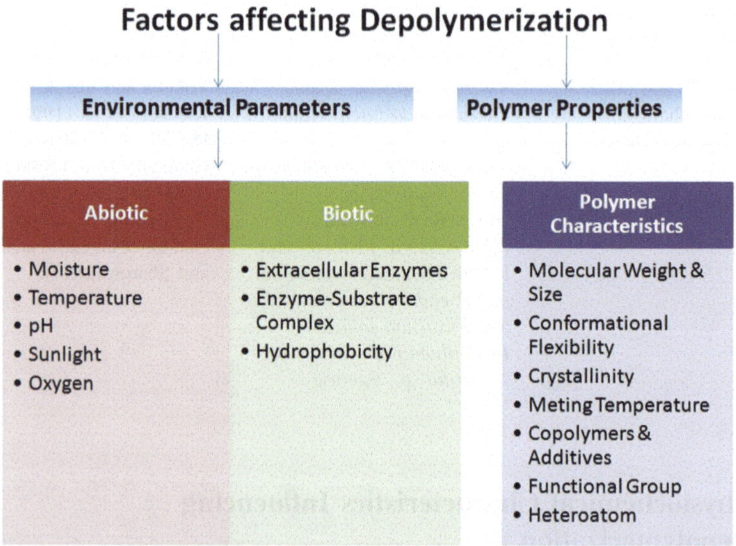

Fig. 3 Factors affecting depolymerization

4.2 Biotic Factors

4.2.1 Production of Extracellular Enzymes

Extracellular enzymes excreted by different microbes may have active sites with different structure/configuration and biodegrade certain polymers better than others. For example, *Aspergillus niger* and *Aspergillus flavus* fungi produce enzymes which are known to better digest 6-12 carbon di-acid-derived aliphatic polyesters as opposed to those from other monomers (Chandra and Rustgi 1997).

4.2.2 Enzyme–Substrate Complex Formation

Enzyme–substrate complex formation is generally explained by either ***lock and key fit*** or by ***induced-fit*** mechanisms (Donald et al. 2006; Koshland 1995). For a reaction to take place and enzyme–substrate complex to be formed, some geometrically similar pattern between enzyme and substrate is required. The lack of these similar geometric patterns between plastic substrate and available enzymes makes the succeeding biodegradation difficult.

4.2.3 Hydrophobicity

Although water solubility does not guarantee biodegradability, the particularly hydrophobic nature of plastics makes them harder to biodegrade. Enzymes and microbes often have complex metabolic pathways for mineralizing such extremely hydrophobic materials (Nakajima-Kambe et al. 1999).

4.3 Polymer Characteristics

4.3.1 Conformational Flexibility

Flexibility in a polymer makes it easily reachable for the microbes. Most polymers have highly cross-linked branched structures, which make formation of an enzyme–substrate complex harder. For example, flexible aliphatic polyesters are comparatively easier to biodegrade than rigid aromatics like polyethylene terephthalate (PET) (Omichi 1992).

4.3.2 Molecular Weight (M_W)

Microorganisms can only digest lower molecular weight segments of polymer and convert them into metabolites. Synthetic polymers with $M_W > 1000$ are found to be less susceptible to biodegradation than those with $M_W < 1000$ (Suzuki et al. 1978).

4.3.3 Degree of Crystallinity

The amorphous part of polymeric chain is prone to enzymatic attack than the crystalline part because

(a) molecules in amorphous part are freely packed, making the enzymatic attack easier and more susceptible. For example, the rate of polylactide acid degradation decreases with increasing crystallinity (Tsuji and Miyauchi 2001).
(b) Amorphous regions show high permeability to O_2. Oxidation rates depend on the reactivity of formed peroxy radicals and dissociation energies of available C–H bonds of polymer substrate. A polymer with unreactive methyl and phenyl groups or with no *H* at all shows resistance to oxidation (Bovey and Winslow 1979).

4.3.4 Melting Temperature (T_m)

Melting temperatures of polyesters is an important parameter for the enzymatic depolymerization. Higher melting temperatures imply lower polymer biodegradation (Tokiwa and Suzuki 1978, 1981; Tokiwa et al. 1979).

4.3.5 Size of the Polymer

In order to avoid mass transfer restrictions and to increase surface area available to microorganism for interaction, the polymer must be reduced in size before subjecting it to biodegradation.

4.3.6 Copolymers and Additives

Naturally occurring polymers are biodegradable but biodegradation of chemically tailored natural polymers is conditioned on type and degree of modification. Adding comonomers in a polymer increases chain irregularity, thereby reducing its crystallinity and providing easy access to microbes and H_2O. Impurities such as catalytic residues, fillers, pigments, and altered products of additives also influence the resistance to degradation. Metals like Mn act as good pro-oxidants in polyolefins making them susceptible for thermo-oxidative degradation. After heat activation in the presence of oxygen, pro-oxidants produce free radicals on hydrocarbon chain which undergo oxidation (Orhan et al. 2004). Furthermore, the pro-oxidant catalyzes chain scissions giving low molecular mass oxidation products containing –OH, –COOH, C=O groups (Jakubowicz 2003). Traces of transition metals present also speed up thermal oxidative processes inducing hydroperoxide decomposition (Zheng et al. 2005, Zhang and Thomas 2011). For example, TiO_2 delustrant makes polyamides more prone for light- and heat-induced oxidation. Making polyolefins blend with other polyester or starch also increases its susceptibility for degradation (Erlandsson et al. 1997).

4.3.7 Role of Heteroatom

It has been observed that polymers that include heteroatoms in the backbones such as polyamines and polyesters show higher susceptibility to degradation than polymers with pure carbon backbones (Zheng et al. 2005). Incorporating heterogroups like oxygen in polymer chains influences bond strength (of neighboring C–H bonds) promoting carbanion formation (e.g., in presence of bases) subsequently making them labile for thermal and biodegradation (Gowariker et al. 2000). Similarly, unsaturation in polymer chains makes polymer prone to oxidation, i.e., natural rubber is more vulnerable to degradation than polyethylene (Seymour 1971). Aromatic polyesters despite of having easily hydrolysable ester bonds have seen to be resistant to degradation. For example, PET despite having ester bonds in its polymer chains still remains non-degradable (in normal conditions) due to its aromatic content (Webb et al. 2012).

4.3.8 Introduction of Functionality

Introducing carbonyl groups like C=O in polyolefins makes them more vulnerable for photodegradation. Increase in the number of chromophores results in increased sites for photon absorption and hence increases the rate of photodegradation. Incorporation of additional chromophores like metal–metal bonds in polymer backbone too induces photodegradability. In such case, metal–metal bond can cleave homolytically when irradiated (Meyer and Caspar 1985).

4.3.9 Chemical Bonding

In thermoplastics, during addition polymerization head-to-head or tail-to-tail addition of monomers results in weak spots making plastic susceptible for degradation. For example, head-to-head linkage in poly (methyl methacrylate) improves its thermal degradation (Holland and Rae 1987). Similarly, branching in polymeric chains too increases thermal degradation (Gowariker et al. 2000).

4.3.10 Mode of Synthesis

Methods of synthesis exhibit an effect on the stability of the polymers (Hendrickson and Connole 1995), e.g., polystyrene (anionic polymerized) shows more photo-stability than its free radically formed equivalent because of the presence of peroxide residue in the latter, which is labile for photodegradation (Pospisil et al. 2006). Similarly, copolymerized polypropylene (PP) is less inclined to photodegradation than PP synthesized via bulk polymerization and by using Ziegler–Natta catalysts (Tang et al. 2005).

5 ASTM Standards for Plastic Biodegradation

The American Society for Testing and Materials (ASTM) has recognized an assortment of scientific and technological tests to assess true biodegradation in numerous plastic articles under varying environments using different analytical methods (Table 4). ASTM (standard D-5488-94d) recognizes biodegradation as "*process which is capable of decomposition of materials into carbon dioxide, methane, inorganic compounds or biomass and water in which the predominant mechanism is the enzymatic action of microorganism that can be measured as*

Table 4 ASTM standards for biodegradation of plastics

Test number	Methods
ASTM G21	Standard practice determining resistance of synthetic polymer materials to fungi
ASTM G29	Standard practice determining algal resistance of plastic films
ASTM D5511 Equivalent to ISO DIS 15985	To determine anaerobic biodegradation, under high solid conditions (total solids > 30%)
ASTM D5247	For determining the aerobic biodegradability of degradable plastics by specific microorganisms
ASTM D5526	To determine anaerobic biodegradation under accelerated landfill conditions
ASTM D5338	To determine aerobic biodegradation under controlled composting conditions
ASTM D6400	List specifications for compostable plastics
ASTM D6002	Standard guide to assess compost ability of environmentally degradable plastics
EN ISO 846	Evaluation of action of microorganism on plastics
ISO 846	Plastics: Determination of behavior under the action of fungi and bacteria. Evaluation by visual examination or measurement of changes in mass or physical properties

standard tests, in a specified period of time, reflecting available disposal conditions." Taking into account the fact that plastics more often than not have multifaceted compositions and are generally degraded via heterogeneous surface mechanism, these tests are chiefly based on concepts used for assessing low molecular weight substances. Some modifications have been done over the years for different environments in which plastics might degrade.

5.1 Test Methods for Study of Biodegradation

These tests assess the changes in the material properties of polymers. Multiple test measures are required to evaluate the biodegradability of polymer for the reasons that

- Leaching of additives, plasticizers, etc., may also result in the observed weight loss and not only polymer degradation;
- Very small change in chemical makeup of polymer could result in high reduction in material strength, e.g., 5% of mineralization can cause 90% decrease of polymer strength;
- CO_2 production may be a result of degradation of low molecular weight part of the polymer, with no degradation of longer chains (Mohan 2011).

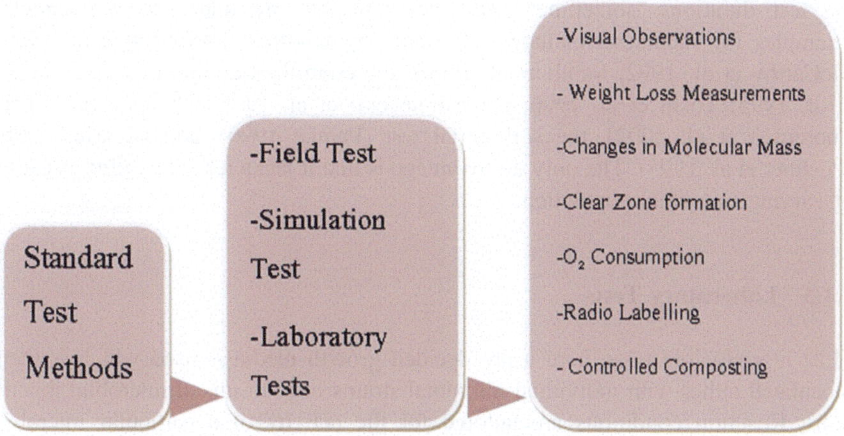

Fig. 4 Test methods used for evaluating polymer degradation

The current test methods are not sufficient and offer a very little flexibility to take into account the wide assortment of advanced polymers (like the ones used in electronics or aviation applications) consumed in different environmental conditions (Gu and Gu 2005). In principle, these tests can be divided into three types (Fig. 4).

5.1.1 Field Test(s)

These tests are carried out by burying the plastic sample in the soil, placing them in water sources like rivers, lakes, or by performing complete composting. Although these tests are easy and widely used, there are some associated disadvantages as well such as (i) the environmental factors like temperature, humidity, pH cannot be well controlled; (ii) analytical prospects to monitor the degradation are limited. In such situations, only possibility is to evaluate disintegration via measuring weight loss or to estimate visible changes in the polymer sample. Complex and open-ended environmental conditions make study of residues, intermediates, degradation pathways, CO_2 evolution, or O_2 consumption complicated.

5.1.2 Simulation Test(s)

As a substitute to field tests, a range of simulation test has been developed in which degradation process is done in a real environment like soil, in H_2O, or in compost, but the contact to the environment is provided in a laboratory reactor. The important abiotic factors such as temperature, pH, humidity that affect the degradation process can be controlled and adjusted. To speed up the degradation and thereby reducing the test durations, sometimes additional nutrients are added to the sample. Examples for such tests include test simulating landfills (McCartin et al. 1990; McCarthy et al. 1992; Smith et al. 1990), the controlled composting test (Pagga et al. 1995; Tosin et al. 1996; Degli-Innocenti et al. 1998; Ohtaki et al. 1998; Tuominen et al. 2002), the soil burial test (Pantke 1990), and aquarium tests (Püchner et al. 1995). The only disadvantage is that it lacks reproducibility because of varying microbial population.

5.1.3 Laboratory Tests

Most reproducible tests, here a pre decided growth media is used which is then inoculated either with individual microbial strains or with mixed microbial inoculums. Reaction conditions are tailored for the activity of a particular microbe. Polymers frequently show higher degradation rates in laboratory tests compared to natural conditions. Examples of such tests are petri dish screen, closed bottle test, rapid detection method, gravimetry, surface hydrolysis, respirometry, and measurement of biogas. These tests offer the advantage of systematic investigations into reaction mechanism but it is still not possible to prove biodegradation in terms of metabolization by microorganisms (Mohan 2011).

5.2 Standard Test Methods

5.2.1 Visual Observations

Visual observations can be made in almost all the tests. Effects used to express degradation comprise of changes in color, roughening of the surface, defragmentation, formation of holes or cracks, or formation of biofilms on the plastic surface. These parameters can be used as indications of microbial attack but they do not essentially prove the biodegradation in terms of metabolism. To get information about the reaction mechanism, more sophisticated observations can be made using various characterizations techniques including but not limited to scanning electron microscopy or atomic force microscopy (Ikada 1999).

5.2.2 Residual Polymer Determination by Weight Loss

The mass loss of test samples is commonly noted in degradation tests (mainly in field and simulation tests), although this also does not give any direct proof for biodegradation. By combining structural analysis of residual material and low molecular weight intermediates, detailed information concerning degradation process can be gathered, especially when using a defined synthetic test medium (Witt et al. 2001).

5.2.3 Change in Molecular Mass and Subsequently in Mechanical Properties

If only small changes in the mass of samples are noted, then the changes in mechanical properties are normally used. Some properties like tensile strength are very receptive to changes in the molecular mass of polymers, which is also an indicator of degradation (Erlandsson et al. 1997).

5.2.4 Clear Zone Formation

It is a simple semi-quantitative method. The polymer size is reduced to fine particles and spread onto an agar medium. Microbes are introduced to the medium. The formation of clear halo around microbial colony is taken as an indication of capability of microbe to depolymerize the polymer. Although this method is typically employed to screen microbes for degradation potential for certain polymers (Nishida and Tokiwa 1993; Abou-Zeid 2001), it can also be utilized to gather semi-quantitative results by analyzing the growth of clear zones (Augusta et al. 1993).

5.2.5 O_2 Consumption or CO_2 Evolution

In aerobic respiration, microbes utilize oxygen and form CO_2. Consequently, the respirometric test, used for estimating the oxygen consumption (Hoffmann et al. 1997), and Sturm test, for estimating the formation of CO_2, are often used as good evaluators to estimate polymer degradation in laboratory test with good accuracy.

5.2.6 Radio Labeling

Net CO_2 and $14CO_2$ evolution measurements are simple, non-destructive and measure ultimate biodegradation. Using this system for biodegradation studies in varying microbial environments exhibits precision and consistency (Sharabi and Bartha 1993). However, labeled materials are quite expensive and not easily available. The licensing and related radioactive waste disposal problems present some major disadvantages.

5.2.7 Controlled Composting Test

In this process, a definite weight of dry plastic is mixed with fixed amount of mature compost and incubated at 58 °C with 65% moisture content. Biodegradation is examined by measuring the amount of material carbon changed into gaseous carbon oxide, i.e., the CO_2 evolved from the polymer–compost mixture minus the CO_2

evolved from the unamended compost (as blank) which is tested in a different reactor (Bellina et al. 1999).

6 Characterization Techniques Used to Study Microbial Depolymerization

The biodegradability of the plastic can be characterized using several parameters such as oxygen uptake rate, carbon dioxide evolution rate, accumulation of biomass, surface changes, and changes in physiochemical properties of plastics (Albertsson and Huang 1995). Numerous analytical techniques have been used to investigate the degree and character of degradation. The examination of changes in physical and chemical properties of plastic both before and after degradation helps to understand reaction mechanisms and/or degradation pathway.

6.1 Dynamic Mechanical Analysis (DMR) Using Universal Testing Systems

Study the mechanical properties of plastics such as elongation at fail, tensile strength, and modulus of the polymer. Physical properties observed for changes are morphology (microcracks), contact angle, density, viscosity, glass transition temperatures, melting temperatures, amorphous part and crystalline part of the polymer, etc.

6.2 Scanning Electron Microscopy (SEM)

SEM helps in studying the surface morphology of the polymer films. The samples are generally have to be sputter-coated with gold or some metal ion before examination. Variable-pressure SEM is helpful for direct observation without using any surface metallization at suitable magnification (Pinzari et al. 2006).

6.3 Thermogravimetric Analysis (TGA)

It observes the changes in mass of sample due to heating. Dynamic thermogravimetry can be used to study the activation energy, rate of decomposition reactions, and temperature at maximum decomposition rate (Carrasco and Pages 1996).

6.4 Fourier Transform Infrared Spectroscopy (FTIR)

Chemical changes produced in polymers during or after microbial degradation processes are analyzed by FTIR (Mohamed et al. 2007; Singh and Sharma 2008; Elashmawi et al. 2008; Milstein et al. 1994). It helps to explain physical and chemical structures, degradation reactions, end group detection, hydrogen bonding, cross-linking behavior of molecules, and copolymer composition in solid- and liquid-form polymers.

6.5 Nuclear Magnetic Resonance Spectroscopy (NMR)

NMR spectroscopy (both 1H and 13C) is the most versatile analytical tool used to understand degradation reaction pathways and formed intermediate compounds (Massardier-Nageotte et al. 2006; Schlemmer et al. 2009; Shah et al. 2008).

6.6 Gel Permeation Chromatography (GPC)

GPC also known as size-exclusion chromatography (SEC) is used to find out molecular mass distribution of polymers and changes in molecular weight after biodegradation (Peng and Shen 1999; Kale et al. 2006; Walter et al. 1995).

6.7 High-Pressure Liquid Chromatography (HPLC)

HPLC is extensively used to separate molecules on the basis of their size. It helps to analyze the metabolic degradation products of xenobiotics, e.g., biotransformation products of styrene, epoxy styrene, phenylacetaldehyde, and 2-phenylethanol were detected by using reverse-phase high-pressure liquid chromatography (Beltrametti et al. 1997; Marconi et al. 1996).

6.8 Gas Chromatography–Mass Spectrometry (GC/MS)

The GC/MS technique comprises a gas chromatograph coupled to a mass spectrometer and is used to identify and quantify residual monomers, trace contaminants, analyzing volatile components, gas mixtures, solvent in resins and coatings, verifying additive levels in polymers and for on-line or at-line monitoring of reactions.

6.9 Biochemical Analysis

The metabolic activity of the cells in the culture and in the biofilm can be determined by various biochemical assays, e.g., by adenosine triphosphate assays, by fluorescein diacetate analysis, and by protein analysis (Orr et al. 2004; Koutny et al. 2006a, b).

6.10 Molecular Techniques

Microbial depolymerization and degradation usually take longer time periods. Once the degradation pathway is known, genetic engineering allows us to (i) reduce the reaction time, (ii) make the plastic more susceptible for microbial attack, (iii) modify regulatory mechanisms, (iv) alter the properties of degradative enzymes, and (v) assemble degradative enzymes from different organisms within a single organism. This can be done by exploiting genetic engineering techniques such as DNA and protein sequencing and production of recombinant organisms, etc.

Some biosynthetic genes, e.g., *phbA* (for 3-ketothiolase), *phbB* (NADPH-dependent acetoacetyl-CoAreductase), and *phbC* (PHB synthase) from acetyl-CoA have been cloned to produce PHA and PHB acids (Slater et al. 1988). These genes are clustered, presumably organized in one operon, and expressed in *Escherichia coli* and in also in different species of *Pseudomonas*. The enzymes implicated in plastic degradation should be characterized, and further genes liable for those enzymes should be worked out. Once these genes are identified, they can be modified to boost their degrading capacity (Aswale and Ade 2008). Efficient degrading microbes should be multiplied at larger scale to decompose the plastic at commercial level.

7 Significance

Efficient technology, low cost, and relatively eco-friendly treatment capable of minimizing and even eliminating synthetic polymers hold a great environmental interest. Microbial population is now being extensively studied for biodegradation of synthetic polymers. Biodegradability tests are essential to assess and minimize the environment impacts and accumulation of plastics. If plastics can be made to degrade more effectively by microbes, it will reduce solid waste problems which raise a plethora of environmental concerns. Microbial enzymes are crucial to biodegradation of plastics. There is a great potential and scope in exploring these microbes and tailoring according to specific needs like to grow in varied environmental conditions, for better utilization of plastics carbon for their energy needs.

8 Future Perspective

The microbial depolymerization of synthetic polymers is a multifaceted phenomenon. Strict nature-like experiments are difficult to carry out in laboratory settings because of number of parameters occurring during the biogeochemical recycling which cannot be entirely replicated and directed in vitro. Information available so far regarding microbial ability for degrading synthetic plastics is build on a few bacteria (representing $< 0.1\%$ of total bacteria) that could grow on culture media. Thus, high diversity of microorganisms present in nature has not been even partially explored yet. Making use of molecular technologies such as genomics, proteomics, transcriptomics, metabolomics makes it feasible to discover novel microorganisms (even non-culturable ones) involved in plastic degradation and investigate properties of microorganisms arising from the interplay of proteins, genes, other macro–micro molecules and the environment.

9 Conclusion

Polymers are generally resistant toward biodegradation. Physicochemical methods of degradation have been more successful than biological ones. However, exploration of microbial depolymerization is a step toward plastics management by sustainable method.

References

Abou-Zeid DM, Müller RJ, Deckwer WD (2001) Degradation of natural and synthetic polyesters under anaerobic conditions. J Biotechnol 86(2):113–126

Abrusci C, Pablos JL, Corrales T, López-Marín J, Marín I, Catalina F (2011) Biodegradation of photo-degraded mulching films based on polyethylenes and stearates of calcium and iron as pro-oxidant additives. Int Biodeterior Biodegradation 65(3):451–459

Akutsu Y, Nakajima-Kambe T, Nomura N, Nakahara T (1998) Purification and properties of a polyester polyurethane-degrading enzyme from *Comamonas acidovorans* TB-35. Appl Environ Microbiol 64(1):62–67

Alariqi SA, Kumar AP, Rao BSM, Singh RP (2006) Biodegradation of γ-sterilised biomedical polyolefins under composting and fungal culture environments. Polym Degrad Stab 91 (5):1105–1116

Albertsson AC, Huang SJ (eds) (1995) Degradable polymers, recycling, and plastics waste management. Marcel Dekker, New york

Albertsson AC, Andersson SO, Karlsson S (1987) The mechanism of biodegradation of polyethylene. Polym Degrad Stab 18(1):73–87

Alexander M (1977) Introduction to soil microbiology 2edn. Wiley, US

Ali MI, Ahmed S, Robson G, Javed I, Ali N, Atiq N, Hameed A (2014) Isolation and molecular characterization of polyvinyl chloride (PVC) plastic degrading fungal isolates. J Basic Microbiol 54(1):18–27

Allen NS, Edge M, Mourelatou D, Wilkinson A, Liauw CM, Parellada MD, Barrio JA, Santa Quiteria VR (2003) Influence of ozone on styrene–ethylene–butylene–styrene (SEBS) copolymer. Polym Degrad Stab 79(2):297–307

Arutchelvi J, Sudhakar M, Arkatkar A, Doble M, Bhaduri S, Uppara PV (2008) Biodegradation of polyethylene and polypropylene. Indian J Biotechnol 7:9–22

Arvanitoyannis I, Biliaderis CG, Ogawa H, Kawasaki N (1998) Biodegradable films made from low-density polyethylene (LDPE), rice starch and potato starch for food packaging applications: Part 1. Carbohyd Polym 36(2–3):89–104

Aswale P, Ade A (2008) Assessment of the biodegradation of polythene. Bioinfolet 5:239

Augusta J, Müller RJ, Widdecke H (1993) A rapid evaluation plate-test for the biodegradability of plastics. Appl Microbiol Biotechnol 39(4–5):673–678

Babu RP, O'connor K, Seeram R (2013) Current progress on bio-based polymers and their future trends. Prog Biomater 2(1):8

Barkay T, Navon-Venezia S, Ron EZ, Rosenberg E (1999) Enhancement of solubilization and biodegradation of polyaromatic hydrocarbons by the bioemulsifier alasan. Appl Environ Microbiol 65(6):2697–2702

Basnett P, Knowles JC, Pishbin F, Smith C, Keshavarz T, Boccaccini AR, Roy I (2012) Novel biodegradable and biocompatible Poly (3-hydroxyoctanoate)/bacterial cellulose composites. Adv Eng Mater 14(6):B330–B343

Bellia G, Tosin M, Floridi G, Degli-Innocenti F (1999) Activated vermiculite, a solid bed for testing biodegradability under composting conditions. Polym Degrad Stab 66(1):65–79

Beltrametti F, Marconi AM, Bestetti G, Colombo C, Galli E, Ruzzi M, Zennaro E (1997) Sequencing and functional analysis of styrene catabolism genes from *Pseudomonas fluorescens* ST. Appl Environ Microbiol 63(6):2232–2239

Benedict CV, Cameron JA, Huang SJ (1983) Polycaprolactone degradation by mixed and pure cultures of bacteria and a yeast. J Appl Polym Sci 28(1):335–342

Bhardwaj H, Gupta R, Tiwari A (2013) Communities of microbial enzymes associated with biodegradation of plastics. J Polym Environ 21(2):575–579

Bonhomme S, Cuer A, Delort AM, Lemaire J, Sancelme M, Scott G (2003) Environmental biodegradation of polyethylene. Polym Degrad Stab 81(3):441–452

Bovey FA, Winslow FH (1979) Macromolecules: an introduction to polymer science. Academic, New York

Calabia BP, Tokiwa Y (2004) Microbial degradation of poly (D-3-hydroxybutyrate) by a new thermophilic Streptomyces isolate. Biotech Lett 26(1):15–19

Carrasco F (1996) Thermogravimetric analysis of polystyrene: influence of sample weight and heating rate on thermal and kinetic parameters. J Appl Polym Sci 61(1):187–197

Caruso G (2015) Plastic degrading microorganisms as a tool for bioremediation of plastic contamination in aquatic environments. J. Pollut Eff Cont 3:112

Chaisu K, Siripholvat V, Chiu CH (2015) New method of rapid and simple colorimetric assay for detecting the enzymatic degradation of poly lactic acid plastic films. Int J Life Sci Biotechnol Pharm 4(1):57–61

Chambliss WG (1983) The forgotten dosage form: enteric-coated tablets. Pharm Technol 7 (9):124–140

Chandra R, Rustgi R (1997) Biodegradation of maleated linear low-density polyethylene and starch blends. Polym Degrad Stab 56(2):185–202

Chandra R, Rustgi R (1998) Biodegradable polymers. Prog Polym Sci 23(7):1273–1335

Chowdhury AA (1963) Poly-β-hydroxybuttersäure Abbauende Bakterien und Exoenzym. Arch Microbiol 47(2):167–200

Das MP, Kumar S (2015) An approach to low-density polyethylene biodegradation by *Bacillus amyloliquefaciens*. 3 Biotech 5(1):81–86

David C, De Kesel C, Lefebvre F, Weiland M (1994) The biodegradation of polymers: recent results. Macromol Mater Eng 216(1):21–35

Degli-Innocenti F, Tosin M, Bastioli C (1998) Evaluation of the biodegradation of starch and cellulose under controlled composting conditions. J Polym Environ 6(4):197–202

Deguchi T, Kakezawa M, Nishida T (1997) Nylon biodegradation by lignin-degrading fungi. Appl Environ Microbiol 63(1):329–331

Deguchi T, Kitaoka Y, Kakezawa M, Nishida T (1998) Purification and characterization of a nylon-degrading enzyme. Appl Environ Microbiol 64(4):1366–1371

Devi RS, Kannan VR, Nivas D, Kannan K, Chandru S, Antony AR (2015) Biodegradation of HDPE by *Aspergillus spp*. from marine ecosystem of Gulf of Mannar, India. Mar Pollut Bull 96(1):32–40

Donald V, Voet Judith G, Pratt Charlotte W (2006) Fundamentals of Biochemistry Life at the Molecular Level. Wiley, New york

Donelli I, Freddi G, Nierstrasz VA, Taddei P (2010) Surface structure and properties of poly-(ethylene terephthalate) hydrolyzed by alkali and cutinase. Polym Degrad Stab 95(9):1542–1550

Duddu MK, Tripura KL, Guntuku G, Divya DS (2015) Biodegradation of low density polyethylene (LDPE) by a new biosurfactant-producing thermophilic *Streptomyces coelicoflavus* NBRC 15399^{T}. Afr J Biotech 14(4):327–340

Edmonds P, Cooney JJ (1968) Microbial growth in a fuel-water system containing polyesterurethane foam. Appl Microbiol 16(2):426

Elashmawi IS, Hakeem NA, Abdelrazek EM (2008) Spectroscopic and thermal studies of PS/PVAc blends. Physica B 403(19):3547–3552

Erlandsson B, Karlsson S, Albertsson AC (1997) The mode of action of corn starch and a pro-oxidant system in LDPE: influence of thermo-oxidation and UV-irradiation on the molecular weight changes. Polym Degrad Stab 55(2):237–245

Expandable Polystyrene Molders Association (2004) Test initiatives highlights EPS performance. EPS Newsline 8:1–8

Flavel BS, Shapter JG, Quinton JS (2006, July). Nanosphere lithography using thermal evaporation of gold. In: International Conference on IEEE Nanoscience and Nanotechnology (ICONN'06)

Flemming HC (1998) Relevance of biofilms for the biodeterioration of surfaces of polymeric materials. Polym Degrad Stab 59(1–3):309–315

Fuhs GW (1961) Der mikrobielle Abbau von Kohlenwasserstoffen. Arch Microbiol 39(4):374–422

Gajanand E, Soni LK, Dixit VK (2014) Biodegradable polymers: a smart strategy for today's crucial needs. Crit Rev Environ Sci Technol 3

Gajendiran A, Krishnamoorthy S, Abraham J (2016) Microbial degradation of low-density polyethylene (LDPE) by *Aspergillus clavatus* strain JASK1 isolated from landfill soil. 3 Biotech 6(1):1–6

Garlotta D (2001) J Polym Environ 9:63

Gautam R, Bassi AS, Yanful EK (2007) A review of biodegradation of synthetic plastic and foams. Appl Biochem Biotechnol 141(1):85–108

Ghosh RN, Ray BC (2004) Optimized print properties on starch blended and surface grafted polyethylene films for biodegradable packaging. J Polym Mater 21(4):425–437

Gimouhopoulos K, Doulia D, Vlyssides A, Georgiou D (2000) Optimization studies of waste plastics-lignite catalytic coliquefaction. Waste Manage Res 18(4):352–357

Gowariker VR, Viswanathan NV, Sreedhar J (2000) Polymer science. New Age International (P) Limited Publishers, New Delhi, India, pp 263–290

Gross RA (1993) Cellulose acetate biodegradability in simulated aerobic composting and anaerobic bioreactors as well as by a bacterial isolate derived from compost. In: Fundamentals of biodegradable materials and packaging, pp 257–279

Gross RA, Gu JD, Eberiel D, McCarthy SP (1995) Laboratory-scale composting test methods to determine polymer biodegradability: model studies on cellulose acetate. J Macromol Sci Part A Pure Appl Chem 32(4):613–628

Grover A, Gupta A, Chandra S, Kumari A, Khurana SP (2015) Polythene and environment. Int J Environ Sci 5(6):1091

Gu JD (2003) Microbiological deterioration and degradation of synthetic polymeric materials: recent research advances. Int Biodeterior Biodegradation 52(2):69–91

Gu JG, Gu JD (2005) Methods currently used in testing microbiological degradation and deterioration of a wide range of polymeric materials with various degree of degradability: a review. J Polym Environ 13(1):65–74

Gu JD, Ford TE, Mitton DB, Mitchell R (2000) Microbial degradation and deterioration of polymeric materials. In: Revie W (ed) The Uhlig Corrosion Handbook, vol 2. Wiley, New York, pp 439–460

Hadad D, Geresh S, Sivan A (2005) Biodegradation of polyethylene by the thermophilic bacterium *Brevibacillus borstelensis*. J Appl Microbiol 98(5):1093–1100

Hazen TC, Dubinsky EA, DeSantis TZ, Andersen GL, Piceno YM, Singh N, Jansson JK, Probst A, Borglin SE, Fortney JL, Stringfellow WT (2010) Deep-sea oil plume enriches indigenous oil-degrading bacteria. Science 330(6001):204–208

Hendrickson L, Connole KB (1995) Review of stabilization of polyolefin insulated conductors. Part I: theory and factors affecting stability. Polym Eng Sci 35(2):211–217

Hermanová S, Šmejkalová P, Merna J, Zarevúcka M (2015) Biodegradation of waste PET based copolyesters in thermophilic anaerobic sludge. Polym Degrad Stab 111:176–184

Heumann S, Eberl A, Pobeheim H, Liebminger S, Fischer-Colbrie G, Almansa E, Cavaco-Paulo A, Gübitz GM (2006) New model substrates for enzymes hydrolysing polyethyleneterephthalate and polyamide fibres. J Biochem Biophys Methods 69(1):89–99

Ho KLG, Pometto AL, Gadea-Rivas A, Briceño JA, Rojas A (1999) Degradation of polylactic acid (PLA) plastic in Costa Rican soil and Iowa state university compost rows. J Polym Environ 7 (4):173–177

Hoang KC, Tseng M, Shu WJ (2007) Degradation of polyethylene succinate (PES) by a new thermophilic Microbispora strain. Biodegradation 18(3):333–342

Hoffmann J, Řezníčèková I, Vaňòková S, Kupec J (1997) Manometric determination of biological degradability of substances poorly soluble in aqueous environments. Int Biodeterior Biodegradation 39(4):327–332

Holland KA, Rae ID (1987) Thermal degradation of polymers. III. Thermal degradation of a compound which models the head-to-head linkage in poly (methyl methacrylate). Aust J Chem 40(4):687–692

Howard GT, Mackie RI, Cann IKO, Ohene-Adjei S, Aboudehen KS, Duos BG, Childers GW (2007) Effect of insertional mutations in the pueA and pueB genes encoding two polyurethanases in *Pseudomonas chlororaphis* contained within a gene cluster. J Appl Microbiol 103(6):2074–2083

Howard GT (2012) Polyurethane biodegradation. In: Microbial degradation of xenobiotics. Berlin Heidelberg, Springer, pp 371–394

Huang JC, Shetty AS, Wang MS (1990) Biodegradable plastics: a review. Adv Polym Technol 10 (1):23–30

Ikada E (1999) Electron microscope observation of biodegradation of polymers. J Environ Polym Degrad 7(4):197–201

Inoue T, Miyazaki M, Kamitani M, Kano J, Saito F (2004) Mechanochemicaldechlorination of polyvinyl chloride by co-grinding with various metal oxides. Adv Powder Technol 15(2):215–225

Ishigaki T, Kawagoshi Y, Ike M, Fujita M (1999) Biodegradation of a polyvinyl alcohol-starch blend plastic film. World J Microbiol Biotechnol 15(3):321–327

Jakubowicz I (2003) Evaluation of degradability of biodegradable polyethylene (PE). Polym Degrad Stab 80(1):39–43

Jarerat A, Tokiwa Y (2001) Degradation of poly (tetramethylene succinate) by thermophilic actinomycetes. Biotech Lett 23(8):647–651

Jen-hou L, Schwartz A (1961) Zum Verhalten von bakteriengemischen gegentiber polyfithylen verschiedenen mittleren Molekulargewichts. Kunststoffe 51:317–320

Kakudo SHINJI, Negoro SEIJI, Urabe ITARU, Okada HIROSUKE (1993) Nylon oligomer degradation gene, nylC, on plasmid pOAD2 from a Flavobacterium strain encodes endo-type

6-aminohexanoate oligomer hydrolase: purification and characterization of the nylC gene product. Appl Environ Microbiol 59(11):3978–3980

Kale G, Auras R, Singh SP (2006) Degradation of commercial biodegradable packages under real composting and ambient exposure conditions. J Polym Environ 14(3):317–334

Kanagawa K, Negoro S, Takada N, Okada H (1989) Plasmid dependence of Pseudomonas sp. strain NK87 enzymes that degrade 6-aminohexanoate-cyclic dimer. J Bacteriol 171 (6):3181–3186

Kanagawa K, Oishi M, Negoro S, Urabe I, Okada H (1993) Characterization of the 6-aminohexanoate-dimer hydrolase from Pseudomonas sp. NK87. Microbiology 139(4):787–795

Kefeli AA, Razumovskii SD, Zaikov GY (1971) Interaction of polyethylene with ozone. Polym Sci USSR 13(4):904–911

Kinoshita S, Negoro S, Muramatsu M, Bisaria VS, Sawada S, Okada H (1977) 6-Aminohexanoic acid cyclic dimer hydrolase. A new cyclic amide hydrolase produced by *Acromobacter guttatus* KI72. Eur J Biochem 80(2):489–495

Kinoshita S, Terada T, Taniguchi T, Takene Y, Masuda S, Matsunaga N, Okada H (1981) Purification and characterization of 6-Aminohexanoic-Acid-Oligomer Hydrolase of *Flavobacterium sp*. KI72. Eur J Biochem 116(3):547–551

Kırbaş Z, Keskin N, Güner A (1999) Biodegradation of polyvinylchloride (PVC) by white rot fungi. Bull Environ Contam Toxicol 63(3):335–342

Kleeberg I, Hetz C, Kroppenstedt RM, Müller RJ, Deckwer WD (1998) Biodegradation of aliphatic-aromatic copolyesters by *Thermomonospora fusca* and other thermophilic compost isolates. Appl Environ Microbiol 64(5):1731–1735

Kleeberg I, Welzel K, VandenHeuvel J, Müller RJ, Deckwer WD (2005) Characterization of a new extracellular hydrolase from *Thermobifida fusca* degrading aliphatic–aromatic copolyesters. Biomacromolecules 6(1):262–270

Koshland DE (1995) The key–lock theory and the induced fit theory. Angew Chem Int Ed Engl 33 (23–24):2375–2378

Koutny M, Lemaire J, Delort AM (2006a) Biodegradation of polyethylene films with prooxidant additives. Chemosphere 64(8):1243–1252

Koutny M, Sancelme M, Dabin C, Pichon N, Delort AM, Lemaire J (2006b) Acquired biodegradability of polyethylenes containing pro-oxidant additives. Polym Degrad Stab 91 (7):1495–1503

Kyrikou I, Briassoulis D (2007) Biodegradation of agricultural plastic films: a critical review. J Polym Environ 15(2):125–150

Lee SY (1996) Bacterial polyhydroxyalkanoates. Biotechnol Bioeng 49(1):1–14

Lenz R (1993) Biodegradable polymers. Biopolymers I 1–40

Liang TW, Jen SN, Nguyen AD, Wang SL (2016) Application of Chitinous materials in production and purification of a Poly (L-lactic acid) depolymerase from Pseudomonas tamsuii TKU015. Polymers 8(3):98

Lim HA, Raku T, Tokiwa Y (2005) Hydrolysis of polyesters by serine proteases. Biotech Lett 27 (7):459–464

Lin YH, Yen HY (2005) Fluidised bed pyrolysis of polypropylene over cracking catalysts for producing hydrocarbons. Polym Degrad Stab 89(1):101–108

Liu C, Zeng JB, Li SL, He YS, Wang YZ (2012) Improvement of biocompatibility and biodegradability of poly (ethylene succinate) by incorporation of poly (ethylene glycol) segments. Polymer 53(2):481–489

Loredo-Treviño A, Gutiérrez-Sánchez G, Rodríguez-Herrera R, Aguilar CN (2012) Microbial enzymes involved in polyurethane biodegradation: a review. J Polym Environ 20(1):258–265

Lugauskas A, Levinskait L, Pečiulyt D (2003) Micromycetes as deterioration agents of polymeric materials. Int Biodeterior Biodegradation 52(4):233–242

Ma A, Wong Q (2013) Identification of esterase in *Aspergillus flavus* during degradation of polyester polyurethane 1. Can Young Sci J 2013(2):24–31

Mahalakshmi V, Siddiq A, Andrew SN (2012) Analysis of polyethylene degrading potentials of microorganisms isolated from compost soil. Int J Pharm Biol Arch 3(5):1190–1196

Marcilla A, Gómez A, Menargues S, García-Martínez J, Cazorla-Amorós D (2003) Catalytic cracking of ethylene-vinyl acetate copolymers: comparison of different zeolites. J Anal Appl Pyrol 68:495–506

Marconi AM, Beltrametti F, Bestetti G, Solinas F, Ruzzi M, Galli E, Zennaro E (1996) Cloning and characterization of styrene catabolism genes from *Pseudomonas fluorescens* ST. Appl Environ Microbiol 62(1):121–127

Masaki K, Kamini NR, Ikeda H, Iefuji H (2005) Cutinase-like enzyme from the yeast *Cryptococcus* sp. strain S-2 hydrolyzes polylactic acid and other biodegradable plastics. Appl Envir Microbiol 71(11):7548–7550

Massardier-Nageotte V, Pestre C, Cruard-Pradet T, Bayard R (2006) Aerobic and anaerobic biodegradability of polymer films and physico-chemical characterization. Polym Degrad Stab 91(3):620–627

McCarthy SP, Gada M, Smith GP, Tolland V, Press B, Eberiel D, Bruell C, Gross R (1992) The accelerated biodegradability of plastic materials in simulated compost and landfill environments. ANTEC 92–Shaping Future 1:816–818

McCartin SM, Press B, Eberiel D, McCarthy SP, Kaplan D, Meyer J (1990) Simulated landfill study on the accelerated biodegradability of plastics materials. Polym Prepr Div Polym Chem Am Chem Soc 440

Mehmood CT, Qazi IA, Hashmi I, Bhargava S, Deepa S (2016) Biodegradation of low density polyethylene (LDPE) modified with dye sensitized titania and starch blend using *Stenotrophomonas pavanii*. Int Biodeterior Biodegradation 113:276–286

Meyer TJ, Caspar JV (1985) Photochemistry of metal-metal bonds. Chem Rev Washington, DC, United States, p 187–218

Milstein O, Gersonde R, Huttermann A, Frund R, Feine HJ, Ludermann HD, Chen MJ, Meister JJ (1994) Infrared and nuclear magnetic resonance evidence of degradation in thermoplastics based on forest products. J Polym Environ 2(2):137–152

Mitchell R (1996) Hazards to space missions from microbial biofilms. Biodeterior Biodegradation 133:3–16

Mohamed A, Gordon SH, Biresaw G (2007) Polycaprolactone/polystyrene bioblends characterized by thermogravimetry, modulated differential scanning calorimetry and infrared photoacoustic spectroscopy. Polym Degrad Stab 92(7):1177–1185

Mohan K (2011) Microbial deterioration and degradation of polymeric materials. J Biochem Technol 2(4):210–215

Mohanty AK, Misra M, Hinrichsen G (2000) Biofibres, biodegradable polymers and biocomposites: an overview. Macromol Mater Eng 276(1):1–24

Motiwalla MJ, Punyarthi PP, Mehta MK, D'Souza JS, Kelkar-Mane V (2013) Studies on degradation efficiency of polycaprolactone by a naturally-occurring bacterium. J Env Bio 34:43–49

Muenmee S, Chiemchaisri W, Chiemchaisri C (2015) Microbial consortium involving biological methane oxidation in relation to the biodegradation of waste plastics in a solid waste disposal open dump site. Int Biodeterior Biodegradation 102:172–181

Muenmee S, Chiemchaisri W, Chiemchaisri C (2016) Enhancement of biodegradation of plastic wastes via methane oxidation in semi-aerobic landfill. Int Biodeterior Biodegradation 113:244–255

Murphy CA, Cameron JA, Huang SJ, Vinopal RT (1996) Fusariumpolycaprolactonedepolymerase is cutinase. Appl Environ Microbiol 62(2):456–460

Murthy KS, Enders NA, Mahjour M, Fawzi MB (1986) A comparative evaluation of aqueous enteric polymers in capsule coating. Pharm Technol 10(10):36–46

Nagai Y, Nakamura D, Miyake T, Ueno H, Matsumoto N, Kaji A, Ohishi F (2005) Photodegradation mechanisms in poly (2, 6-butylenenaphthalate-co-tetramethyleneglycol) (PBN–PTMG). I: influence of the PTMG content. Polym Degrad Stab 88(2):251–255

Nagarajan V, Singh M, Kane H, Khalili M, Bramucci M (2006) Degradation of a terephthalate-containing polyester by a thermophilic microbial consortium. J Polym Environ 14(3):281–287

Nakajima-Kambe T, Shigeno-Akutsu Y, Nomura N, Onuma F, Nakahara T (1999) Microbial degradation of polyurethane, polyester polyurethanes and polyether polyurethanes. Appl Microbiol Biotechnol 51(2):134–140

Nakamiya K, Ooi T, Kinoshita S (1997) Non-heme hydroquinone peroxidase from *Azotobacter beijerinckii* HM121. J Ferment Bioeng 84(1):14–21

Narayan R (1993) Biodegradation of polymeric materials (anthropogenic macromolecules) during composting. In: Science and engineering of composting: design, environmental, microbiological and utilization aspects. Washington, OH, Renaissance Publishers, pp 339–62

National Research Council (1993) In Situ bioremediation when does it work?. National Academy Press, Washington, DC, p 224

Negoro SEIJI, Kakudo SHINJI, Urabe I, Okada HIROSUKE (1992) A new nylon oligomer degradation gene (nylC) on plasmid pOAD2 from a *Flavobacterium sp*. J Bacteriol 174 (24):7948–7953

Nishida H, Tokiwa Y (1993) Distribution of poly (β-hydroxybutyrate) and poly (ε-caprolactone) aerobic degrading microorganisms in different environments. J Polym Environ 1(3):227–233

Nishida H, Suzuki S, Tokiwa Y (1998) Distribution of poly (β-propiolactone) aerobic degrading microorganisms in different environments. J Environ Polym Degrad 6(1):43–58

Nomura N, Deguchi T, Shigeno-Akutsu Y, Nakajima-Kambe T, Nakahara T (2001) Gene Structures and Catalytic Mechanisms of Microbial Enzymes Able to Blodegrade the Synthetic Solid Polymers Nylon and Polyester Polyurethaoe. Biotech Gene Eng Rev 18(1):125–147

Nowak B, Pająk J, Łabużek S, Rymarz G, Talik E (2011) Biodegradation of poly (ethylene terephthalate) modified with polyester "Bionolle®" by *Penicillium funiculosum*. Polimery 56 (1):35–44

Nwachukwu S, Obidi O, Odocha C (2010) Occurrence and recalcitrance of polyethylene bag waste in Nigerian soils. Afr J Biotech 9(37):6096–6104

Oda Y, Asari H, Urakami T, Tonomura K (1995) Microbial degradation of poly (3-hydroxybutyrate) and polycaprolactone by filamentous fungi. J Ferment Bioeng 80 (3):265–269

Odusanya SA, Nkwogu JV, Alu N, Udo GE, Ajao JA, Osinkolu GA, Uzomah AC (2013) Preliminary studies on microbial degradation of plastics used in packaging potable water in Nigeria. Niger Food J 31(2):63–72

Ohtaki A, Sato N, Nakasaki K (1998) Biodegradation of poly-ε-caprolactone under controlled composting conditions. Polym Degrad Stab 61(3):499–505

Omichi H (1992) Handbook of Polymer Degradation. In: Hamid SH, Amin MB, Maadhah AG (eds) Marcel Dekker, New york, pp 335–344

Orhan Y, Hrenovic J, Buyukgungor H (2004) Biodegradation of plastic compost bags under controlled soil conditions. Acta Chimica Slovenica 51(3):579–588

Orr IG, Hadar Y, Sivan A (2004) Colonization, biofilm formation and biodegradation of polyethylene by a strain of *Rhodococcus ruber*. Appl Microbiol Biotechnol 65(1):97–104

Pagga U, Beimborn DB, Boelens J, De Wilde B (1995) Determination of the aerobic biodegradability of polymeric material in a laboratory controlled composting test. Chemosphere 31(11–12):4475–4487

Pantke M, Seal K (1990) An interlaboratory investigation into the biodeterioration testing of plastics, with special reference to polyurethanes. Part 2: soil burial experiments. Material und Organismen 25(2):87–98

Park H, Park K, Shalaby WS (2011) Biodegradable hydrogels for drug delivery. CRC Press

Peng X, Shen J (1999) Preparation and biodegradability of polystyrene having pyridinium group in the main chain. Eur Polymer J 35(9):1599–1605

Pinzari F, Pasquariello G, De Mico A (2006 April) Biodeterioration of paper: a SEM study of fungal spoilage reproduced under controlled conditions. Macromolecular Symposia 238(1):57–66, WILEY-VCH Verlag

Pometto AL, Lee BT, Johnson KE (1992) Production of an extracellular polyethylene-degrading enzyme (s) by Streptomyces species. Appl Environ Microbiol 58(2):731–733
Pospíšil J, Pilař J, Billingham NC, Marek A, Horak Z, Nešpůrek S (2006) Factors affecting accelerated testing of polymer photostability. Polym Degrad Stab 91(3):417–422
Potts JE (1978) In: Grayson M (ed) Kirk-Othmer, Encyclopedia of Chemical Technology, Wiley-Interscience, New York, pp 626–668
Pramila R, Padmavathy K, Ramesh KV, Mahalakshmi K (2012) *Brevibacillus parabrevis, Acinetobacter baumannii* and *Pseudomonas citronellolis*-Potential candidates for biodegradation of low density polyethylene (LDPE). Afr J Bacteriol Res 4(1):9–14
Pranamuda H, Tokiwa Y, Tanaka H (1995) Microbial degradation of an aliphatic polyester with a high melting point, poly (tetramethylene succinate). Appl Environ Microbiol 61(5):1828–1832
Pranamuda H, Tokiwa Y, Tanaka H (1997) Polylactide degradation by an *Amycolatopsis* sp. Appl Environ Microbiol 63(4):1637–1640
Prema S, Uma MDP (2013) Degradation of poly lactide plastic by mesophilic bacteria isolated from compost. Int J Res Purif Appl Microbiol 3(4):121–126
Puechner P, Mueller WR, Bardtke D (1995) Assessing the biodegradation potential of polymers in screening-and long-term test systems. J Environ Polym Degrad 3(3):133–143
Qiu Z, Ikehara T, Nishi T (2003) Crystallization behaviour of biodegradable poly (ethylene succinate) from the amorphous state. Polymer 44(18):5429–5437
Ramis X, Cadenato A, Salla JM, Morancho JM, Valles A, Contat L, Ribes A (2004) Thermal degradation of polypropylene/starch-based materials with enhanced biodegradability. Polym Degrad Stab 86(3):483–491
Restrepo-Flórez JM, Bassi A, Thompson MR (2014) Microbial degradation and deterioration of polyethylene—a review. Int Biodeterior Biodegradation 88:83–90
Ronkvist ÅM, Xie W, Lu W, Gross RA (2009) Cutinase-catalyzed hydrolysis of poly (ethylene terephthalate). Macromolecules 42(14):5128–5138
Rutkowska M, Heimowska A, Krasowska K, Janik H (2002) Biodegradability of polyethylene starch blends in sea water. Pol J Environ Stud 11(3):267–272
Salgado CL, Sanchez E, Zavaglia CA, Granja PL (2012) Biocompatibility and biodegradation of polycaprolactone-sebacic acid blended gels. J Biomed Mater Res, Part A 100(1):243–251
Sanchez JG, Tsuchii A, Tokiwa Y (2000) Degradation of polycaprolactone at 50 °C by a thermotolerant *Aspergillus sp*. Biotech Lett 22(10):849–853
Sand W (2003) Microbial life in geothermal waters. Geothermics 32(4):655–667
Santo M, Weitsman R, Sivan A (2013) The role of the copper-binding enzyme–laccase–in the biodegradation of polyethylene by the actinomycete *Rhodococcus ruber*. Int Biodeterior Biodegradation 84:204–210
Schink B, Janssen PH, Frings J (1992) Microbial degradation of natural and of new synthetic polymers. FEMS Microbiol Rev 9(2–4):311–316
Schlemmer D, Sales MJ, Resck IS (2009) Degradation of different polystyrene/thermoplastic starch blends buried in soil. Carbohyd Polym 75(1):58–62
Schmidt-Naake G, Drache M, Weber M (2002) Combination of mechanochemical degradation of polymers with controlled free-radical polymerization. Macromol Chem Phys 203(15):2232–2238
Scott G (1984) Mechano-chemical degradation and stabilization of polymers. Polym Eng Sci 24 (13):1007–1020
Sekiguchi T, Ebisui A, Nomura K, Watanabe T, Enoki M, Kanehiro H (2009) Biodegradability evaluation of several fibers soaked into deep sea and isolation of the biodegradable plastic degrading bacteria from deepocean water. Nippon Suisan Gakkaishi 75:1011–1018
Seretoudi G, Bikiaris D, Panayiotou C (2002) Synthesis, characterization and biodegradability of poly (ethylene succinate)/poly (ε-caprolactone) block copolymers. Polymer 43(20):5405–5415
Seymour Raymond B (1971) Additives for polymers, introduction to polymer chemistry. McGraw-Hill Book Company, New York, pp 268–271
Shah AA, Hasan F, Hameed A, Ahmed S (2008) Biological degradation of plastics: a comprehensive review. Biotechnol Adv 26(3):246–265

Shahnawaz M, Sangale MK, Ade AB (2016) Rhizosphere of Avicennia marina. Environ Sci Pollut Res 23(14):14621–14635

Sharabi NED, Bartha RICHARD (1993) Testing of some assumptions about biodegradability in soil as measured by carbon dioxide evolution. Appl Environ Microbiol 59(4):1201–1205

Sharon C, Sharon M (2017) Studies on biodegradation of polyethylene terephthalate: a synthetic polymer. J Microbiol Biotechnol Res 2(2):248–257

Sheik S, Chandrashekar KR, Swaroop K, Somashekarappa HM (2015) Biodegradation of gamma irradiated low density polyethylene and polypropylene by endophytic fungi. Int Biodeterior Biodegradation 105:21–29

Shigeno-Akutsu Y, Nakajima-Kambe T, Nomura N, Nakahara T (1999) Purification and properties of culture-broth-secreted esterase from the polyurethane degrader *Comamonas acidovorans* TB-35. J Biosci Bioeng 88(5):484–487

Shimao M (2001) Biodegradation of plastics. Curr Opin Biotechnol 12(3):242–247

Singh B, Sharma N (2008) Mechanistic implications of plastic degradation. Polym Degrad Stab 93 (3):561–584

Slater SC, Voige WH, Dennis DE (1988) Cloning and expression in *Escherichia coli* of the *Alcaligenes eutrophus* H16 poly-beta-hydroxybutyrate biosynthetic pathway. J Bacteriol 170 (10):4431–4436

Smith GP, Press B, Eberiel D, McCarthy SP, Gross RA, Kaplan DL (1990) Accelerated in-laboratory test to evaluate the degradation of plastics in landfill environments. In: Polymeric Materials Science and Engineering, Proceedings of the ACS Division of Polymeric Materials Science and Engineering, vol 63, pp 862–866

Stevens ES (2003) What makes green plastics green? Biocycle 44(3):24–27

Sukhumaporn S, Tokuyama S, Kitpreechavanich V (2012) Poly (*L*-Lactide)- degrading enzyme production by *Actinomadura keratinilytica*T16-1 in 3 L airlift bioreactor and its degradation ability for biological. J Microbiol Biotechnol 22(1):92–99

Suzuki J, Hukushima K, Suzuki S (1978) Effect of ozone treatment upon biodegradability of water-soluble polymers. Environ Sci Technol 12(10):1180–1183

Takamoto T, Uyama H, Kobayashi S (2001) Lipase-catalyzed degradation of polyester in supercritical carbon dioxide. Macromol Biosci 1(5):215–218

Tang L, Wu Q, Qu B (2005) The effects of chemical structure and synthesis method on photodegradation of polypropylene. J Appl Polym Sci 95(2):270–279

Tezuka Y, Ishii N, Kasuya KI, Mitomo H (2004) Degradation of poly (ethylene succinate) by mesophilic bacteria. Polym Degrad Stab 84(1):115–121

Thomas BT, Olanrewaju-Kehinde DSK, Popoola OD, James ES (2015) Degradation of plastic and polythene materials by some selected microorganisms isolated from soil. World Appl Sci J 33 (12):1888–1891

Tokiwa Y, Calabia BP (2006) Biodegradability and biodegradation of poly (lactide). Appl Microbiol Biotechnol 72(2):244–251

Tokiwa Y, Suzuki T (1974) Degradation of poly-ethylene glycol adipate by a fungus. J Ferment Technol

Tokiwa Y, Suzuki T (1977a) Purification and some properties of polyethylene adipate-degrading enzyme produced by Penicillium sp. Strain 14–3. Agric Biol Chem 41(2):265–274

Tokiwa Y, Suzuki T (1977b) Hydrolysis of polyesters by lipases. Nature 270(5632):76–78

Tokiwa Y, Suzuki T (1978) Hydrolysis of polyesters by *Rhizopus delemar* lipase. Agric Biol Chem 42(5):1071–1072

Tokiwa Y, Suzuki T (1981) Hydrolysis of copolyesters containing aromatic and aliphatic ester blocks by lipase. J Appl Polym Sci 26(2):441–448

Tokiwa Y, Suzuki T, Ando T (1979) Synthesis of copolyamide–esters and some aspects involved in their hydrolysis by lipase. J Appl Polym Sci 24(7):1701–1711

Tokiwa Y, Calabia BP, Ugwu CU, Aiba S (2009) Biodegradability of plastics. Int J Mol Sci 10 (9):3722–3742

Tomita K, Ikeda N, Ueno A (2003) Isolation and characterization of a thermophilic bacterium, Geobacillus thermocatenulatus, degrading nylon 12 and nylon 66. Biotech Lett 25(20):1743–1746

Tosin M, Degli-Innocenti F, Bastioli C (1996) Effect of the composting substrate on biodegradation of solid materials under controlled composting conditions. J Polym Environ 4(1):55–63

Tribedi P, Sil AK (2013) Founder effect uncovers a new axis in polyethylene succinate bioremediation during biostimulation. FEMS Microbiol Lett 346(2):113–120

Tribedi P, Sil AK (2014) Cell surface hydrophobicity: a key component in the degradation of polyethylene succinate by *Pseudomonas sp.* AKS2. J Appl Microbiol 116(2):295–303

Tribedi P, Gupta AD, Sil AK (2015) Adaptation of Pseudomonas sp. AKS2 in biofilm on low-density polyethylene surface: an effective strategy for efficient survival and polymer degradation. Bioresources Bioprocess 2(1):14

Tseng M, Hoang KC, Yang MK, Yang SF, Chu WS (2007) Polyester-degrading thermophilicactinomycetes isolated from different environment in Taiwan. Biodegradation 18(5):579–583

Tsuji H, Miyauchi S (2001) Poly (L-lactide): VI Effects of crystallinity on enzymatic hydrolysis of poly (L-lactide) without free amorphous region. Polym Degrad Stab 71(3):415–424

Tuominen J, Kylmä J, Kapanen A, Venelampi O, Itävaara M, Seppälä J (2002) Biodegradation of lactic acid based polymers under controlled composting conditions and evaluation of the ecotoxicological impact. Biomacromol 3(3):445–455

Walter T, Augusta J, Müller RJ, Widdecke H, Klein J (1995) Enzymatic degradation of a model polyester by lipase from *Rhizopus delemar*. Enzym Microb Technol 17(3):218–224

Webb HK, Crawford RJ, Sawabe T, Ivanova EP (2009) Poly (ethylene terephthalate) polymer surfaces as a substrate for bacterial attachment and biofilm formation. Microbes Environ 24 (1):39–42

Webb HK, Arnott J, Crawford RJ, Ivanova EP (2012) Plastic degradation and its environmental implications with special reference to poly (ethylene terephthalate). Polymers 5(1):1–18

Wei L, Liang S, McDonald AG (2015) Thermophysical properties and biodegradation behavior of green composites made from polyhydroxybutyrate and potato peel waste fermentation residue. Ind Crops Prod 69:91–103

Williams PT, Bagri R (2004) Hydrocarbon gases and oils from the recycling of polystyrene waste by catalytic pyrolysis. Int J Energy Res 28(1):31–44

Witt U, Einig T, Yamamoto M, Kleeberg I, Deckwer WD, Müller RJ (2001) Biodegradation of aliphatic–aromatic copolyesters: evaluation of the final biodegradability and ecotoxicological impact of degradation intermediates. Chemosphere 44(2):289–299

Woodruff MA, Hutmacher DW (2010) The return of a forgotten polymer—polycaprolactone in the 21st century. Prog Polym Sci 35(10):1217–1256

Yagi H, Ninomiya F, Funabashi M, Kunioka M (2009) Anaerobic biodegradation tests of poly (lactic acid) and polycaprolactone using new evaluation system for methane fermentation in anaerobic sludge. Polym Degrad Stab 94(9):1397–1404

Yoon MG, Jeon HJ, Kim MN (2012) Biodegradation of polyethylene by a soil bacterium and AlkB cloned recombinant cell. J Bioremed Biodegrad 3(4):1–8

Zafar U (2013) Ph.D. Thesis. The University of Manchester, Manchester

Zafar U, Houlden A, Robson GD (2013) Fungal communities associated with the biodegradation of polyester polyurethane buried under compost at different temperatures. Appl Environ Microbiol 79(23):7313–7324

Zhang M, Thomas NL (2011) Blending polylactic acid with polyhydroxybutyrate: the effect on thermal, mechanical, and biodegradation properties. Adv Polym Technol 30(2):67–79

Zhang J, Wang X, Gong J, Gu Z (2004) A study on the biodegradability of polyethylene terephthalate fiber and diethylene glycol terephthalate. J Appl Polym Sci 93(3):1089–1096

Zheng Y, Yanful EK, Bassi AS (2005) A review of plastic waste biodegradation. Crit Rev Biotechnol 25(4):243–250

Author Biographies

Ms. Anvita Sheel is currently perusing Ph.D under the guidance of Prof. Deepak Pant in the subject of Environmental Science from Central University of Himachal Pradesh, Dharamshala, Himachal Pradesh, India. She received post graduation degree in Environment Science (Gold Medalist) from Panjab University, Chandigarh in 2014. She received first position (poster presentation) in National Seminar on Emerging Trends in Environmental Science and Technology-2017 and 2nd position in 8th Chandigarh Science Congress (oral presentation). She was also awarded with Certificate of Excellence in Rio+21 IYWC India Certification Program.

Dr. Deepak Pant is currently the Dean and Head of Environmental Science, Central University of Himachal Pradesh, Dharamshala. He received the Silver Jubilee Research Fellowship Award in 2003 from Kumaun University, Nainital, India; the UCOST Young Scientist Award 2009, INSA Fellow (2010), DST-SERC, Visiting Fellow (2010) and DST-SERC Young Scientist Award (2011). He has 5 patents in the area of waste management using green techniques and published 10 books, 45 research papers in various national and international journals. He recently received Visitor's Award 2017 from the President of India for his innovations in the area of waste management.

Chapter 5
Bioremediation Techniques for E-waste Management

Deepak Pant, Anand Giri and Varun Dhiman

Abstract Bioremediation or microbial cooperation can improve the e-waste management process in a greener way. Every management strategy is concentrated upon the organic and inorganic portion of the e-waste. Organic part consists of variety of thermo and thermosetting plastic with the presence of halogenated material. Microbes are involved in the process of dehalogenation in many ways. Microbes can manage the leaching of inorganic portion of e-waste which consists of both metallic and nonmetallic components.

Keywords E-waste management · Microbial participation · Bioleaching Green technology

Abbreviations

WEEE	Waste of electrical and electronic equipment
E-waste	Electronic waste
DNA	Deoxyribose nucleic acid
CRT	Cathode ray tube
PCBs	Polychlorinated biphenyls
NADH	Nicotinamide adenine dinucleotide
HXRF	Handheld X-ray fluorescence analysis
IC	Ion Chromatography
ICP	Inductively coupled plasma mass spectrometry
OES	*Optical Emission Spectroscopy*, Mt: Million Ton

D. Pant (✉) · A. Giri · V. Dhiman
Central University of Himachal Pradesh, Dharamshala 176215, Himachal Pradesh, India
e-mail: dpant2003@gmail.com

S. J. Varjani et al. (eds.), *Waste Bioremediation*, Energy, Environment, and Sustainability, https://doi.org/10.1007/978-981-10-7413-4_5

1 Introduction

E-waste includes waste electronic gadgets, electronic appliances such as television, mobile phones, computers while traditional goods such as refrigerators, air conditioning unit, and radio included in WEEE. Physical and chemical characteristics of e-waste are different as compared to municipal and household waste because it has varieties of components that require advance technical procedures for its proper management and scientific disposal.

E-waste production is 20–50 Mt (around 1–3% of municipal solid waste) in 2009, and this amount is increasing day by day (Robinson 2009). The metals present in e-waste recovered using chemical, biological, and hybrid leaching techniques play an important role in metal leaching process.

Chemical leaching using cyanide leaching, halide leaching, thiourea leaching, and thiosulfate leaching is the main leaching technique for the recovery of metals from electronic waste (Table 1). Chemical leaching involves the formation of different complexes. Inorganic acids like sulfuric acid (H_2SO_4), hydrochloric acid (HCl), solution of H_2SO_4 and HNO_3, sodium hypochlorite with acid or alkali can also be used for the recovery of metals. EDTA is one of the strong chelating agents used for metal extraction process from different ores or electronic waste. Releasing of toxic gases, high energy consumption, costly, and non-environmentally friendly nature were the major limitation of chemical leaching methods.

Biological leaching as the name indicated involves the use of various biological or natural agents for the purpose of metal leaching from e-waste scrap. These natural agents have the capability of transforming metals present in electronic waste. Biological leaching includes basic, associate, and precious metals. For extracting gold from printed electronic circuits by bioleaching (Faramarzi et al. 2004; Brandl 2008), *Chromobacterium violaceum is* proved to be more efficient in the mobilization of gold. Its bioleaching includes small pieces of printed circuit boards (5 × 10 mm treat with *C. violaceum* or *Pseudomonas fluorescens,* which dissolved Au in the form of dicyanoaurate [Au (CN) $^{2-}$]. *C. violaceum* involvement significantly improves Au recovery with copper removal from the waste (Pham and Ting 2009). Mobilization of metal through microbes is initiated through the secreted complexing agents with various oxidation and reduction reactions. Various involved mineral composition phases are as follows: FeO(OH), $Mn_3O_3(OH)_6$, M_2SiO_4 (M; Fe, Zn, Al), Cu_9S_5, Cu_5FeS_4, $ZnFe_2O_4$, $FeSiO_4$, $FeCr_2O_4$, Fe_3O_4, Al_2O_3, MnO, Ni/Mo/P/Al_2O_3, FeS_2, MoS_2 and WS_2) AS_2S_3, $CuFeS_2$, Cu_2S, FeS, Fe_7S_8, MnS_2, PbS, ZnS (Ilyas and Lee 2015). Proton-induced and ligand-induced reactions are involved for the leaching of metal oxides from these materials.

The interactions between microbial metabolisms and materials during the bioleaching process may have efficient integrated biometallurgical processes for metals recovery from any sources (Morin et al. 2006). The hybrid combination of both chemical and biological process can be applied to overcome the above limitation of bioleaching through the treatment of substrate with biological leaching approach followed by chemical processing or vice versa (Ilyas and Lee 2015).

Table 1 Chemistry in chemical leaching

Reagent	Metal	Chemistry	References
Sulfuric acid	Copper (Cu), zinc (Zn)	$ZnS + H_2SO_4 + 0.5O_2 \rightarrow ZnSO_4 + SO + H_2O$ $CuFeS_2 + 2H_2SO_4 + O_2$ $CuSO_4 + FeSO_4 + 2SO + 2H_2O$	Akcil et al. (2015)
Cyanide (CN^-)	Gold (Au), silver (Ag), palladium (Pd), and platinum (Pt)	$4Au + 8CN^- \rightarrow 4Au(CN)^{2-} + 4e$ $O_2 + 2H_2O + 4e \rightarrow 4OH^-$	Dorin and Woods (1991)
Thiourea	Gold (Au), silver (Ag)	$Au + 2CS(NH_2)^2 + Fe^{3+} \rightarrow Au(CN(NH_2)_2)^{3+} + Fe^{3+}$	Jing-ying et al. (2012)
Thiosulphate leaching	Copper (Cu), gold (Au)	$2Cu(NH_3)_4^{2-} + 8S_2O_3^{2-} \rightarrow 2Cu(S_2O_3)_3^{5-} + S_4O_6^{2-} + 8NH_3$ $2Cu(S_2O_3)_3^{5-} + 8NH_3 + 0.5O_2 + H_2O \rightarrow (Cu(NH_3)_4^{2+} + 6S_2O_3^2 + 2OH^-$	Chu et al. (2003), Akcil et al. (2015)
Aqua regia	Gold (Au)	$2Au + 9HCl + 3HNO_3 \rightarrow 2AuCl_3 + 3NOCl + 6H_2O$	Sheng and Etsell (2007)
H_2O_2	Copper (Cu)	$Cu^0 + H_2O_2 \rightarrow CuSO_4 + 2H_2O$	Akcil et al. (2015)
EDTA, oxalate, citrate, and tartrate	Cr (chromium), Cu (copper), and Zn (zinc), Pb (lead)	–	Peters (1999), Hong et al. (2000), Elliott and Shastri (1999), Pant et al. (2012)

Recycling of electronic waste by using microorganisms is an important method of valuable material recovery from electronic waste. Bioleaching is a promising technology for extraction of valuable metals from e-waste by using microorganisms and their metabolic products (Brandl et al. 2008; Mishra et al. 2008; Hong and Valix 2014). Microorganisms, mainly acidophilic group of bacteria like *Acidithiobacillus ferrooxidans, Acidithiobacillus thiooxidans, Leptospirillum ferrooxidans, and Sulfolobus* sp., play an important role in bioleaching of heavy metals from the wastes (Mishra and Rhee 2010). In the bioleaching process, microorganism transforms solid metallic compounds to its soluble and extractable form. Many chemolithoautotrophs bacteria, e.g., *A. ferrooxidans and a. thiooxidans,* use CO_2 as carbon source, Fe^{2+} and reduced S as an energy source for growth and play important role in bioleaching applications due to their ability to biooxidation and bioleaching reactions (Brandl et al. 2001). Mainly, three major groups of microbes like autotrophic bacteria (e.g. *Thiobacilli* sp.), heterotrophic bacteria (e.g. *Pseudomonas* sp., *Bacillus* sp.), and heterotrophic fungi (e.g. *Aspergillus* sp., *Penicillium* spp.) are involved in bioleaching of metals from any electronic waste or their ores (Schinner and Burgstaller 1989; Pant et al. 2012). Acidophilic microbes *A. ferrooxidans, A. thiooxidans, L. ferrooxidans,* etc, can tolerate high metal ion concentration (Lee et al. 2012).

1.1 Limitation of Leaching Techniques

There are several disadvantages of various chemical and biological techniques for e-waste management (Table 2). Future perspective of sustainable development depends upon the use of new techniques which must be suitable from economic as

Table 2 Advantages and disadvantages of different leaching techniques

Process	Advantages	Disadvantages
Biological leaching	• Simple and inexpensive • Ideal for low grade sulfide ores • No emission of gaseous pollutants • More economic • An eco-friendly technique • No need for high temperature and pressure	• The process is slow and causes delay in the process • Once started, the process cannot be stopped quickly • Low yield • Require large open area for treatment • No process control • High risk of contamination • Inconsistent yield as microbes cannot grow uniformly
Chemical leaching	• Take less time as compared to other techniques • High yield of mineral • High energy demand • Process control • High leaching efficiency	• Involve the use of organic solvents and acids that are harmful in nature • Expensive • Non-environment friendly • Potential for acid leaks

well as environmental point of view. The field of metal extraction involves the using of various processes that have been continuously expanding its applications for the discovery of new eco-friendly, economic, and efficient technique for metal recovery.

2 Mechanisms of Bioleaching

Various mechanisms involved in bioleaching are acidolysis, complexolysis, redoxolysis, and bioaccumulation for metal recovery (Bosshard et al. 1996). The leaching of copper from PCB shreds by *A. ferrooxidansis* involves:

(i) Fe^{3+} can be formed by *A. ferrooxidans* from Fe^{2+} through following reaction (1).

$$\mathbf{Fe^{2+} + O_2 + H^+ \overset{\textit{A.ferrooxidansis}}{\rightarrow} Fe^{3+} + H_2O} \quad (1)$$

(ii) Fe^{3+} formed by *A. ferrooxidans* in reaction (1) oxidizes the elemental copper to Cu^{2+} through following chemical reaction (2), (Bard et al. 1985).

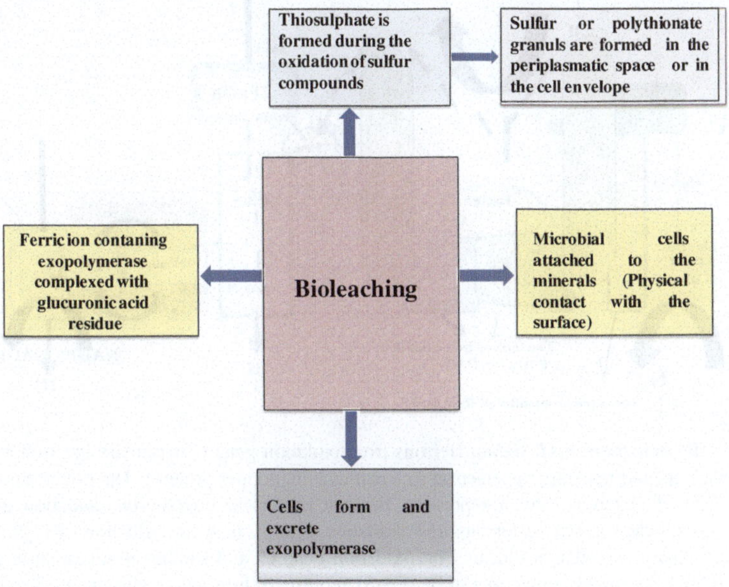

Fig. 1 Metal bioleaching

$$Cu^0 + \mathbf{Fe}^{3+}_{(\mathbf{Aq})} \rightarrow \mathbf{Fe}^{2+}_{(\mathbf{Aq})} + Cu^{2+}_{(\mathbf{Aq})} \quad (2)$$

Microorganism can also oxidize metal sulfide by direct and indirect mechanism. In direct mechanism, the source of electron is direct minerals, and in the indirect mechanism, oxidation of reduced metals occurs through ferric iron (**Fe^{3+}**) (Brandl 2008). Following equation explains both direct and indirect mechanism of bioleaching of pyrite.

$$2FeS_2 + 7O_2 + 2H_2O + 2H_2SO_4 \xrightarrow{\textit{Thiobacilli}} 2FeSO_4 + 2H_2SO_4 \quad (3)$$

$$4FeS_2 + O_2 + 2H_2SO_4 \xrightarrow{\textit{T.Ferrooxidans, L.Ferrooxidans}} 2Fe_2(SO_4)_3 + 2H_2O \quad (4)$$

Another mechanism of metal leaching proposed by Sand et al. (1995, 1999) is presented in Fig. 1.

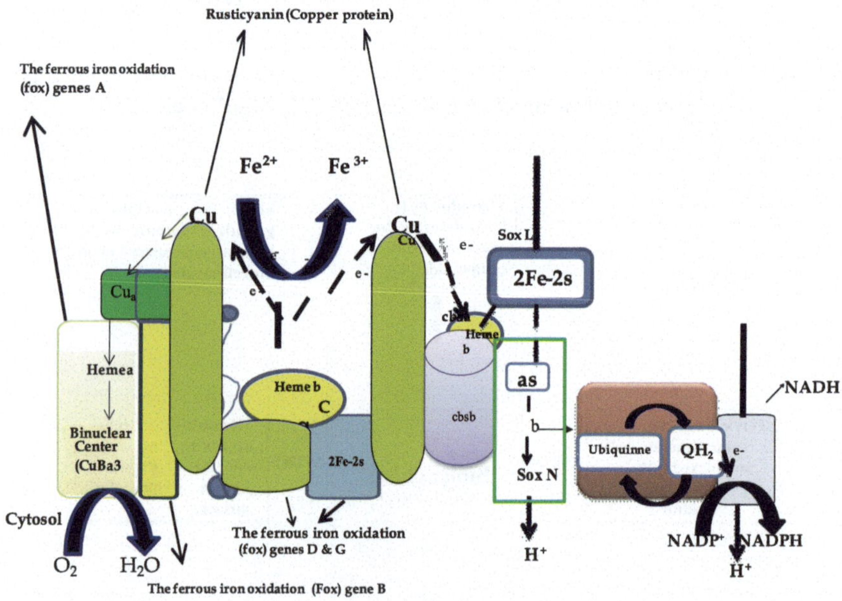

Fig. 2 Fe(II) oxidation in *M. sedula* (ferrous iron oxidation gene C is primary electron acceptor from metal ions and transfers the electron to a rusticyanin, copper protein). The dotted line shows the direction of electron flow; **a** represents heme a center in ferrous iron oxidation gene A; (CuBa3) is binuclear center in ferrous iron oxidation gene A (site for reduction of O_2 to H_2O); **b** represents heme b center in ferrous iron oxidation gene C; and sulphur iron oxidation gene N (soxN), (CuA) is copper center in FoxB, (Fe–S) represents iron sulfur clusters in ferrous iron oxidation gene G and sulphur iron oxidation gene (SoxL); (Q) and (QH2) represents ubiquinone and hydroquinone) (Auernik et al. 2008; Kozubal et al. 2011; Wheaton et al. 2015).

Biooxidation of Fe(II) and molecular properties of associated protein-like oxidases and associated electron transfer proteins have been most widely and extensively studied in mesophilic bacterium *A. ferrooxidans and A. ambivalens* (Bandeiras et al. 2009). Some researcher explained a hypothetical models for Fe(II) oxidation by the helping of ferrous iron oxidation (fox) genes cluster in the *M. sedula* species (Aurnik and Kelly 2008; Kozubal et al. 2011) electron acceptor from Fe(II) are ferrous iron oxidation gene FoxCD), cytochrome b, and shuttled by a multi-copper oxidase. Ferrous iron oxidation gene G (FoxG) can form complex with ferrous iron oxidation gene D (FoxD). The way of electron transfer involving multi-copper oxidase to cytochrome complex through quinone pool and finally to nicotinamide adenine dinucleotide-hydrogen dehydrogenase (NADH dehydrogenase) (Fig. 2).

Bioleaching accounts for 10% of the copper production worldwide. Importance of bioleaching as a biotechnological application involves the biology of *A. ferrooxidans* and associated bioleaching microorganisms (Valdés et al. 2008). *A. ferrooxidans* is mainly used in industrial recovery of copper by bioleaching or biomining methods using energy from the oxidation of iron- and sulfur-containing minerals for the growth and generates reverse electron flow from Fe(II) to nicotinamide adenine dinucleotide-hydrogen (NADH) (Levican et al. 2002; Yarzabal et al. 2004; Brasseur et al. 2004; Bruscella et al. 2007; Valdés et al. 2008). It can also obtain energy by the oxidation of reduced sulfur compounds, hydrogen, and formate. *A. ferrooxidans* obtain the necessary carbon for metabolic processes from carbon dioxide in the environment (Valdés et al. 2008); this process falls under the chemolithoautotrophic metabolism (Fig. 3).

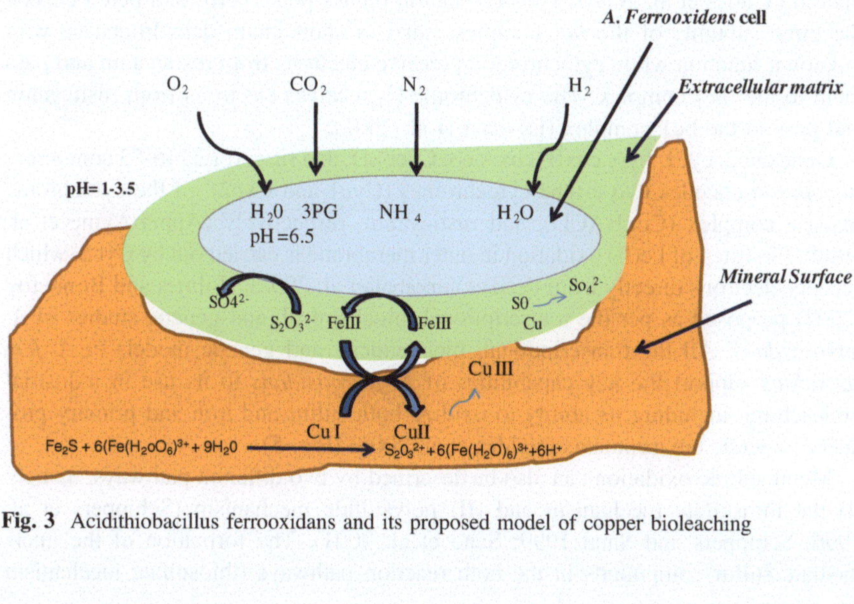

Fig. 3 Acidithiobacillus ferrooxidans and its proposed model of copper bioleaching

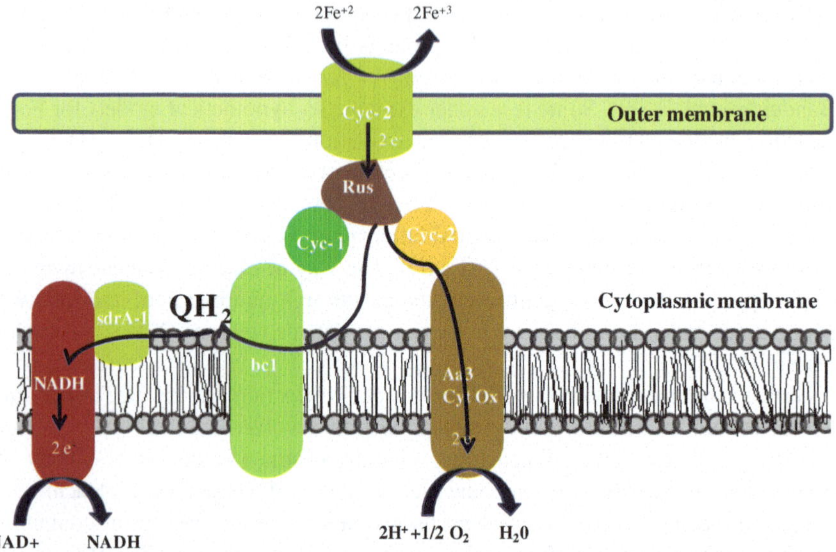

Fig. 4 Enzymes and electron transfer proteins involved in the oxidation of ferrous iron (Levican et al. 2002)

As per the genome-based models, analysis of *A. ferrooxidans* identifies the role as main components in electron transport chain and mainly involved in the oxidation of ferrous iron (Fig. 4). DNA sequence analysis and bioinformatic interpretations of *A. ferrooxidans* identified two transcriptional units, the *petI* and *rus* operon (Valdés et al. 2008). The *petI* operon (genes petA, petB, and petC) encode the three subunits of the bc_1 complex, sdrA is short chain dehydrogenase with unknown function while cytochrome c_4 receive electrons from rusticyanin and pass them to the bc_1 complex. The cytochrome C_4 receives electrons from rusticyanin and pass to the bc1 complex (Levican et al. 2002).

Gene cyc2, cyc1, hyp, coxB, coxA coxC, coxD, and rus; AFE3146-53 containing rus *operon* encodes two c-type cytochromes (Cyc1 and Cyc2), of the cytochrome oxidase complex (CoxBACD), and rusticyanin, respectively (Appia-Ayme et al. 1999). First step of Fe(II) oxidation in outer membrane is carried out by Cyc2, which accepts electrons directly from Fe(II) (Yarzábal et al. 2002). Holmes and Bonnefoy (2007) proposed as per the transcriptional, biochemical, and genetic studies of *A. ferrooxidans*. All the transcriptional, biochemical, and genetic models of *A. ferrooxidans* support the key capabilities of *A. ferrooxidans* to its use in industrial bioleaching, including its ability to oxidize both sulfur and iron and primary producer in acidic environments (Valdés et al. 2008) (Fig. 5).

Metal sulfide oxidation can also be described by two different pathways, namely (I) the thiosulfate mechanism and (II) polysulfide mechanism (Schippers et al. 1996; Schippers and Sand 1999; Sand et al. 2001). The formation of the intermediate sulfur compounds in the both reaction pathways (thiosulfate mechanism

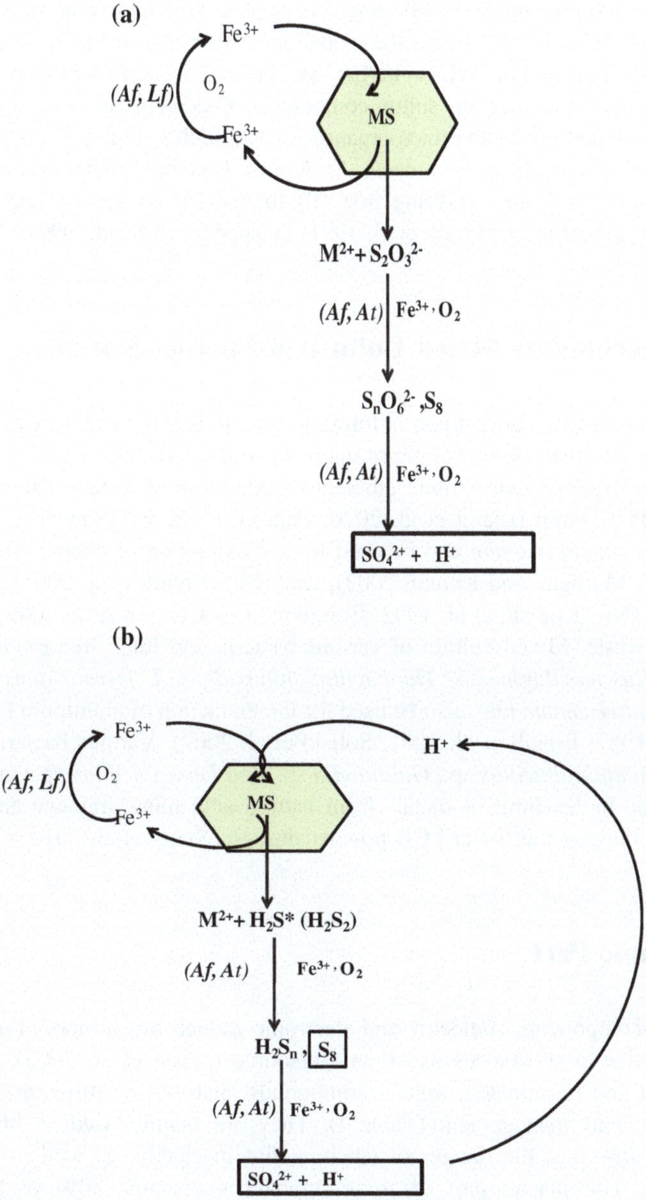

Fig. 5 Comparison of the thiosulfate (**a**) and polysulfide (**b**) mechanisms (from Schippers and Sand 1999, modified)

and (II) polysulfide mechanism) depends on the pH of the geochemical environment, mineralogy of the metal sulfide, and the presence of different oxidants (Schippers 2004). Chemical dissolution of the metal sulfides by the microorganisms formed intermediate sulfur compounds. Microorganisms play a crucial role in the oxidation of Fe(II) to Fe(III) ions and catalyze the oxidation of intermediate sulfur compounds to sulfuric acid. Oxidation of Fe(II) to Fe(III) ions serves as oxidants for the metal sulfides for most of the intermediate sulfur compounds. Oxidation of elemental sulfur is exclusively carried out by the microorganisms because this sulfur species is inert to a biotic oxidation in acidic environments. Acidophilic bacteria *A. Ferrooxidans (Af) and L. ferrooxidans (Lf)* are catalyzing iron (II) to iron(III) conversion and associated recycling of ions (Fig. 5) (Pronk et al. 1991; Schippers and Sand 1999).

3 Bioleaching by Mixed Culture of Microorganism

Mixed culture of microorganism is found to be efficient for metal extraction from electronic waste than single culture of microorganism, like culture of *A. thiooxidans* and *A. ferrooxidans, that* is more efficient in extraction of metals (Cu, Ni, Zn and Pb) from PCB scrap (Liang et al. 2010; Pant et al. 2012). Fungus *A. niger* and *Penicillium simplicissimum* can be used for the extraction of copper (Cu) (Brandl et al. 2001; Mulligan and Kamali 2003), lead (Pb) (Brandl et al. 2001), zinc (Zn), and nickel (Ni) (Konishi et al. 1992; Brandl et al. 2001; Ren et al. 2009) from any electronic waste. Mixed culture of various bacteria and fungi like *Bacillus circulans, Bacillus mucilaginosus, Thiobacillus thiooxidans, T. ferrooxidans, A. niger,* and *P. Simplicissimum* may also be used for the extraction of aluminum from its ore (Groudev 1987; Brandl et al. 2001; Solisio et al. 2002). Various bacterial consortiums of (*Acidithiobacillus* sp., *Gallionella* sp., and *Leptospirillum* sp.) may also be used for the bioleaching of metals from natural acid mine drainage and showed higher leaching rate at lower PCB powder dosage (Xiang et al. 2010) (Table 3).

4 Organic Part

Electronic components of electric and electronic gadgets are composed of different fibers, polymeric substances as organic resource (Alaee et al. 2003). Some are chlorinated and brominated organic compounds analyzed by different techniques for organic part management (Table 4). They are mainly used as fire-resistant agents for lowering the danger of catching fire in electric as well as electronic appliances. The plastic part of these appliances contains 30% of flame and fire-resistant agents. They are efficient and suitable for different types of plastics (Bientinesi and Petarca 2009).

There are management issues with the halogenated plastics in electronic waste. Several methods were suggested to counter its management issues. Dehalogenation and waste regeneration from old electronic material are described in a patent no. cn101220173a (Chen et al. 2008), and another method describes the high temperature treatment of the waste to transform it in diverse kind of bromine

Table 3 Involvement of different microbial species for bioleaching

	Species involved	Leaching agent	Condition/ pH range	References
Bacterial species	*Acetobacter methanolicus*	Gluconate	Acidophilic	Glombitza et al. (1988)
	Acidianusbrierleyi	Sulfuric acid	Acidophilic	
				Barrett (1993), Olson (1994)
	Acidophillum cryptum	Organic acids	2.0–6.0	Rossi (1990)
	Arthrobacter sp.	–	–	Bosecker (1993)
	Bacillus megaterium	Citrate, amino acids	–	Krebs et al. (1996)
	Chromobacterium violaceum	Cyanide	–	Olson (1994)
	Corynebacterium sp.	Ferric iron, sulfuric acid		
			–	Müller et al. (1995)
	Crenothrix sp.		5.5–6.2	
	Leptotrix sp.		5.8–7.8	Rossi (1990)
	Leptospirillum ferrooxidans		–	Rossi (1990), Olson (1994)
	Metallosphaera sedula		Acidophilic	Barrett (1993), Rossi (1990)
		Sulfuric acid		
	Thermothrixthiopara sp.		Neutral	Brierley (1982)
	Thiobacillus acidophilus		1.5–6.0	Rossi (1990)
	Thiobacillus albertis		2.0–4.5	
	Thiobacillus capsulatus		–	
	Thiobacillus concretivorus		0.5–6.0	Rossi (1990)
	Thiobacillus cuprinus		–	Huber and Stetter (1990)
	Thiobacillus denitrificans		5.0–7.0	Rossi (1990)
	Thiobacillus ferrooxidans		1.4–6.0	Glombitza et al. (1988), Olson (1994)
	Thiobacillus intermedius		1.9–7.0	Rossi (1990)
	Thiobacillus neapolitanus		3.0–8.5	
	Thiobacillus perometabolis		2.6–6.8	
	Thiobacillus prosperus		–	Huber et al. (1989)
	Thiobacillus rubellus		–	Barret (1993), Rossi (1990)
	Thiobacillus tepidarius		–	Hughes and Poole(1989)
	Thiobacillus thiocynoxidans		–	Jain and Tyagi (1992)
	Thiobacillus thioparus		4.5–10.0	Rossi (1990)

(continued)

Table 3 (continued)

	Species involved	Leaching agent	Condition/ pH range	References
Archaea species	*Sulfobacillus thermosulfidooxidans*	Ferric iron	Extremely acidophilic	Barrett (1993)
	Sulfobacillus thermosulfidooxidans			
	Sulfobacillus thermosulfidooxidans sub. *asporogenes*			
	Sulfolobus acidocaldarius		0.9–5.8	Olson (1994), Rossi (1990)
	Sulfolobus ambivalens		–	
	Sulfolobus solfataricus		–	
	Sulfolobus thermosulfidooxidans		–	
	Sulfolobus brierleyi		–	Brierley (1982)
	Sulfolobus yellowstonii		–	Olson (1994)
Fungal species	*Actinomucor* sp.	Oxalate, malate, pyruvate, oxalacetate	–	
	Alternaria sp.			
	Aspergillus amstelodami			
	Aspergillus clavatus	Aspartate		
	Aspergillus ficuum	Oxalate		Strasser et al. (1994)
	Aspergillus fumigatus	–		
	Aspergillus niger	Oxalate, citrate, gluconate, lactate		Bosshard et al. (1996), Sayer et al. (1995)
	Aspergillus ochraceus	Citrate, glutamate		
	Candida sp.	–		
	Cerostamella sp.	–		
	Cladosporium resinae	–		
	Coriolus versicolor	–		
	Cunninghamiella sp.	–		
	Fusarium sp.	–		
	Gleophyllum trabeum	Oxalate		Strasser et al. (1994)
	Mucor sp.	Fumarate, gluconate		
	Paecilomyces variotii	Malate		
	Penicillium cyclopium			
	Penicillium funiculosum	Citrate, glutamate		
	Penicillium notatum	–		
	Penicillium ochrochloron	Oxalate		
	Penicillium oxalicum			Strasser et al. (1994)
	Penicillium simplicissimum			

(continued)

Table 3 (continued)

	Species involved	Leaching agent	Condition/ pH range	References
		Citrate, oxalate, gluconate		
	Penicillium spinulosum	Oxalate		
	Penicillium variotii	–		
	Phanerochaete chrysosporium	–		
	Pichia sp.	–		

Table 4 Percentage analysis of chlorine and bromine part in WEEE

Equipments	Chlorine (%)	Bromine (%)	Analytical technique	References
Small WEEE	–	0.53	HXRF	Dimitrakakis et al. (2009)
Modest WEEE	1	0.6	X-ray fluorescence, ICP-OES	
Television and computer screen	0.01–1.3	0.02–4.4	Semiquantitative XRF	
Fridge	1.3	–	IC	
Landline phone	2.1	0.6	IC	Caballero et al. (2016)

compounds (Hornung et al. 2006). Another two processes: (i) staged gasification and (ii) co-combustion were compared by researchers in 2009 (Bientinesi and Petarca 2009). After the experiment, they conclude that the co-combustion produces less emission of HCl, HBr, NOx than the staged gasification. But the later process is more energy efficient method and saves resources and lowers carbon dioxide emissions. Its environmental impact is less than the previous method. Therefore, performance ability of staged gasification is much better as compared to co-combustion and also results in high Br yield (Bientinesi and Petarca 2009). Above study concludes that WEEE thermal high-temperature treatment for energy and metal recovery is much efficient and effective process to deal with such type of waste. Following conclusions are being made on the basis of studies carried on the process of high-temperature thermal degradation process:

(1) The degradation product obtained is highly enriched with free halogens and hydrogen halides when bench scale oxidation reactor is used (Yang et al. 2011).
(2) In the process of pyrolysis, alkali metals have effective role in decreasing Br percentage and VOCs in the resultant product.
(3) Starting temperature is 400 °C, and for complete degradation, a temperature of 1000 °C is required.

Apart from that, fungi (*Pestalotiopsis microspora*), a plastic eating fungi, has been recently discovered, and the enzymes (serine hydrolase) can breakdown the plastics (Mohammed et al. 2013). This may also help in remediating e-waste from environment.

5 Factors Influencing Bioleaching

Metal bioleaching is influenced by (Fig. 6) a series of different factors such as physicochemical parameters (temperature, pH, redox potential, water potential, nutrient availability, mass transfer, other gases content like carbon dioxide, etc.). Efficiency of leaching (Brandl 2008) depends on bioleaching environment (microbial diversity, population density, microbial activity, metal tolerance, spatial distribution of microorganism, etc.) and the properties of metals to be leached (metal type, composition, surface area, hydrophobicity, etc.). In the case of sphalerite (Zn ore) oxidation by *T. ferrooxidans*, different parameters like activities of microbial cell, source of energy, temperature, and particle size of metal strongly influence metal leaching (Ballester et al. 1990). The presence of organic compounds, surface active agents and solvents inhibited the pyrite oxidation of *T. ferrooxidans* (Bacelar-nicolau and Johnson 1999). Sometimes, other metals like copper, nickel, uranium, or thorium can also influence the iron(II) oxidation by *T. ferrooxidans* (Leduc et al. 1997). Industrial biocides such as tetra-n-butyltin, isothiazolinones, 2,2b-dihydroxy-5,5b-dichlorophenylmethane (dichlorophen) decrease the leaching rate of manganese oxides by heterotrophic microorganisms (Arief and Madgwick 1992).

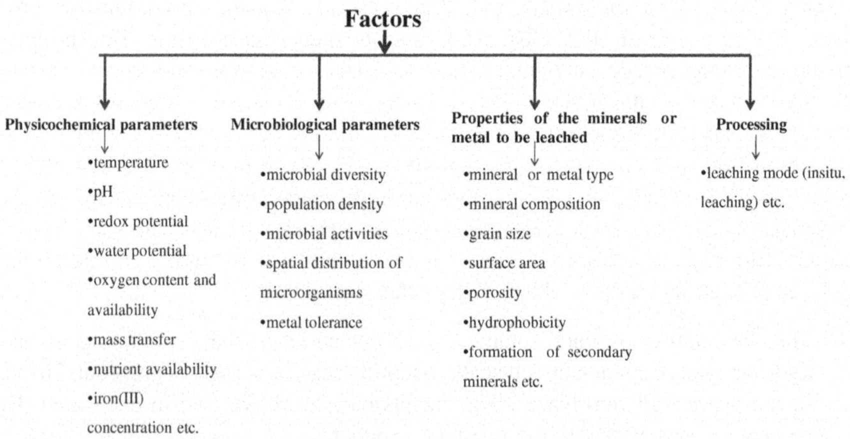

Fig. 6 Factors influencing the bioleaching

6 Hybrid Technique for Metal Extraction

Biological leaching (Fig. 7) of metals from any electronic waste or metals ore is a cost-effective technique, but it is time-consuming because microbial growth is a complex process involving numerous anabolic and catabolic reactions within the microbial cells (Maier 2000), and several environmental issues in chemical leaching hybrid technique involving combination of chemical (eco-friendly chemicals) and biological leaching may be the best and efficient techniques for metal extraction (Table 5; Pant et al. 2012). In hybrid technique, various combinations of ligands with associated microbes can be used for metal extraction. For the extraction of Zn, Cd, Pb, and Cu from electronic waste or their ores, EDTA (Wasay et al. 1998) can reasonably enhance the efficiency of metal extraction with either *A. ferrooxidan* (Cheikh et al. 2010) or *Shewanella putrefaciens* (Dobbin et al. 1995). Various organic acids like citric, tartaric, and oxalic acids secreted from fungi (*A. niger, Penicillium bilaiae,* and other *Penicillium* sp.) act as a chelating agents for the extraction of Cu, Cd, Zn, and Pb (Arwidsson and Allard 2010; Wasay et al. 1998;

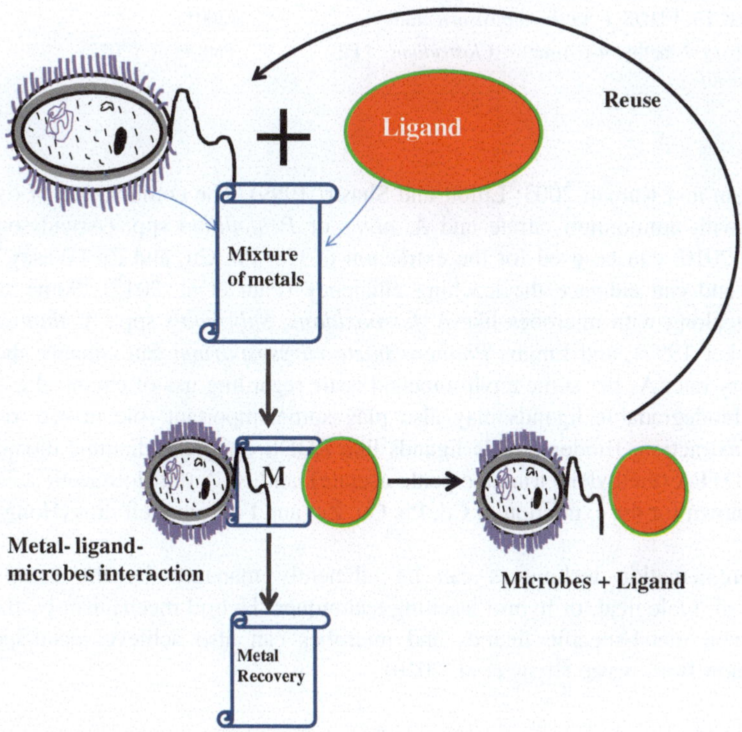

Fig. 7 Flow diagram of hybrid technique

Table 5 Hybrid combination for the extraction of metals

Ligands and microbial combination	Recovered metals	References
EDTA + *A. ferrooxidans* and *Shewanella putrefaciens*	Zn and Pb	Cheikh et al. (2010), Wasay et al. (1998), Satroutdinov et al. (2000), Dobbin et al. (1995), Pant et al. (2012)
Citrate + *Aspergillus niger, Penicillium* sp.	Cd, Cu, Pb	Arwidsson and Allard (2010), Mulligan and Kamali (2003), Pant et al. (2012)
Tartrate + *Aspergillus niger, Penicilliumbilaiae, Penicillium* sp.	Cd, Pb and Zn	Arwidsson and Allard (2010), Pant et al. (2012), Wasay et al. (1998)
Oxalate + *Aspergillus niger, Penicillium* sp.	Zn	Elliott and Shastri (1999), Arwidsson and Allard (2010)
Oxalate + ammonium citrate + *Aspergillus niger, Penicillium* sp.	Cd, Pb, Cu and Zn	Wasay et al. (1998), Arwidsson and Allard (2010), Pant et al. (2012)
DTPA, + *Candida albicans*; NTA (nitrilotriacetic acid) + *Acidithiobacilluscaldus* strain BC13; EDDS + *Amycolatopsisorientalis*	Cr, Pb, Cu, Zn	Hong et al. (2000), Zwicker et al. (1997), Sohnle et al. (2001)
(3-hydroxy-2-methyl-4-pyrone) + *Clostridium beijerinckii*	Fe	Pant et al. (2012)

Mulligan and Kamali 2003; Elliott and Shastri 1999). The combination of oxalate along with ammonium citrate and *A. niger* or *Penicillium* spp. (Arwidsson and Allard 2010) can be used for the extraction of Cd, Pb, Cu, and Zn (Wasay et al. 1998) and can enhance the leaching efficiency (Pant et al. 2012). Some sulfide ligands along with microbes like *A. ferrooxidans*, *Sulfobolus* spp., *A. thiooxidans* (Bosecker 1997), and fungus *Phanerochaete chrysosporium* can enhance the rate recovery rate. As per some environmental issue regarding use of chemical ligands, some biodegradable ligands may also play some important role in eco-friendly metal extraction. Biodegradable ligands like EDDS (ethylene diamine di-succinic acid), DTPA (diethylene triamine penta-acetate), and NTA (nitrilotriacetic acid) are well known for the extraction of Cd, Pb, Cu, Zn, and Fe from their ores (Hong et al. 2000).

Bioremediation techniques can be efficiently managed by the compatible chemical, biological, or hybrid leaching techniques. Hybrid mechanism by the use of certain metal-specific ligands and microbes can also achieve metal-specific extraction from waste (Ilyas et al. 2010).

References

Akcil A, Erust C, Gahan CS, Ozgun M, Sahin M, Tuncuk A (2015) Precious metal recovery from waste printed circuit boards using cyanide and non-cyanide lixiviants–a review. Waste Manag 45:258–271

Alaee M et al. (2003) An overview of commercially used brominated flame retardants, their applications, their use patterns in different countries/regions and possible modes of release. Environ Int 29 (6):683–689

Appia-Ayme C, Guiliani N, Ratouchniak J, Bonnefoy V (1999) Characterization of an Operon Encoding Two c-Type Cytochromes, an aa3-Type Cytochrome Oxidase, and Rusticyanin in *Thiobacillus ferrooxidans ATCC 33020.* Appl Environ Microbiol 65(11):4781–4787

Arief YY, Madgwick JC (1992) The effect of biocides on microaerobic bacterial biodegradation of manganese oxide. Biorecovery 2(2):95–106

Arwidsson Z, Allard B (2010) Remediation of metal-contaminated soil by organic metabolites from fungi II—metal redistribution. Water Air Soil Pollut 207(1–4):5–18

Auernik KS, Kelly RM (2008) Identification of components of electron transport chains in the extremely thermoacidophilic crenarchaeon Metallosphaera sedula through iron and sulfur compound oxidation transcriptomes. Appl Environ Microbiol 74(24):7723–7732

Bacelar-Nicolau P, Johnson DB (1999) Leaching of pyrite by acidophilic heterotrophic iron-oxidizing bacteria in pure and mixed cultures. Appl Environ Microbiol 65(2):585–590

Ballester A, Gonzalez F, Blazquez ML, Mier JL (1990) The influence of various ions in the bioleaching of metal sulphides. Hydrometallurgy 23(2–3):221–235

Bandeiras TM, Refojo PN, Todorovic S, Murgida DH, Hildebrandt P, Bauer C, Pereira MM, Kletzin A, Teixeira M (2009) The cytochrome ba complex from the thermoacidophilic crenarchaeote *Acidianus ambivalens.* is an analog of bc 1 complexes. Biochim et Biophys Acta (BBA)-Bioenerg 1787(1):37–45

Bard AJ, Parsons R, Jordan J (1985) Standard potentials in aqueous solution, vol 6. CRC press

Barrett J (1993) Metal extraction by bacterial oxidation of minerals. Horwood

Bientinesi M, Petarca L (2009) Comparative environmental analysis of waste brominated plastic thermal treatments. Waste Manag 29(3):1095–1102

Bosecker K (1993) Bioleaching of silicate manganese ores. Geomicrobiol J 11(3–4):195–203

Bosecker K (1997) Bioleaching: metal solubilization by microorganisms. FEMS Microbiol Rev 20 (3–4):591–604

Bosshard PP, Bachofen R, Brandl H (1996) Metal leaching of fly ash from municipal waste incineration by *Aspergillus niger.* Environ Sci Technol 30(10):3066–3070

Brandl H (2008) Microbial leaching of metals. Biotechnol Set, Second Ed, pp 191–224

Brandl H, Lehmann S, Faramarzi MA, Martinelli D (2008) Biomobilization of silver, gold, and platinum from solid waste materials by HCN-forming microorganisms. Hydrometallurgy 94(1–4):14–17

Brandl H, Bosshard R, Wegmann M (2001) Computer-munching microbes: metal leaching from electronic scrap by bacteria and fungi. Hydrometallurgy 59(2):319–326

Brasseur G, Levicán G, Bonnefoy V, Holmes D, Jedlicki E, Lemesle-Meunier D (2004) Apparent redundancy of electron transfer pathways via bc 1 complexes and terminal oxidases in the extremophilic chemolithoautotrophic *Acidithiobacillus ferrooxidans.* Biochim et Biophys Acta (BBA)-Bioenerg, 1656(2), 114–126

Brierley CL (1982) Microbiological mining. Sci Am 247:44–53

Bruscella P, Appia-Ayme C, Levican G, Ratouchniak J, Jedlicki E, Holmes DS, Bonnefoy V (2007) Differential expression of two bc1 complexes in the strict acidophilic chemolithoautotrophic bacterium *Acidithiobacillus ferrooxidans* suggests a model for their respective roles in iron or sulfur oxidation. Microbiol 153(1):102–110

Caballero BM, De Marco I, Adrados A, López-Urionabarrenechea A, Solar J, Gastelu N (2016) Possibilities and limits of pyrolysis for recycling plastic rich waste streams rejected from phones recycling plants. Waste Manag 57:226–234

Cheikh M, Magnin JP, Gondrexon N, Willison J, Hassen A (2010) Zinc and lead leaching from contaminated industrial waste sludges using coupled processes. Environ Technol 31(14):1577–1585

Chen L, Guan G, Wang B, Peng S, Zhou W (2008) Apparatus and method fordehalogenation regeneration of waste and old electric plastics, Patent Publication No.CN101220173 (A)

Chu CK, Breuer PL, Jeffrey MI (2003) The impact of thiosulfate oxidation products on the oxidation of gold in ammonia thiosulfate solutions. Miner Eng 16(3):265–271

Dimitrakakis E, Janz A, Bilitewski B, Gidarakos E (2009) Determination of heavy metals and halogens in plastics from electric and electronic waste. Waste Manag 29(10):2700–2706

Dobbin PS, Powell AK, McEwan AG, Richardson DJ (1995) The influence of chelating agents upon the dissimilatory reduction of Fe (III) by *Shewanella putrefaciens*. Biometals 8(2):163–173

Dorin R, Woods R (1991) Determination of leaching rates of precious metals by electrochemical techniques. J Appl Electrochem 21(5):419–424

Elliott HA, Shastri NL (1999) Extractive decontamination of metal-polluted soils using oxalate. Water Air Soil Pollut 110(3):335–346

Faramarzi MA, Stagars M, Pensini E, Krebs W, Brandl H (2004) Metal solubilization from metal-containing solid materials by cyanogenic *Chromobacterium violaceum*. J Biotechnol 113 (1):321–326

Glombitza F, Iske U, Bullmann M (1988) Mikrobielle laugung von seltenen erdelementen und spurenelementen. BioEngineering 4:37–43

Groudev SN (1987) Use of heterotrophic microorganisms in mineral biotechnology. Eng Life Sci 7(4):299–306

Holmes DS, Bonnefoy V (2007) Genetic and bioinformatic insights into iron and sulfur oxidation mechanisms of bioleaching organisms. In: Biomining. Springer Berlin, Heidelberg, pp 281–307

Hong KJ, Tokunaga S, Kajiuchi T (2000) Extraction of heavy metals from MSW incinerator fly ashes by chelating agents. J Hazard Mater 75(1):57–73

Hong Y, Valix M (2014) Bioleaching of electronic waste using acidophilic sulfur oxidising bacteria. J Clean Prod 65:465–472

Hornung A et al. (2006) Method for treating waste materials containing halogen. U.S. Patent No. 7,060,242. 13 Jun. 2006

Huber H, Stetter KO (1989) Thiobacillus prosperus sp. nov., represents a new group of halotolerant metal-mobilizing bacteria isolated from a marine geothermal field. Arch Microbiol 151(6):479–485

Huber H, Stetter KO (1990) *Thiobacillus cuprinus* sp. nov., a novel facultatively organotrophic metal-mobilizing bacterium. Appl Environ Microbiol 56(2):315–322

Hughes MN, Poole RK (1989) Metals and Micro-organisms

Ilyas S, Ruan C, Bhatti HN, Ghauri MA, Anwar MA (2010) Column bioleaching of metals fromelectronic scrap. *Hydrometallurgy* 101(3):135–140

Ilyas S, Lee JC (2015) Hybrid leaching: an emerging trend in bioprocessing of secondary resources. In: Microbiology for minerals, metals, materials and the environment, CRC Press, pp 359–382

Jain DK, Tyagi RD (1992) Leaching of heavy metals from anaerobic sewage sludge by sulfur-oxidizing bacteria. Enzyme and microbial technology 14(5):376–383

Jing-ying L, Xiu-Li X, Wen-quan L (2012) Thiourea leaching gold and silver from the printed circuit boards of waste mobile phones. Waste Manag 32(6):1209–1212

Konishi Y, Kubo H, Asai S (1992) Bioleaching of zinc sulfide concentrate by *Thiobacillus ferrooxidans*. Biotechnol Bioeng 39(1):66–74

Kozubal MA, Dlakić M, Macur RE, Inskeep WP (2011) Terminal oxidase diversity and function in "*Metallosphaera yellowstonensis*": gene expression and protein modeling suggest mechanisms of Fe (II)-oxidation in the Sulfolobales. Appl Environ Microbiol

Krebs W, Bosshard PP, Brandl H, Bachofen R (1996) March. From waste to resource: metal recovery from solid waste incineration residues by microorganisms. In: Spring Meeting of the

Vereinigung für Allgemeine und Angewandte Mikrobiologie (VAAM), Bayreuth. Biospektrum Suppl (p 98)

Leduc LG, Ferroni GD, Trevors JT (1997) Resistance to heavy metals in different strains of *Thiobacillus ferrooxidans*. World J Microbiol Biotechnol 13(4):453–455

Lee JC, Pandey BD (2012) Bio-processing of solid wastes and secondary resources for metal extraction–a review. Waste Manag 32(1):3–18

Levicán G, Bruscella P, Guacunano M, Inostroza C, Bonnefoy V, Holmes DS, Jedlicki E (2002) Characterization of the petI and res operons of *Acidithiobacillus ferrooxidans*. J Bacteriol 184 (5):1498–1501

Liang G, Mo Y, Zhou Q (2010) Novel strategies of bioleaching metals from printed circuit boards (PCBs) in mixed cultivation of two acidophiles. Enzym Microb Technol 47(7):322–326

Maier RM, Pepper I, Gerba C (2000) Bacterial growth. Environ microbiol, pp 43–59

Mishra D, Kim DJ, Ralph DE, Ahn JG, Rhee YH (2008) Bioleaching of metals from spent lithiumion secondary batteries using *Acidithiobacillus ferrooxidans*. Waste Manage 28(2):333–338

Mishra D, Rhee YH (2010) Current research trends of microbiological leaching for metal recovery from industrial wastes. Curr Res Technol Educ Topics Appl Microbiol Microb Biotechnol 2:1289–1292

Mohammed A, Oyeleke SB, Ndamitso MM, Achife CE, Badeggi UM (2013) The Impact of Electronic Waste Disposal and Possible Microbial and Plant Control. Int J Eng Sci 2(10):35–42

Morin D, Lips A, Pinches T, Huisman J, Frias C, Norberg A, Forssberg E (2006) BioMinE–Integrated project for the development of biotechnology for metal-bearing materials in Europe. Hydrometallurgy 83(1):69–76

Müller B, Burgstaller W, Strasser H, Zanella A, Schinner F (1995) Leaching of zinc from an industrial filter dust with *Penicillium*, *Pseudomonas* and *Corynebacterium*: Citric acid is the leaching agent rather than amino acids. J Ind Microbiol Biotechnol 14(3):208–212

Mulligan CN, Kamali M (2003) Bioleaching of copper and other metals from low-grade oxidized mining ores by Aspergillus niger. J Chem Technol Biotechnol 78(5):497–503

Olson GJ (1994) Microbial oxidation of gold ores and gold bioleaching. FEMS Microbiol Lett 119 (1–2):1–6

Pant D, Joshi D, Upreti MK, Kotnala RK (2012) Chemical and biological extraction of metals present in E waste: a hybrid technology. Waste Manag 32(5):979–990

Peters RW (1999) Chelant extraction of heavy metals from contaminated soils. J Hazard Mater 66 (1):151–210

Pham VA, Ting YP (2009) Gold bioleaching of electronic waste by cyanogenic bacteria and its enhancement with bio-oxidation. In: Advanced materials research, vol 71. Trans Tech Publications, pp 661–664

Pronk JT, Meijer WM, Hazeu W, Van Dijken JP, Bos P, Kuenen JG (1991) Growth of *Thiobacillus ferrooxidans* on formic acid. Appl Environ Microbiol 57(7):2057–2062

Ren WX, Li PJ, Geng Y, Li XJ (2009) Biological leaching of heavy metals from a contaminated soil by *Aspergillus niger*. J Hazard Mater 167(1):164–169

Robinson BH (2009) E-waste: an assessment of global production and environmental impacts. Sci Total Environ 408(2):183–191

Rossi G (1990) Biohydrometallurgy. McGraw-Hill

Sand W, Gehrke T, Jozsa PG, Schippers A (1999) Direct versus indirect bioleaching. Process Metall 9:27–49

Sand W, Gehrke T, Jozsa PG, Schippers A (2001) (Bio) chemistry of bacterial leaching—direct vs. indirect bioleaching. Hydrometallurgy 59(2):159–175

Sand W, Gerke T, Hallmann R, Schippers A (1995) Sulfur chemistry, biofilm, and the (in) direct attack mechanism—a critical evaluation of bacterial leaching. Appl Microbiol Biotechnol 43 (6):961–966

Satroutdinov AD, Dedyukhina EG, Chistyakova TYI, Witschel M, Minkevich IG, Eroshin VK, Egli T (2000) Degradation of metal—EDTA complexes by resting cells of the bacterial strain DSM 9103. Environ Sci Technol 34(9):1715–1720

Sayer JA, Raggett SL, Gadd GM (1995) Solubilization of insoluble metal compounds by soil fungi: development of a screening method for solubilizing ability and metal tolerance. Mycol Res 99(8):987–993

Schinner F, Burgstaller W (1989) Extraction of zinc from industrial waste by a *Penicillium* sp. Appl Environ Microbiol 55(5):1153–1156

Schippers A, Sand W (1999) Bacterial leaching of metal sulfides proceeds by two indirect mechanisms via thiosulfate or via polysulfides and sulfur. Appl Environ Microbiol 65(1):319–321

Schippers A (2004) Biogeochemistry of metal sulfide oxidation in mining environments, sediments, and soils. Geol Soc America Spec Pap 379:49–62

Schippers A, Jozsa P, Sand W (1996) Sulfur chemistry in bacterial leaching of pyrite. Appl Environ Microbiol 62(9):3424–3431

Sheng PP, Etsell TH (2007) Recovery of gold from computer circuit board scrap using aqua regia. Waste Manage Res 25(4):380–383

Sohnle PG, Hahn BL, Karmarkar R (2001) Effect of metals on *Candida albicans* growth in the presence of chemical chelators and human abscess fluid. J Lab Clin Med 137(4):284–289

Solisio C, Lodi A (2002) Bioleaching of zinc and aluminium from industrial waste sludges by means of Thiobacillus ferrooxidans. Waste Manag 22(6):667–675

Strasser H, Burgstaller W, Schinner F (1994) High-yield production of oxalic acid for metal leaching processes by *Aspergillus niger*. FEMS Microbiol Lett 119(3):365–370

Valdés J, Pedroso I, Quatrini R, Dodson RJ, Tettelin H, Blake R, Eisen JA, Holmes DS (2008) *Acidithiobacillus ferrooxidans* metabolism: from genome sequence to industrial applications. BMC Genom 9(1):597

Wasay SA, Barrington SF, Tokunaga S (1998) Remediation of soils polluted by heavy metals using salts of organic acids and chelating agents. Environ Technol 19(4):369–379

Wheaton G, Counts J, Mukherjee A, Kruh J, Kelly R (2015) The confluence of heavy metal biooxidation and heavy metal resistance: implications for bioleaching by extreme *thermoacidophiles*. Minerals 5(3):397–451

Xiang Y, Wu P, Zhu N, Zhang T, Liu W, Wu J, Li P (2010) Bioleaching of copper from waste printed circuit boards by bacterial consortium enriched from acid mine drainage. J Hazard Mater 184(1):812–818

Yang H, Liu J, Yang J (2011) Leaching copper from shredded particles of waste printed circuit boards. J Hazard Mater 187(1):393–400

Yarzabal A, Appia-Ayme C, Ratouchniak J, Bonnefoy V (2004) Regulation of the expression of the Acidithiobacillus ferrooxidans rus operon encoding two cytochromes c, a cytochrome oxidase and rusticyanin. Microbiol 150(7):2113–2123

Yarzábal A, Brasseur G, Ratouchniak J, Lund K, Lemesle-Meunier D, DeMoss JA, Bonnefoy V (2002) The high-molecular-weight cytochrome c Cyc2 of *Acidithiobacillus ferrooxidans* is an outer membrane protein. J Bacteriol 184(1):313–317

Zwicker N, Theobald U, Zähner H, Fiedler HP (1997) Optimization of fermentation conditions for the production of ethylene-diamine-disuccinic acid by Amycolatopsis orientalis. J Ind Microbiol Biotechnol 19(4):280–285

Author Biographies

Dr. Deepak Pant is currently the Dean and Head of Environmental Science, Central University of Himachal Pradesh, Dharamshala. He received the Silver Jubilee Research Fellowship Award in 2003 from Kumaun University, Nainital, India; the UCOST Young Scientist Award 2009, INSA Fellow (2010), DST-SERC, Visiting Fellow (2010) and DST-SERC Young Scientist Award (2011). He has 5 patents in the area of waste management using green techniques and published 10 books, 45 research papers in various national and international journals. He recently received Visitor's Award 2017 from the President of India for his innovations in the area of waste management.

Mr. Anand Giri currently perusing his Ph.D under the supervision of Prof. Deepak Pant in the subject of Environmental Science from Central University of Himachal Pradesh, Dharamshala, Himachal Pradesh, India. He did his post graduation in Environmental Science from G.B. Pant University of Agriculture and Technology Pantnagar, Uttarakhand in year of 2012.

Mr. Varun Dhiman currently perusing his Ph.D under the supervision of Prof. Deepak Pant in the subject of Environmental Sciences from Central University of Himachal Pradesh, Dharamshala, Himachal Pradesh, India. He did his post graduation in Environmental Science from Central University of Himachal Pradesh (CUHP), Himachal Pradesh in year of 2014.

Chapter 6
A Review on Bioremediation Potential of Vetiver Grass

Jegannathan Kenthorai Raman and Edgard Gnansounou

Abstract Wastes generation and contamination are increasing drastically due to industrial activities and population growth. Many approaches are in place such as chemical and physical treatments to tackle these wastes and contamination. However, other means of remediation are necessary to overcome the challenges in wastes treatment. In the last decades, research in bioremediation has increased dramatically owing to its potential application in cleaning up all kinds of waste from soil and other contaminated areas. Numerous plants and microorganisms prove to be efficient for bioremediation for various kinds of wastes such as metal, nuclear, mineral, fuels, and chemicals. Among them, vetiver grass could play pivotal role as bioremediation agent for numerous waste categories. Originally being a native from India, vetiver perennial grass has spread to the tropical and subtropical parts of the world due to its application in various fields, such as soil conservation, fragrance, energy fuel, handcrafts, hedge formation, medicinal preparations, and carbon sequestration. The unique property of vetiver plant is their ability to grow in harsh conditions and absorb contaminants leading to remediation. Therefore, several research studies reported on bioremediation potential of vetiver grass among other potential application. In this study, a review of these studies is carried out to highlight the benefits and the limitation of vetiver in bioremediation across industries and to provide some insights on how to incorporate vetiver grass in industries, municipalities, and households in the current and future wastes remediation programs that demand a sustainable approach.

Keywords Bioremediation · Contamination · Heavy metals · Chemicals
Soil remediation · Vetiver grass · Water contamination

J. K. Raman · E. Gnansounou (✉)
Bioenergy and Energy Planning Research Group (BPE),
Ecole Polytechnique Fédérale de Lausanne (EPFL), GC A3 444, Station 18,
1015 Lausanne, Switzerland
e-mail: edgard.gnansounou@epfl.ch

S. J. Varjani et al. (eds.), *Waste Bioremediation*, Energy, Environment, and Sustainability, https://doi.org/10.1007/978-981-10-7413-4_6

1 Introduction

The contamination level in soil and water has drastically increased several folds due to population growth, modernization, and industrial revolution, and it is to increase further. High-, medium-, and low-level toxic chemicals reach soil and water from several industries, households, municipals, and hospital and agriculture areas despite several toxic and waste treatment and management measures. It is more prevalent in developing countries due to limited management measures, awareness, and regulation (Kahn Danielle et al. 2009). Wastewater treatment used to tackle various wastes is chemical treatment and biological treatment (aerobic and anaerobic). Waste management measures are landfilling, composting, recycling, and incineration.

Bioremediation is an emerging and effective waste treatment or management procedure that uses biological resources to degrade and detoxify waste pollutants. Bioremediation is considered as the best sustainable way to treat toxic, household waste, and contaminated soil and water. Within bioremediation, the treatment of waste using plants is termed as phytoremediation, and treatment using both plants and microbes is termed as rhizoremediation (Kuiper et al. 2004). Bioremediation works through the principle of biogeochemical cycle that could be applied for soil, surface water, groundwater, sediments, ecosystem cleanup and restoration (Prasad 2011). Bioremediation of various wastes uses several microorganism and plant variants depending on their ability to degrade or accumulate certain type of waste chemical. Bioremediation is in use since ancient times to recover soil and water from contamination, e.g., Lotus and vetiver (Mishra and Clark 2013; Lavania 2008; Prasad 2011).

Bioremediation use numerous plants and microbes. Vetiver (*Vetiveria zizanioides* (L.) Nash), a perennial grass variety, native to India and grown in several parts of the world, has its application in soil and water conservation, disaster risk management, perfumery, medicine, households, energy and bioremediation. Vetiver has taken a predominant place in bioremediation research for mining spills, oil spills, landfills and garbage dumps treatment, agrochemicals and pesticides removal, heavy metal absorption, water treatment and purification, removal of effluents, nuclear waste, wetland application (Islam et al. 2008; Maiti and Kumar 2015). The unique characteristics of vetiver such as growing in harsh conditions, deep long roots, fleshy leaves, root aroma, soil agglomeration due to high root penetration, metal adsorption capacity, and withstanding extreme weather conditions being both hydrophyte and xerophyte make vetiver an excellent candidate for bioremediation (Dalton et al. 1996; Truong et al. 2010). Vetiver bioremediation has been the topic of study in several journals, reports, and news. In this chapter, the objective was to make a review of such literature to give an updated status on vetiver bioremediation in different industries and to provide few policy recommendations for successful implementation along with the limitation in the vetiver bioremediation system.

2 Vetiver Bioremediation

The vetiver system for bioremediation has its application potential in most of the industries to manage the soil and water contamination. This section is the review of the corresponding studies related to each industry to highlight the importance of vetiver plant in bioremediation.

2.1 Mining Industry

Mining operation in natural resources is inevitable to supply the mineral needs for the growing population and modernization in the developing and developed countries. However, these operation leaves behind a huge scar in the land creating a large soil dump with heavy metals that correspondence to 35% of the soil contaminant in Europe (EEA 2015) for instance that abolishes almost all the possibility to use that land. In addition, it also contaminates the surrounding area such as lakes, rivers, and fertile land as the minerals runoff in the rain due to erosion (Navarro et al. 2008). Recovery of such sites is important to decrease the contamination level, well-being of the species living around the mines and to create a pleasant environment. Several measures are available to accomplish this target, among that vetiver cultivation in such dump soil could be one of the cheapest options due to its cultivation easiness, mineral absorption, and soil erosion control capacity. Recently, Banerjee et al. (2016) tested dump soil from Odisha, India, containing heavy metals from iron ore mine with vetiver to study the mineral removal capacity. The pot experiments planted with vetiver grass showed that vetiver plant could remove heavy metals (Fe, Ni, Cr, Mn, Al, Cu, and Zn) from iron ore soil after 3 months. The roots accumulated more heavy metals compared to the shoot. Heavy metals in the soil stimulated malondialdehyde and proline as the tolerant factor during the plant growth due to stress and several enzymes that favor metal accumulation in roots and shoot (Banerjee et al. 2016). Similarly, Shu et al. (2002) examined the heavy metals (lead, copper, and zinc) removal capacity of vetiver and three other types of grass from contaminated soils. Vetiver grown in glasshouse using contaminated soil showed the highest accumulation of Pb, Cu, and Zn after 7 months in roots and shoots compared to other grasses. Study on the impact of root oil yield on heavy metal contamination leads to conclusion that oil yield decrease with increase in heavy metals concentration mainly due to low root growth because of the heavy metal stress. In line with this work, Pang et al. (2003) and Danh et al. (2010) demonstrated the capability of vetiver cultivation for heavy metals removal and oil production in mining contaminated soil. Similarly, Antiochia et al. (2007) reported the use of vetiver for remediation of soil contaminated with Pd and Zn.

2.2 Chemical Industry

The increased economic growth in developing countries will lead to higher chemical generation and use contributing to a third of global chemical consumption by 2020 (UNEP 2017). According to cleantech service (cleantechsg.com), annually around 10 million tons of toxic chemicals are released into the environment (cleantechsg.com). Management and treatment of these chemical wastes is vital for the good health condition of all living being. Bioremediation of chemical waste is economical among the waste treatment options. Several research findings were reported in the last decade on bioremediation of chemical waste, and vetiver has played a predominant role in chemical bioremediation. Research on bioremediation of phenol using vetiver was reported by Singh et al. (2008a) showing the results of complete removal of phenol from a solution with 50–100 mg/L and 89, 76 and 70% removal from 200, 500, and 1000 mg/L, respectively. Inherent production of peroxidase and hydrogen peroxide during vetiver plant growth in phenol environmental were found to be responsible for removing phenol. On the other hand, phenol inhibited vetiver biomass growth (Singh et al. 2008a). In another study, Ho et al. (2012) reported that use of endophytic bacterium *Achromobacter xylosoxidans* strain improves vetiver biomass growth while degrading phenol. Similar to phenol, cyanide remediation using vetiver was reported by Saeb et al. (2015) with a capacity of 8 mg/kg and the cyanide removal capacity increased with increasing pH, water quantity and growth time. Cyanide accumulation was higher in vetiver roots compared to that of shoots.

Explosive chemicals such as 2,4,6-trinitrotoluene (TNT) are carcinogen and pose a serious threat. TNT migration from explosive site and wastewater site to river and lakes is a concern. Similar to other chemicals, treatment with vetiver grass uptake TNT as well (Makris et al. 2007; Das et al. 2010). Makris et al. 2007 studied the TNT uptake capacity of vetiver and reported 1.03 mg/g in soil-pot experiments. Whereas, Das et al. (2010) reported a higher uptake of TNT (40 mg/g) using vetiver-urea system, where urea enhances the TNT uptake by acting as a chaotropic agent. Moreover, Liu et al. (2010) had also reported on cadmium accumulation using vetiver bioremediation. Treatment at 30 mg/L concentration, cadmium accumulation of up to 2232 and 93 mg/kg Cd dry weight were detected in roots and shoots, respectively.

2.3 Nuclear Industry

Radioactive compounds contaminants are generated from nuclear power plant, nuclear fuel mill, nuclear weapon testing, and from nuclear disasters. Significant amount of these compounds end up in soil and water. They cause serious health hazard to living beings by entering the food chain. Bioremediation of nuclear waste is also possible using vetiver cultivation (Shaw and Bell 1991; Singh et al. 2008b).

Radionuclides such as ^{90}strontium and ^{137}cesium were tested using vetiver plant for their bioremediation potential. Vetiver grown on solution spiked with ^{90}Sr and ^{137}Cs individually for 168 h showed 94 and 61% removal of ^{90}Sr and of ^{137}Cs, respectively. However, when they were grown on combined solution, the removal reduced to 91 and 59%. The accumulation was higher in shoots for ^{90}Sr compared to the roots, and the opposite behavior was observed for ^{137}Cs. Moreover, vetiver plant exposed to low-level nuclear waste for 15 days was able to reduce the waste level below the detection limit (Singh et al. 2008b). Similarly, Mahmood et al. (2014) also reported vetiver remediation to treat cesium-contaminated soil with 95.58% removal capacity. In addition, vetiver has the capacity to uptake uranium as well. According to the report by Hung et al. (2010), up to 126.45 mg/kg plant biomass of uranium could accumulate in vetiver grass when grown in soil contaminated with uranium at a concentration of 250 mg/kg.

2.4 *Pharmaceutical Industry*

Pharmaceutical industry is a major source for contamination, and the trend is increasing at alarming rate (Depledge 2011). The source of pharmaceutical pollutants is from pharmaceutical production firms, disposal of unused and expired drugs from households and hospitals (Wu et al. 2012; Wojcieszynska 2015). These drugs pose a serious danger to aquatic organism and human life. In addition, they also increase the antibiotic resistance of several microorganisms in the environment (Fent et al. 2006; Pruden et al. 2006; Luo et al. 2014; Sengupta et al. 2016). Several micro-pollutants contained in pharmaceutical waste reach the soil and water. Based on the frequency of use, naproxen is classified under micro-pollutant. It is a non-steroidal anti-inflammatory general medicine drug taken for inflammation and pain. Marsidi et al. (2016) reported the capacity of vetiver to treat wastewater and soil containing naproxen at concentrations ranging from 50, 100, 300 and 500 mg/L. After 56 days of treatment, 81.5, 64.5, 36.4, and 60.9% of naproxen were removed, respectively. In addition, the phytotoxicity results showed that bacterial activity around the roots also played a role in naproxen degradation. Similarly, the contamination of antibiotic in water could be reduced using vetiver. After 60 days growth in contaminated hydroponic system at a concentration of 5, 10, and 15 mg/L, complete removal of tetracycline was observed within 40 days (Datta et al. 2013). The metabolic profiling of vetiver treated with tetracycline reveals that biosynthesis of secondary metabolites TCA cycle, glyoxylate metabolism tryptophan metabolism, and inositol phosphate metabolism pathways has an impact due to tetracycline, and the tetracycline detoxification process is mediated by amide hydrolysis and GST-mediated conjugation (Sengupta et al. 2016).

2.5 Food Industry

Food is the primary need of all living beings and its demand directed the industrial revolution throughout the world. However, as the development toward modern agriculture technologies to improve the feed output and to fill the food needs of the growing population, several fertilizers, pesticides, herbicides, and fungicides were introduced. Despite gaining several benefits from these chemicals, environmental contamination occurred that lead to ill effects for aquatic, human, and other living organisms. This trend will continue to rise in future and several measures are being taken to control contamination from food industry (Aktar et al. 2009; Pretty and Bharucha 2014). Agriculture and food processing industries are the backbone of food industry that caters the nutrition needs (protein, carbohydrate, and lipids) through cereals, vegetables, meat, and oil.

Fertilizer and pesticide runoff to lakes and rivers is the main concern associated with agriculture practices. Bioremediation could be one of the strategies to clean up the waterways and soil. Several reports on vetiver used as a bioremediation agent is available in the literature. Atrazine is used a common herbicide for controlling grass weeds and broadleaf, and it is one of the general chemicals detected in water and soil (Lin et al. 2008; Fan and Song 2014). Vetiver's ability to take herbicide atrazine was the topic of research reported by Gupta et al. (2004), Schwitzguébel et al. (2006), and Marcacci et al. (2006). Vetiver grown in water contaminated with atrazine herbicides, deethylatrazine, and deisopropylatrazine (compounds formed due to degradation of atrazine by the action of bacteria in soil) showed a reduction of 5.2, 4.9, and 4.8 mmol/L of transpired water, respectively. Experiments conducted by Marcacci et al. (2006) on the mechanism of atrazine resistance in vetiver confirmed that atrazine conjugation to glutathione catalyzed by glutathione-S-transferases was the dominant pathway to detoxification. Inorganic arsenic was used in pesticides in the previous decades and several arsenic pesticides were banned in early 80s and 90s. However, the pesticide contamination spreads across and persists in many regions around the world (Pandey and Singh 2015). Greenhouse experiments on vetiver grown in arsenic contaminated soil showed good arsenic tolerance up to 225 mg/kg. Maximum arsenic removal (10.6%) was observed in 45 mg/kg arsenic soil contaminated and reduced to 0.6 and 4.5% at 450 and 225 mg/kg (Datta et al. 2011). Persistent organic pollutant such as organochlorine pesticides is a concern when present in adequate measures. These pollutants can be easily magnified over the food chain (Burgess 2013; Ye et al. 2014). After washing organochlorine with oil and natural polymers to remove majority of the pesticides, vetiver cultivation in washed soil further degrades these pollutants and enriches the soil with nutrients (Ye et al. 2013). Similar to other pesticides mentioned above, 2,4-bis(Isopropylamino)-6-methylthio-s-triazine (prometryn) is also a common pesticide used in USA, China, India, and other parts of the world for broadleaf weeds and annual grasses control in the cultivation of vegetables, cotton, wheat, and rice. Greenhouse hydroponic experiments on application of vetiver in prometryn pesticide removal for 65 days revealed that vetiver could reduce the half-life of this pesticide to 11.5 days in three-phase uptake following first-order kinetics. Vetiver roots rapidly removed prometryn from the solution and translocated to the shoots in first 8 days followed by a slower removal in the second phase (between 14 and 30 days) and a more gradual removal in third phase till 67 days (Sun et al. 2016a, b). Endosulfan is a hazardous organochlorine insecticide used in several crop and vegetable cultivations. In Africa, endosulfan is commonly used in the cotton fields and alarming rate of this insecticide is found in the air, water, and soil (Abaga et al. 2014; Dousset et al. 2016). The influence of vetiver on endosulfan-contaminated soil was studied by Abaga et al. (2014). The research results from the pot experiments revealed that half-lives of endosulfan could be reduced significantly and endosulfan could be completely removed after six months. On the other report, addition of rhizospheric microbes enhances the uptake of endosulfan in treatment of contaminated soil with vetiver (Singh et al. 2016). Similar to pesticides and insecticides, fertilizers are applied to the cultivation field and gardens. Excess amount of fertilizers in the soil contaminates the environment through runoff and erosion. Boron is an essential requirement for plant grown that is required in low quantity due to its high toxicity. Boron toxicity is reported in several countries, and bioremediation technique is the cost-effective and environmental friendly technique to treat boron-polluted locations (Rámila et al. 2016). Application of vetiver to treat boron-contaminated sites has been reported by Angin et al. (2008), Smolcz and Cortés (2015) and Xin and Huang (2017). In addition, vetiver cultivation toward successful onsite obsolete pesticide contamination bioremediation in Africa was reported by Harmsen et al. (2009).

Palm industry is one of the major industries that supply cheap and essential oil needs for food in many parts of the world. However, processing of palm for its oil generates high amount of secondary effluent that contaminates the soil and water around the mill. Several conventional measures are available to treat such effluent; however, they are not cost effective. Bioremediation using vetiver has been identified as one of the cost-effective approaches to reduce the BOD and COD from palm oil mill secondary effluent (Darajeh et al. 2016). Similarly, vetiver application in remediation of wastewater from coconut husk retting area has shown promising results in reducing biological oxygen demand, total dissolved solid, electrical conductivity, chemical oxygen demand, dissolved CO_2, H_2S, fluoride, chloride, hardness, phosphate, iron, polyphenols, nitrate, and coliform, and increasing the pH and oxygen (Girija et al. 2011).

2.6 Textile Industry

Textile, being the second basic need of a human being, is the second largest polluter in the world accounting to more than 8000 chemicals use that are toxic and persist in the environment such as rich dyes, fixing agents, detergents, solvents, bleaches, and heavy metals (Scott 2015). Bioremediation of chemicals from the textile industry through vetiver is reported in various studies. Charoenlarp et al. (2016)

used floating platform technique to study the textile effluent treatment using vetiver grass. This technique was effective in removing 36, 77, 88, and 43% of COD, BOD, SS, and color removal, respectively. The vetiver grass growth was good in textile effluent with 95.5% survival rate and the shoot reached 134 cm in height and the root to 24.62 cm.

2.7 *Energy Industry*

As one of the essential need in modern era, energy production generates a significant amount of pollutants released to air, water, and soil. Among different energy supply available, the largest contribution for pollution is from fossil energy that is mainly responsible for climate change. Most of the waste from fossil oil processing is dealt with utmost care to control contamination; however, complete control is not achievable, and the contamination is large due natural seeps followed by urban runoff, accidental oil spills, and extraction of oil (NRC 2003). Bioremediation using vetiver in oil-contaminated site is a promising technique to degrade the oil components in the soil. Experiments conducted in non-remediated and vetiver-remediated crude oil spilled Nigerian soil for jute mallow cultivation showed that vetiver was effective in hydrocarbon and heavy metal accumulation, decreasing the soil acidic pH and improving soil microbial community favoring jute mallow to attain maximum growth compared to non-remediated soil (Charles et al. 2013). Similar results were reported by Bangash et al. (2016), Li et al. (2006), and Xia (2004) for the study on effect of diesel, benzo[a]pyrene, and oil shale-contaminated soil on vetiver growth, respectively. Likewise, 90 days vetiver pot culture experiments conducted in real field oil sludge contaminated soil concluded that vetiver system helps to restore soil quality and augmentation of microbial consortium. In addition, vetiver cultivation is more effective for oil sludge contaminated sites remediation (Dhote et al. 2017) and fly ash remediation (Ghosh et al. 2015). On the other hand, greenhouse study results from vetiver cultivation in Venezuelan heavy crude oil in soil show that vetiver could tolerate mild petroleum contamination up to 5% and could be used to improve slightly polluted soil (Brandt et al. 2006). In addition to the above-mentioned industries, vetiver has also been applied in compost (Vázquez et al. 2013; Bakhshoodeh et al. 2017), biomethanation (Wietlisbach et al. 2016), municipal wastewater treatment (Kumar and Prasad 2015; Badejo et al. 2017), and tannery industry (Srisatit et al. 2003).

3 Policy Suggestions and Limitation

The review results and highlights on vetiver bioremediation to reduce contamination in soil and water in the above sections clearly support the advantages that could be gained from vetiver use. These advantages reassure to frame policy recommendations to harness the full benefit of this bioremediation system through proper and regulated implementation along with addressing all the associated limitations.

Policy suggestions

- Most of the results on vetiver bioremediation in mitigating contamination are from laboratory research. Therefore, policies and projects on testing vetiver bioremediation in real-time large application have to be drawn to know the effect of scale in this application.
- Most of the studies are from developing and least developed countries. Therefore, more policies should support such kind of studies in the developed countries in collaboration with developing countries.
- In several parts of the world, vetiver is cultivated as a commercial crop for its root oil. However, root removal from unregulated cultivation has devastating effect on the environment. Therefore, policies on vetiver use in bioremediation should be regulated to avoid roots removal.
- Environment agencies in all countries should recognize the benefits of vetiver remediation and provide tax incentives to encourage this contamination mitigation tool.
- Recommendation to open vetiver research stations and nursery to teach the cultivation mechanism to people and to provide the necessary saplings.
- Provide instructions across industries to implement vetiver bioremediation in the contaminated sites around the industry and to have a target plan for every five years.
- Conduct public vetiver cultivation drive to introduce vetiver in farm areas through public–private (fertilizer and pesticide companies) partnership.
- Policies to encourage research activities to improve economic returns through combine vetiver bioremediation and energy from vetiver grass shoots such as biofuels (Pandey and Singh 2015; UNEP 2015; Kumar and Prasad 2015; Schwitzguébel et al. 2006).
- Vetiver cultivation not only favors bioremediation, it also provides carbon sequestering that mitigates greenhouse gas. Therefore, providing incentive for greenhouse gas reduction could be an additional benefit to encourage vetiver cultivation for bioremediation.
- Conduct more awareness programs through TV, radio, concerts, street arts, sports event, public cleaning events, and social media to dissipate the benefits of vetiver as a bioremediation tool in urban and rural areas.

Limitations

- The benefit of vetiver in bioremediation is proved mostly in research and pilot scale studies. However, the output results from large-scale studies are uncertain.
- The detoxification of contaminants is often slow and there is no complete decomposition. In such case, the toxic compounds accumulated in the plant do not favor the environment while using the shoots or roots for other application and it could enter the food chain if the cattle graze the shoots (Aken 2008).
- Vetiver is effective for soil remediation with medium to low toxic chemicals; therefore, more improvement is needed to tackle contamination involving high concentration and more variety of chemicals such as heavy and extra-heavy oil contamination (Infante et al. 2010).
- At present, vetiver is cultivated in several parts of the world for its root to extract oil. Root removal could trigger soil erosion and create adverse effect to the environment. Therefore, controlling misuse of vetiver from remediation soil could be complex issue.

4 Conclusion

Vetiver is a highly resilient and versatile plant that is used in various applications including soil erosion control, water conservation, bioremediation, fragrance oil production, household and energy applications. Among the applications, vetiver application in bioremediation is vastly studied. The review of all these studies confirms that vetiver is a promising agent in bioremediation for removal and degradation of several toxic contaminants in soil and water. However, several policy measures that are given as a suggestion here are required for successful implement of this bioremediation system in large scale that needs to overcome several limitations through rigorous research, collaboration from developed and developing countries and large-scale test trials. In addition, growing vetiver for bioremediation and harvesting the shoots for energy purpose could generate economic returns and employment. Therefore, government should draft the suitable policies measures to encourage vetiver cultivation for bioremediation and other applications.

References

Abaga NO, Dousset S, Munier-Lamy C, Billet D (2014) Effectiveness of vetiver grass (*Vetiveria zizanioides L. Nash*) for phytoremediation of endosulfan in two cotton soils from Burkina Faso. Int J Phytoremediation 16:95–108

Aken BV (2008) Transgenic plants for phytoremediation: helping nature to clean up environmental pollution. Trends Biotechnol 26:225–227

Aktar MW, Sengupta D, Chowdhury A (2009) Impact of pesticides use in agriculture: their benefits and hazards. Interdiscip Toxicol 2:1–12

Angin I, Turan M, Ketterings QM, Cakici A (2008) Humic acid addition enhances B and Pb phytoextraction by Vetiver grass (*Vetiveria zizanioides (L.) Nash*). Water Air Soil Pollut 188:335–343

Antiochia R, Campanella L, Ghezzi P, Movassaghi K (2007) The use of vetiver for remediation of heavy metal soil contamination. Anal Bioanal Chem 388:947–956

Badejo AA, Omole DO, Ndambuki JM, Kupolati WK (2017) Municipal wastewater treatment using sequential activated sludge reactor and vegetated submerged bed constructed wetland planted with *Vetiveria zizanioides*. Ecol Eng 99:525–529

Bakhshoodeh R, Alavi N, Majlesi M, Paydary P (2017) Compost leachate treatment by a pilot-scale subsurface horizontal flow constructed wetland. Ecol Eng 105:7–14

Banerjee R, Goswami P, Pathak K, Mukherjee A (2016) Vetiver grass: An environment clean-up tool for heavy metal contaminated iron ore mine-soil. Ecol Eng 90:25–34

Bangash N, Saleem AR, Rashid A, Dawson L (2016) Effect of diesel contamination on the physico-chemical characteristics of soil and growth of vetiver grass. Soil Environ 35(1)

Brandt R, Merkl N, Schultze-Kraft R, Infante C, Broll G (2006) Potential of vetiver (*Vetiveria zizanioides (L.) Nash*) for phytoremediation of petroleum hydrocarbon-contaminated soils in Venezuela. Int J Phytoremediation 8:273–284

Burgess LC (2013) Organic pollutants in soil. In: Brevik EC, Burgess LC (eds) Soils and human health. CRC Press, FL, pp 83–106

Charles UI, Edem D, Nkereruwem JM (2013) Application of phyto-remediation (sunflower and vetiver grass) on crude oil spilled soil cultivated to jute mallow (*Corchorus olitorius L.*). Resource and Environ 3:169–175

Charoenlarp K, Surakul K, Winitkhetkamnoun P, Kanthupthim P, Panbumrung P, Udorn S (2016) Textile wastewater treatment using vetiver grass cultivated with floating platform technique. RMUTKJ 10:51–57

Dalton PA, Smith RJ, Truong PNV (1996) Vetiver grass hedges for erosion control on a cropped flood plain: hedge hydraulics. Agric Water Manag 31:91–104

Danh LT, Truong P, Mammucari R, Foster N (2010) Economic incentive for applying vetiver grass to remediate lead, copper and zinc contaminated soils. Int J Phytoremediation 13:47–60

Darajeh N, Idris A, Masoumi HR, Nourani A, Truong P, Sairi NA (2016) Modeling BOD and COD removal from Palm Oil Mill Secondary Effluent in floating wetland by *Chrysopogon zizanioides* (L.) using response surface methodology. J Environ Manage 181:343–352

Das P, Datta R, Makris KC, Sarkar D (2010) Vetiver grass is capable of removing TNT from soil in the presence of urea. Environ Pollut 158:1980–1983

Datta R, Das P, Smith S, Punamiya P, Ramanathan DM, Reddy R, Sarkar D (2013) Phytoremediation potential of vetiver grass [*Chrysopogon zizanioides (L.)*] for tetracycline. Int J Phytoremediation 15:343–351

Datta R, Quispe MA, Sarka, D (2011) Greenhouse study on the phytoremediation potential of vetiver grass, Chrysopogon zizanioides L., in arsenic-contaminated soils. Bull Env Contam Toxicol 86:124–128

Depledge M (2011) Pharmaceuticals: reduce drug waste in the environment. Nature 478:36

Dhote M, Kumar A, Jajoo A, Juwarkar A (2017) Assessment of hydrocarbon degradation potentials in plant-microbe interaction system with oil sludge contamination: a sustainable solution. Int J Phytoremediation. (Accepted manuscript)

Dousset S, Abaga NO, Billet D (2016) Vetiver grass and micropollutant leaching through structured soil columns under outdoor conditions. Pedosphere 26:522–532

EEA (2015) Progress in management of contaminated sites. European Environment Agency, Denmark

Fan X, Song F (2014). Bioremediation of atrazine: recent advances and promises. J soils sediments 14:1727–1737

Fent K, Weston AA, Carminada D (2006) Ecotoxicology of human pharmaceuticals. Aquatic Toxicol 76:122–159

Ghosh M, Paul J, Jana A, De A, Mukherjee A (2015) Use of the grass, *Vetiveria zizanioides* (L.) nash for detoxification and phytoremediation of soils contaminated with fly ash from thermal power plants. Ecol Eng 74:258–265

Girija N, Pillai SS, Koshy M (2011) Potential of vetiver for phytoremediation of waste in retting area. Ecoscan 1:267–273

Gupta S, Schulin R, Kanekar P, Pakniker KM, Schwitzguebel JP, Raghu K, Kathrin W (2004) In-situ removal of Atrazine using synergy between Plant roots and rhizospheric organisms. Environ Biotechnol ESEB 1:331

Harmsen J, Ammati M, Davies M, Sylla CH, Sidibe T, Traore HK, Diallo A, Demba AS (2009) An African approach for risk reduction of soil contaminated by obsolete pesticides. In: Tenth International In Situ and On-Site Bioremediation Symposium. Baltimore, May 5–8

Ho YN, Mathew DC, Hsiao SC, Shih CH, Chien MF, Chiang HM, Huang CC (2012) Selection and application of endophytic bacterium *Achromobacter xylosoxidans* strain F3B for improving phytoremediation of phenolic pollutants. J Hazard Mater 219:43–49

Hung LV, Maslov OD, Nhan DD, My TT, Ho PK (2010) Uranium uptake of *Vetiveria zizanioides* (L.) Nash. JINR Commun. E18-2010-71

Infante C, Morales F, Ehrmann U, Hernández-Valencia I, León N (2010) Hydrocarbon bioremediation and phytoremediation in tropical soils: Venezuelan study case. Trends in Bioremediation and Phytoremediation. Reserarch Signpost. Kerala, India, pp 429–451

Islam MP, Bhuiyan MK, Hossain MZ (2008). Vetiver grass as a potential resource for rural development in Bangladesh. Agric Eng Int CIGR J 1276:1–8

Kahn Danielle J, Kaseva ME, Mbuligwe SE (2009) Hazardous wastes issues in developing countries. Hazardous waste Manage 11:112

Kuiper I, Lagendijk EL, Bloemberg GV, Lugtenberg BJ (2004) Rhizoremediation: a beneficial plant-microbe interaction. MPMI 17:6–15

Kumar A, Prasad R (2015) Production of renewable energy and waste water management from vetiver grass. Management of water, energy and bio-resources in the era of climate change emerging issues and challenges. Springer International Publishing, Berlin, pp 169–181

Lavania UC (2008) Vetiver in India: historical perspective and prospective for development of specific genotypes for environmental or industrial application. In: Truong P (ed) Ist Indian Vetiver Workshop–Vetiver System for Environment Protection and National Disaster Management. Cochin, India, pp 40–47

Li H, Luo YM, Song J, Wu LH, Christie P (2006) Degradation of benzo [a] pyrene in an experimentally contaminated paddy soil by vetiver grass (*Vetiveria zizanioides*). Environ Geochem Health 28:183–188

Lin CH, Lerch RN, Garrett HE, George MF (2008) Bioremediation of atrazine-contaminated soil by forage grasses: transformation, uptake, and detoxification. J Environ Qual 37:196–206

Liu Y, Zeng G, Wang X, Chen B, Song H, Xu L (2010) Cadmium accumulation in *Vetiveria zizanioides* and its effects on growth, physiological and biochemical characters. Bioresource Technol 101:6297–6303

Luo Y, Guo W, Ngo HH, Nghiem LD, Hai FI, Zhang J, Ziang S, Wang XC (2014) A review on the occurrence of micropollutants in the aquatic environment and their fate and removal during wastewater treatment. Sci Total Environ 473–474:619–641

Mahmood ZU, Abdullah NH, Rahim KA, Mohamed N, Muhamad NA, Wahid AN, Desa ND, Ibrahim MZ, Othman NA, Anuar AA, Ishak K (2014) Uptake evaluation of caesium by glasshouse grown grasses for radiophytoremediation of contaminated soil. Research and Development Seminar, Bangi (Malaysia)

Maiti SK, Kumar A (2015) Energy plantation, medicinal and aromatic on contaminated soil. In: Prasad MNV (ed) Bioremediation and Bioeconomy. Elsevier, pp 29–47

Makris KC, Shakya KM, Datta R, Sarkar D, Pachanoor D (2007) High uptake of 2, 4, 6-trinitrotoluene by vetiver grass–potential for phytoremediation? Environ Pollut 146:1–4

Marcacci S, Raveton M, Ravanel P, Schwitzguébel JP (2006) Conjugation of atrazine in vetiver (*Chrysopogon zizanioides* Nash) grown in hydroponics. Environ Exp Bot 56:205–215

Marsidi N, Nye CK, Abdullah SR, Abu Hassan H, Halmi MI (2016) Phytoremediation of naproxen in waste water using *vetiver zizaniodes*. J Eng Sci Technol 11:1086–1097

Mishra A, Clark JH (2013) Green materials for sustainable water remediation and treatment. RSC Publishing, Cambridge

Navarro MC, Pérez-Sirvent C, Martínez-Sánchez MJ, Vidal J, Tovar PJ, Bech J (2008) Abandoned mine sites as a source of contamination by heavy metals: a case study in a semi-arid zone. J Geochem Explor 96:183–193

NRC (2003) Oil in the sea III: inputs, fates and effects. The National Research Council (NRC) committee on oil in the sea: inputs, fates, and effects. The National Academies Press, Washington

Pandey VC, Singh N (2015) Aromatic plants versus arsenic hazards in soils. J Geochem Explor 157:77–80

Pang J, Chan GS, Zhang J, Liang J, Wong MH (2003) Physiological aspects of vetiver grass for rehabilitation in abandoned metalliferous mine wastes. Chemosphere 52:1559–1570

Prasad MNV (2011) A state-of-the-art report on bioremediation, its applications to contaminated sites in India. Ministry Environ Forests. New Delhi. http://www.moef.nic.in/downloads/public-information/BioremediationBook.pdf

Pretty J, Bharucha ZP (2014) Sustainable intensification in agricultural systems. Ann Bot 114:1571–1596

Pruden A, Pei R, Storteboom H, Carlson KH (2006) Antibiotic resistance genes as emerging contaminants: studies in Northern Colorado. Environ Sci Technol 40:7445–7450

Saeb K, Khadami R, Khoramnejadian S, Abdollahi E (2015) Use of vetiver (*Vetiveria zizanoides*) in remediation of cyanide soil contamination. J Biol Today's World 4:150–155

Schwitzguébel JP, Meyer J, Kidd P (2006) Pesticides removal using plants: phytodegradation versus phytostimulation. In: Phytoremediation rhizoremediation, pp 179–198. Springer, Netherlands

Scott A (2015) Cutting out textile pollution. Chem Eng News 93:18–19

Sengupta A, Sarkar D, Das P, Panja S, Parikh C, Ramanathan D, Bagley S, Datta R (2016) Tetracycline uptake and metabolism by vetiver grass (*Chrysopogon zizanioides* L. Nash). Environ Sci Pollut Res 23:24880–24889

Shaw G, Bell JN (1991) Competitive effects of potassium and ammonium on caesium uptake kinetics in wheat. J Environ Radioact 13:283–296

Shu WS, Xia HP, Zhang ZQ, Lan CY, Wong MH (2002) Use of vetiver and three other grasses for revegetation of Pb/Zn mine tailings: field experiment. Int J Phytoremediation 4:47–57

Singh S, Melo JS, Eapen S, D'souza SF (2008a) Potential of vetiver (*Vetiveria zizanoides* L. Nash) for phytoremediation of phenol. Ecotoxicol Environl Saf 71:671–676

Singh S, Eapen S, Thorat V, Kaushik CP, Raj K, D'souza SF (2008b) Phytoremediation of ^{137}cesium and ^{90}strontium from solutions and low-level nuclear waste by *Vetiveria zizanoides*. Ecotoxicol Environ Saf 69:306–311

Singh V, Singh P, Singh N (2016) Synergistic influence of *Vetiveria zizanioides* and selected rhizospheric microbial strains on remediation of endosulfan contaminated soil. Ecotoxicol 25:1327–1337

Smolcz SU, Cortés VG (2015) Remediation of boron contaminated water and soil with vetiver phytoremediation technology in Northern Chile. In: 6th International Conference on Vetiver (ICV6). Da Nang, Vietnam, May 5–8

Srisatit T, Sengsai W (2003) Chromium removal efficiency by *Vetiveria zizanioides* and *Vetiveria nemoralis* in constructed wetlands for tannery post–treatment wastewater. In: Proceedings of the Third International Conference on Vetiver and Exhibition. Guangzhou, China, Oct 6

Sun S, Li Y, Lv P, Punamiya P, Sarkar D, Dan Y, Ma J, Zheng Y (2016a) Determination of prometryn in vetiver grass and water using gas chromatography–nitrogen chemiluminescence detection. J Chromatogr Sci 54:97–102

Sun SX, Li YM, Zheng Y, Hua Y, Datta R, Dan YM, Lv P, Sarkar D (2016b) Uptake of 2, 4-bis (Isopropylamino)-6-methylthio-s-triazine by vetiver grass (*Chrysopogon zizanioides* L.) from hydroponic media. Bull Environ Contam Toxicol 96:550–555

Truong PN, Foong YK, Guthrie M, Hung YT (2010) Phytoremediation of heavy metal contaminated soils and water using vetiver grass environmental bioengineering. Springer, Netherlands, pp 233–275
UNEP (2015) Vetiver Briquette: Feasibility Report. Carbon Roots International, Inc., Haiti, pp 1–38
UNEP (2017) The chemicals and waste subprogramme. United Nations Environment Programme. Nairobi, Kenya. http://www.unep.org/chemicalsandwaste/who-we-are/overview
Vázquez MA, De la Varga D, Plana R, Soto M (2013) Vertical flow constructed wetland treating high strength wastewater from swine slurry composting. Ecol Eng 50:37–43
Wietlisbach SO, Ram K, Kothurkar NK, Nair R, Harigovind S (2016) Performance of a vertical subsurface flow constructed wetland in treating biomethanation effluent. In: Global Humanitarian Technology Conference (GHTC). IEEE, pp 847–853 Oct 13
Wojcieszynska D, Domaradzka D, Hupert-Kocurek K, Guzik U (2015) Bacterial degradation of naproxen—undisclosed pollutant in the environment. J Environ Manage 145:157–161
Wu S, Zhang L, Chen J (2012) Paracetamol in the environment and its degradation by microorganism. Appl Microbiol Biotechnol 96:875–884
Xia HP (2004) Ecological rehabilitation and phytoremediation with four grasses in oil shale mined land. Chemosphere 54:345–353
Xin J, Huang B (2017) Comparison of boron uptake, translocation, and accumulation in reed, cattail, and vetiver: an extremely boron-tolerant plant, vetiver. Plant Soil 1–9
Ye M, Sun M, Yang X, Wei H, Song Y, Xin J (2013) Remediation of organochlorine pesticides (OCPs) contaminated soil by successive hydroxypropyl-β-cyclodextrin and peanut oil enhanced soil washing–nutrient addition: a laboratory evaluation. J Soils Sediments 13:403–412
Ye M, Sun M, Liu Z, Ni N, Chen Y, Gu C, Kengara FO, Li H, Jiang X (2014) Evaluation of enhanced soil washing process and phytoremediation with maize oil, carboxymethyl-β-cyclodextrin, and vetiver grass for the recovery of organochlorine pesticides and heavy metals from a pesticide factory site. J Environ Manage 141:161–168

Author Biographies

Dr. Jegannathan Kenthorai Raman is a Scientist at Bioenergy and Energy Planning Research Group, EPFL, Switzerland. Currently his research work is centered on experimental and life cycle assessment aspects of biorefinery schemes, industrial biotech products and bioprocess engineering.

Prof. Edgard Gnansounou is a Professor of modelling and planning of energy systems at the Swiss Federal Institute of Technology Lausanne (EPFL) where he is Director of the Bioenergy and Energy Planning Research Group. His current research works comprise techno-economic and environmental assessment of bio-refinery schemes based on conversion of agricultural residues and other renewable energy systems.

Chapter 7
Organic Waste and Pollutants Reduction Through Composting

V. Sudharsan Varma, Shanmugaprakash Muthusamy and Karthik Rajendran

Abstract Management of organic solid waste through composting is the sustainable option to prevent leachate and greenhouse gas emissions after disposal/landfilling. Composting reduces a maximum of 65% of the initial volume and also recovers nutrient-rich end product. During composting, the biodegradable organic carbon, micropollutants, and nuisance gases from the organic fractions are biologically transformed into a stabilized product. Millions of indigenous microbial populations act to degrade the waste by releasing off high temperature and gases. The process is largely influenced by the aeration rate, moisture content, C/N ratio, temperature, particle size and volume of the waste material composted. The active thermophilic phase of the composting determines the rate of waste degradation and organic matter transformation during the process. During composting, the waste material undergoes three different temperature phases: (a) initial moderate temperature (less than 40 °C) for a couple of days, (b) the thermophilic temperature (over 40 °C) for few days to several weeks, and finally, (c) the cooling and maturation phase. During the process, the organic carbon and nitrogen are metabolized to moisture, CO_2, and other nitrogen gases, significantly reducing the availability of exchangeable carbon and heavy metals in the compost. Furthermore, it also increases the bioavailability of essential plant nutrients total nitrogen (N), ammonium (NH_4), phosphate (P_2O_5), and potash (K_2O), and other micronutrients. The release of heavy metals and potential gases during composting is unavoidable, but it

V. S. Varma
Agricultural Research Organization, Newe Ya'ar Research Center,
Institute of Soil, Water and Environmental Sciences, 30095 Ramat Yishay, Israel

S. Muthusamy
Downstream Processing Laboratory, Department of Biotechnology,
Kumaraguru College of Technology, Coimbatore 641049, India

K. Rajendran (✉)
MaREI Center, Environmental Research Institute,
University College Cork, Cork, Ireland
e-mail: Karthik.1988@gmail.com; K.Rajendran@ucc.ie

K. Rajendran
School of Engineering, University College Cork, Cork, Ireland

S. J. Varjani et al. (eds.), *Waste Bioremediation*, Energy, Environment, and Sustainability, https://doi.org/10.1007/978-981-10-7413-4_7

can be controlled by adding appropriate bulking agents such as sawdust, dry leaves, wood chips, and optimizing the above-mentioned process parameters. This chapter focusses especially on organic wastes and pollutants reduction through composting. Also, the major influencing factors and process parameters for effective composting of organic waste and treatment of pollutants gases emitted during the process are discussed.

Keywords Organic waste · Composting · Degradation · Micro-macro pollutant transformation · Process parameters · Stability and maturity

1 Introduction

The global wastes are increasing at an alarming rate and it is important to address the issue at the earliest. There are different methods to treat the waste depending on its composition, moisture content, the market for a treated product, tipping fees, etc. (Onwosi 2017). Some of the waste management strategies include biochemical methods such as anaerobic digestion and composting. Other methods include thermochemical processes such as combustion and incineration. Composting has gained adequate popularity in developing countries due to its conversion of waste to value-added product fertilizer (Qian et al. 2014). Composting is a reliable waste treatment option due to its stability of the process and produced products which are also sanitized. In addition, it is a continuous process which shrinks the volume of waste to 40–60% of its original capacity. Compositing is influenced by several factors such as temperature, pH, particle size, moisture content, aeration, and electrical conductivity (Juarez 2015).

Composting transforms organic wastes to compost which has essential minerals and nutrients such as nitrogen, phosphorus, and potassium which plays a vital role in agriculture and plant growth. About 22 million tons of food waste was composted in the USA in 2013 which resulted in 2.74 million tons of carbon dioxide equivalent emission reduction. Composting has other benefits including soil amendments, diverting wastes from landfilling (Wei 2017). During composting, several environmental issues may arise such as the formation of volatile organic carbon emissions, toxic gases, bio-aerosols resulting in health risks. Managing to compost efficiently is very important to avoid such malfunction.

This chapter focusses on various aspects of composting including the conditions for composting, phases in a composting process, factors which influence them, the type of feedstocks which could be used, types of composters available, what are the micropollutants form during composting, how to treat VOC emission during composting, and finally, a section on economics and profitability aspects of composting.

2 Composting

The solid waste management practices and disposal options in developing countries are still a major concern due to financial constraints, technology advancements and technical qualified people. In such cases, recycling, landfilling, composting, and biogas production seems to be the viable options in most of the developing countries. However, composting could be considered as the economically suitable alternative for the treatment of organic solid waste with the limited resources.

Composting can be generally described as "natural decomposition of biodegradable organic matter under aerobic conditions by the action of bacteria, fungi, and other organisms with higher thermophilic temperatures to produce finally a stable and nutrient-rich end product, compost." The by-products formed during the degradation process serve as food for the other populations. The complex molecules will be broken into simpler forms for easier assimilation by the indigenous microorganisms (Fig. 1).

2.1 *Optimum Conditions for Composting*

Nutritious microbial food: Organic waste-containing water (moisture content between 40–60%) and added nutrients (nitrogen, phosphorous, sulfur) present organic matter content in waste serves as a source of carbon, nutrients, and energy for the metabolic reactions during bioremediation process. Micronutrients in addition to N, P, and S, many other micronutrients are needed to a lower

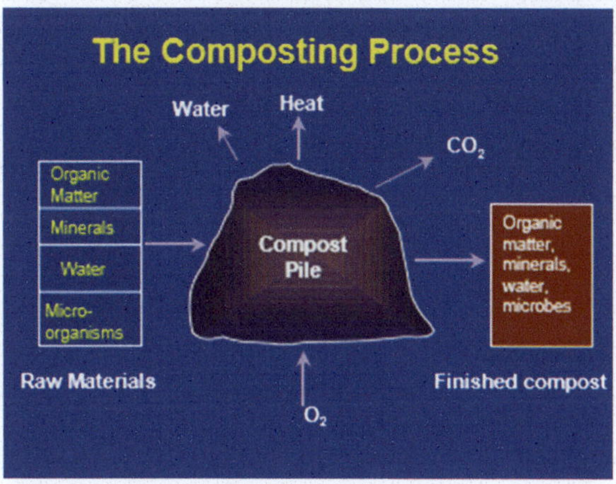

Fig. 1 Composting process (Rynk et al. 1992)

concentration such as K, Ca, Mg, Fe, Ni, and others (Chen et al. 2011). To maintain aerobic conditions in the compost system, it requires 2–5 kg of O_2/kg of the organic matter which is degraded. The pH of the system could in the range of 5.5–8.5.

Enzymes: These are selective chemical compounds that are used to reduce the complex substrate molecules to smaller forms for higher utilization of the microorganisms. Few important enzymes include cellulases, B-glucosidases, urease, phosphatases, and arylsulphatase that are majorly involved in the substrate degradation.

2.1.1 Phases of Composting

The composting process has three major temperature phases, the initial ambient phase, thermophilic phase, and the final cooling phase (Cooperband 2002).

First Phase: During the initial period for the first few days of composting, the temperature seems to increase to 40–50 °C. This temperature is sufficient enough to degrade sugars and other easily biodegradable substances, by the microbial populations (Varma and Kalamdhad 2014a, b).

Second phase: This may occur over extended periods of time, temperatures between 40 and 65 °C prevails. Cellulose and other complex polymers are degraded during this thermophilic phase. However, lignin is consider to degrade slowly even at this higher temperatures. The plant pathogens, weeds, and other contaminants are highly affected during this phase. In few cases, it is important to control the temperature around 60–65 °C, where most of the important microorganisms would be killed above 65 °C. Therefore, turning or aeration to the compost system would prevent such loss.

Third phase: The final of the composting is observed with the reduced temperature, with reduced organic matter degradation. The compost would be in the curing phase with the microflora similar to that of the soil. The compost looks like dark colour and has an earthy smell odour.

2.1.2 Factors Influencing the Composting Process

1. Nutritious microbial food—It should contain the high amount of C, N, P, S, and trace elements of minerals
2. Suitable moisture content—The ideal moisture content is to 45–60% by its weight
3. Optimum pH and Temperature—The pH range occurs between 6.5 and 8.0 and the temperature lies between 54 and 60 °C
4. Oxygen demand—The ideal range for oxygen concentration is >10%
5. Carbon and Nitrogen Ratio—The acceptable C:N ratio is 25 or 35:1
6. Bulk Density—The compost must possess 1000 lbs./cu yd

7. Particle size—The size should be less than <1 inch but it may vary according to the raw materials
8. Bedding materials—Greater porosity will allow for proper aeration.

2.1.3 Various Types of Raw Materials

The raw materials were chosen based on the composting methods and the optimal condition of the composting environment. The major factors influencing the successful compost include:

- The initial feedstock materials
- The size of the input feedstock materials
- The indigenous microorganisms of the composting process.

The raw materials for composting can be split into two categories: carbon sources and nitrogen sources (Tables 1 and 2).

Table 1 Costs of several types of equipment for different capacities in a composting facility plant

Equipment used	Capacity					
	1000 Cu Y		5000 Cu Y		15,000 Cu Y	
	Total cost	Cost per Cu Y	Total cost	Cost per Cu Y	Total cost	Cost per Cu Y
Small loader (40HP) 1/3-yard bucket	$1423	$1.42	$6398	$1.28	$17,276	$1.15
Tractor (85HP) 1-yard bucket	$1116	$1.12	$4800	$0.96	$11,669	$0.78
Front loader (135 HP) 3-yard bucket	$3062	$3.06	$11,365	$2.27	$21,135	$1.41
Windrow turner (small 40HP)	$2326	$2.33	$2885	$0.58	$4205	$0.28
Windrow turner (large 100HP)	$4383	$4.38	$4551	$0.91	$4996	$0.33
Windrow turner (medium 80HP)	$17,360	$17.36	$17,491	$3.50	$17,797	$1.19

Table 2 Carbon sources and nitrogen sources

Sources	Characteristics	Examples
Nitrogen	Wet, high in nitrogen, not very rigid, decompose quickly, high bulk density	Livestock manure, grass, food waste
Carbon	Dryer, decompose slowly, low in nitrogen, low bulk density, somewhat rigid	Corn stalks, leaves, sawdust, wood chips

The initial carbon (C) to nitrogen (N) ratio, i.e., C/N plays an important role in the degradation process.

Estimation of carbon content in the feedstock: (Boraste et al. 2009)

$$\%\text{carbon} = \%\text{volatile solids}/1.8$$

where

$$\%\text{volatile solids} = 100 - \%\text{ash (material incinerated at 500 °C.)}$$

Estimation of moisture content in the feedstock:

$$\text{Moisture content} = 100 \times (\text{wet weight}\ \ n\ \text{dry weight})/\text{wet weight}$$

The optimum C/N range for starting the process would be from 25:1 to 35:1.

$$\text{Bulk Density} = \text{Weight of the material} \times 40$$

where

$$\text{Weight of the material} = \text{Weight of the bucket with material} - \text{Weight of the empty bucket}$$

Based on the Carbon:Nitrogen and Moisture content, the commonly used feedstocks, which are chosen for the composting process and give in Table 3.

2.2 *Types of Composting Methods*

Composting methodology is classified mainly into two categories (Misra and Roy 2003). They are

1. Traditional methods
2. Rapid composting methods.

2.2.1 Traditional Methods

Traditional methods involve mounding materials in the form of piles or as pits that decomposes overtime with the addition of mixing. Traditional methods could further have broken down to aerobic and anaerobic methods based on the supply of oxygen.

Table 3 Types of various feedstock and its composition (Rynt et al. 1992)

Sources	Feedstock	Moisture content	C:N	Bulk density
Carbon	Hay	8–10	15–30	–
	Corn Stalks	12	60–70	32
	Corn silage	655–68	40	–
	Straw	5–20	40–150	50–400
	Fall leaves	–	30–80	100–300
	Saw dust	20–60	200–700	350–450
	Brush, wood chips	–	100–500	–
	Bark (paper mill waste)	–	100–130	–
	News paper	3–8	400–800	200–250
	Cardboard	8	500	250
	Mixed paper	–	150–200	–
Nitrogen	Dairy manure	80	5–25	1400
	Poultry manure	20–40	5–15	1500
	Hog manure	65–80	10–20	–
	Cull potatoes	70–80	18	1500
	Vegetable wastes	–	10–20	–
	Coffee grounds	–	20	–
	Grass clippings	–	15–25	–
	Sewage sludge	–	9–25	–

Aerobic Decomposition

- **Indian Method (Indore)**

In late 1920s, this method was developed in India that uses different raw materials including garden wastes, leftover from plants, cattle manure, and other organics that are stalked as piles. Hard woods were first laid and crushed which were followed by other plant residues. It is necessary that the plant residues do not go more than 10%. Green materials are allowed to shrivel for 2–3 days to remove excess moisture before piled up. Each material is stacked and spread for about 15 cm thick to a pile height of 1.5 m. The pile is cut in between 20 and 25 kg that casts as a bedding. Later, the bedding along with manure and urine is placed into the pits to proceed with composting process.

Pit Method

The site should be placed near to cattle and water source in addition it should be at a higher level. The dimensions of the pit should be 2 m wide and between 1 and 1.5 m in height. Cattle manure should be applied as layers that have a thickness between 10 and 15 cm. Slurry is mixed with manure in the ratio of 4.5:3.5. Around 4.5 kg of inoculums from other composting facility was used that is sprinkled over the substrate to wet it. The composting pit is stacked up in layers consecutively for not more than a week. Compressing should be avoided to ensure enough aeration.

Turning: The substrate/feedstock is rotated three times during composting process:

First turning—15 days after filling the pit,
Second turning—after 15 days of first turning
Third turning—after 1 month of second turning

Heap Method

To withstand rain and harsh weather conditions composting is prepared in the form of piles and shelters the site with a shed. The pile dimension is as follows: 2 m (Length) × 2 m (Width) × 1.5 m (Height). When it goes to the top, sides are narrowed and the top has a width of 0.5 m, i.e., in the form of trapezoid. Carbonaceous rich materials including hay, leaves, sawdust, or straw and nitrogenous materials such as manure or sewage sludge were added in the ratio of 2:1. Repetition of the above-mentioned process for a height of 1.5 m is necessary and wetting the material is needed to dampen it; however, it should not be soggy or oversoaked. To maintain heat inside the compost, the heap is covered with soil or hay that is turned every 6 or 12 weeks once. The overall composting process takes about four months to complete that can be used thereafter.

Anaerobic Decomposition

- ***Indian Method (Bangalore)***

Acharya (1939) developed this method in Bangalore, India for areas with low rainfall that uses night soil for preparing compost. Some of the problems from the Indore method including unfavorable weather patterns, adequate turning, nutrient losses due to external factors were overcome in this method. However, it takes longer time than the Indore method, which is a disadvantage of this method (Misra and Roy 2003).

Pit preparation: The pit is prepared depending on the land, type of feedstock to be composted where the trench could be made in different sizes. In this method, the site shall be chosen based on the Indore method and it is preferable to have slope walls to prevent water logging.

Pit filling: Night soil and organic wastes are stacked alternatively with a 15-20 cm thickness of each feedstock. The contents in the pit shall remain for about three months without turning or watering. Most of the material during this period will break down, reduce in its volume and adding refuse on top as alternate layers helps is loss of moisture. The initial aerobic composting that happens for about 8–10 days followed by anaerobic decomposition that takes about 6–8 months which is a lot slower than the aerobic composting to get the final product.

- **Passive Composting of Manure Piles**

Passive composting was used to treat animal manure that heaps the material, and when given a long time, the material decomposes with little mixing. Only placing manure in the piles is not enough for aerobic composting where other actions including placing bedding material, managing moisture level and porous structure maintenance ensures it. The microbes govern the degradation under anaerobic conditions. Adverse effects linked to anaerobic decomposition are due to low operating temperatures, slower process decomposition, release of toxic hydrogen sulfide, and other odor compounds. Cattle manure management relies on bedding that creates a dried with porous mixture which elevated the C:N ratio. Equal volume of manure and bedding ensures appropriate porosity is reached for composting. When the amount of bedding is low, dry bedding is added. Horse manure could be composted in heaps without bedding while swine, dairy, and poultry require bedding materials. Passive composting is equivalent to windrow methods with less frequent turning requirements. Each pile needs passive air movement less than 6 ft height and 12 ft in width. Passive composting requires low aeration that comes with a disadvantage of odor problems.

- ***Chinese Rural Composting***

Pit Method

In pit method, composting is done in rectangular or circular shape in a corner of the field. Substrates such as swine manure, weeds, straw are commonly used in this method. The first layer is covered with manure or water hyacinth for about 15 cm in thickness while the second layers consist of straw followed by animal manure in the third layer. This cycle is repeated until the pit is full and mud is added to the top. A water layer is created for about 4 cm in height to provide anaerobic conditions that helps in reducing nitrogen content of the pit. Generally, the first turnover is done after 30 days of filling where superphosphate is added and mixed thoroughly. The second turnover is performed after 60 days, i.e., 30 days after first turnover subsequently after three weeks. Approximately, eight tons of compost could be generated in a period of 3 months.

High-Temperature Compost

High-temperature compost is prepared using night soil, sewage, dung, and plant residues at a ratio of 1:4. As mentioned in other methods, materials are heaped in alternate layers and required amount of water is added. For aeration, 3-cm mud plaster and bamboo poles are inserted. They could be removed after a couple of days that leaves the aeration in the heap. The temperature inside the heap increases between 60 and 70 °C in a period of five days. First turnover is done in two weeks while moisture is made up using water or excreta and the whole composting process gets complete in about two months.

- ***Turned Windrows***

Turned windrows are placed in long and narrow piles that undergoe mixing frequently. Turning the materials ensures complete mixing of the materials and increases aeration (Atalia et al. 2015). The dimensions of the windrows are as follows:

Height—1 m for high-dense material while 30 cm for low-dense materials.
Width—Between 10 and 20 ft.

Windrow size is determined using the turning size, shape, and shape. Aeration in windrows occurs mainly through active or passive airflow through convection or gaseous diffusion. The porosity of the windrow determines the rate of air exchanged within. When windrow is large, anaerobic zones form that releases odor when windrow is turned. Sometimes, in small windrows, it is difficult to achieve heat requirements and may not achieve the required temperature that is enough to evaporate moisture and destroy pathogens.

- ***Passively Aerated Windrows***

The advantage of the passively aerated piles is to avoid the turning of the compost materials. The air is supplied to the materials by perforated pipes at the bottom of each pile. The both sides of the pipe are kept open for the flow of air, through which the piles are aerated (Taiwo and Oso 2004).

2.3 Application of Effective Microorganism

2.3.1 Em-Based Quick Compost Production Process

It was developed at 1999 and the dimension is about 6(l) × 4(w) × 3(h) ft. The feedstock materials include 2 parts of cow dung, 1 part of rice husk, 1 part of rice husk/charcoal, and 1 part of rice bran.

- Accelerator—33 L of EM solution or *Trichoderma* solution per pit.

EM solution = 10 ml of EM + 40 ml of molasses + 950 ml of distilled water, incubating it for 5–7 days depending on the temperature. Then, it is added to 98 L of water to make up to 100 L. This reduces the composting period from 3 to 1 month. The 0.5 ft layer of the mixture was made in the pit and sprinkle accelerator over. The process should be repeated till the pit is fully occupied. The materials should be mixed after three weeks for aerobic conditions and the materials will be ready to use after two weeks (Boraste et al. 2009).

2.3.2 Use of Cellulolytic Culture

- **IBS rapid composting technology**

This process involves the initial growth of the compost fungus activator and the actual composting.

Compost fungus activator: The fungus (*Trichoderma harzianum*) is initially grown in the mixture of sawdust and ipil-ipil leaves, which is utilized as compost fungus activator (CFA) (Misra et al. 2003).

Composting process: Rice straw, weeds, grasses are used as raw materials for this process. Better aeration and the large surface area should be provided for the good microbial action. Substrates should be moistened with water either by soaking (small volume) or sprinkling (large volume). The C:N ratio is at 4:1. Some possible combinations of the mixture include 3:1 ratio of rice straw and ipil-ipil leaves, 4:1 ratio of rice straw and chicken manure, and finally, 4:1 ratio of grass and (legume materials + manure).

The bedding of the composting materials should be raised about 30 cm from the bottom and the substrates are loosely packed to get better aeration. During piling, CFA is added to the substrate to enhance the decomposition process. The heap is covered completely to avoid loss of heat, evaporation of water and ammonia volatilization. The composting piles start heated up within initial three days, and the thermophilic temperatures should be maintained at 50 °C or higher. The piles should be turned around 5–7 days, and thereafter once every two weeks. The size of the pile gradually starts decreasing based on the degradation of the raw materials.

2.4 *Lime Pre-treatment for Organic Matter with High Lignin Content*

Lime ($Ca[OH]_2$) and oxygen (O_2) could be effectively used to degrade high lignin biomass. Generally, the enzymatic hydrolysis of lime-treated biomass is restricted by the physical characterization of the biomass from the treatment. These include acetylation, lignification, and crystallization. However, the lime treatment helps in removing the lignin and hemicellulose, thereby increasing the crystallinity index.

2.5 *Advantages and Disadvantages of Composting*

Composting has several advantages for the environment in addition to reducing the quantity of waste reaching the landfills. The process reduces toxins such as fuels and pesticides from the waste reaching water sources and plants. It also reduces the emissions of methane and other potential gases from the waste. Further, the compost contains rich organic matter and other beneficial inorganic nutrients. In addition, composting

- Mitigates climate change (global warming)
- Reduces greenhouse gases
- Reduces water pollution.

Few inconveniences of composting include the attraction of animals which can be prevented by proper maintenance. While, the gas emissions from composting could be misunderstood as pollution, while its actually earthy odor.

3 Bioremediation of Micropollutants

Bioremediation of micropollutants during composting is to merely transform the hazardous nature of the parental compound to the less harmful end product. The bioremediation potential of contaminated resources during composting is highly influenced by process parameters such as time, temperature, microbial flora, and the type of methodology adopted. Hence, the compost process parameters for reducing the contamination levels of micropollutants should be controlled effectively, as compared to that of the regular composting of green waste. The intensity of the microbial action within the compost is facilitated highly by the moisture content, warm temperature, aerobic conditions, and the nutrient balance for effective transformation of the micropollutants. The elevated thermophilic temperatures during composting enable higher reaction rate compared to the ambient temperature in soils. In addition, the solubility and the mass transfer rates of the contaminants at thermophilic temperatures are generally higher, thereby making it more available for the microbial metabolism. Hence, the combination of these factors allows the microbes to attack both the highly degradable organic matter and toxically contaminates at higher significant rates.

The potential of biodegrading toxic organic contents during composting has been successful for many pesticides and herbicides. These pesticides include wide range of chemical compounds including herbicides, insecticides, fungicides, and rodenticides, which are extensively used for agricultural purposes. Even most of the pesticides are readily biodegradable (Porto et al. 2011); however, organochlorine pesticides could be more resistant to microbial degradation. During composting, reduction of micropollutants toxicity levels occurs via many ways, i.e., some are decayed into simpler molecules, or they form bonds with other compounds or volatilize as gas. In most of the cases, mineralization, humification, or leachate during composting have a major effect on reducing micropollutants. Mineralization is the effective breakdown of organic compounds into their mineral and organic constituents. The organic constituents further lead to CO_2 and water, where the mineral constituents of the micropollutants reach the soil medium as non-toxic molecules by the water formed during composting. Humic and fulvic acids are considered to play a major role in binding many metabolites of mineralized pesticides in compost (Hayar et al. 1997). pH of the compost is also effective in bonding the acidic and ionic pesticides to the organic matter by H-bonding when the pH < pK_a of that particular compound.

Biologically, microorganisms release selective enzymes that readily break down the parental compound. Chemical nature of the particular compound determines the water solubility and bioavailability, by which the microbes will assimilate water-soluble compounds. Many hydrolytic enzymes, i.e., esterase and phosphatase help in breaking the liable bonds making them more susceptible to degradation. Many other common enzymes include mono- and dioxygenases, which introduce one or two oxygen atoms to the compound to increase its bioavailability. Many of the enzymes act extracellularly to degrade pesticides during composting. The extracellular enzymes are effective in binding to the target pesticides making into a possibly less toxic and less hazardous form during composting. Fungi are more effective in degrading recalcitrant polymeric organic matter, as the bacteria will be depleted during the later stages of composting. Genera of *Trichoderma, Gliocladium, Penicillium, Rhizoctonia solani, Cunninghamella echinulate, Botrytis cinerea, Phanerochaete chrysosporium, Mortierella sp., Mucor sp., Alternaria sp., Phoma cf., Eupyrena Basidiomycete strain,* and *Mortierella sp.* are effective in degrading pesticides. The degradation pattern of several pesticides during composting has been evaluated. These compounds include atrazine, carbetamide, chlorpyrifos, diainon, isofenphos, pendimethalin, simazine, tertbutyl, azoxystrobin, propiconazole (Lemmon and Pylypiw Jr 1992; Barriuso et al. 1997; Buyuksonmez et al. 1999; Kupper et al. 2008; Lalander et al. 2016).

Heavy metals reach the environment by many different ways through the use of chemicals via low concentrated diffuse either in direct or indirect applications. The application of compost with high heavy metal concentration also increases the contamination in soil medium. In addition, emissions from the combustion processes, motor vehicles, oil spills, or dumping are also the primary sources of contamination (Kastner and Miltner 2016). Millions of sites are contaminated with organic pollutants worldwide and it is increasing every year. Therefore, the cheapest and the effective remediation would be through composting. The total metal concentration of the compost/soil could be considered as the useful pollution indicator. However, it does not provide the information about the ionic form of the heavy metal and the solubility nature, which is considered to be more harmful to the living organisms (Venkateswaran et al. 2007; Jiwan and Kalamdhad 2016). The chemical speciation of heavy metals during composting has been categorized into five different factions based on the mobility and bioavailability of the metals upon the extraction method. These include exchangeable fraction (EF), carbonate fraction (CF), a reducible fraction (RF), an oxidizable fraction (OF), and residual fraction (RF) (Wong and Selvam 2006; Jiwan and Kalamdhad 2012). The fractionate forms of each and every metal have different properties, i.e., EF is easily affected by the changes in the water's ionic content; CF could be affected by the pH (acidic conditions); RF remains highly unstable in anoxic conditions; OF is likely to undergo changes during oxidizing conditions; finally, RF is permanently entrapped in a crystal structure that is less harmful (Katoh et al. 2016). During composting, several million indigenous microbial populations feed on the substrate resulting in the organic and inorganic matter transformation. Temperature, pH, moisture content, and ammonia volatilization are considered to be the major factors affecting the ionic form of heavy metals (Fang and Wong 2000). The initial changes in pH of the compost due to organic matter degradation and ammonia volatilization at higher temperatures affect the exchangeable and carbonate fraction. The reducible and organic fractions of the heavy metals are affected by the oxic and anoxic conditions, due to acetic acid and ammonia. The humic and fulvic acids production during the maturation phase of compost entraps the heavy metals as residual fractions. The residual fractions of heavy metals are more stable and are not easily available for plant uptake. However, the increase in heavy metal canceration during composting could be considered due to the net loss of dry mass (Varma and Kalamdhad 2014a, b). The reduction and transformation of Cu, Zn, Pb, Ni, Cu, Mn, Cr, Cd, and many other heavy metals during different types of composting have been well studied (Amir et al. 2005; Wong and Selvam 2006; Chiang et al. 2007; Singh and Kalamdhad 2013; Liu et al. 2017).

4 Treatment of VOC Emissions During Composting

During composting, the majority of the size and volume reduction of the mass occurs through the volatilization of gases. The potential of greenhouse gases from the compost operations includes CO_2, CH_4, N_2O, and VOCs, respectively. CO_2 emissions from the composting facilities are unavoidable and usually not highlighted for the GHG contribution, due to the biogenic origin of carbon. When the organic residues are decomposed aerobically and with optimum C/N ratio and other controlled process parameters, methane emissions and N_2O could be minimized, which has a higher contribution to the GHGs. But many of the VOCs are of environmental concern which should be treated to minimize the odor nuisance and hazardous nature. The compost gases emitted from the compost mass could be characterized based on the high flow rates; however, as low pollutant concentration, depending on the size of the composting facility and methodology adopted. But, the contribution of volatile organic compounds (VOCs) can be a higher percentage in terms of odorant nuisance and greenhouse gas potential even at lower emission range. The emissions of most of the VOCs occur due to insufficient aeration and incomplete degradation of the organic residues during the process. In this context, biofiltration and treatment of VOCs during composting has been discussed.

4.1 Biofiltration

Biofiltration is a simple biological technique, which is economical and environmentally viable option for the treatment of odors. The filter media used for the biofilter is mostly the organic residue or matured compost, which adsorbs and also biologically degrades the odorous compounds. Sometimes, the soil, peat, chipped brush and bark blended with a biologically inert material such as gravel to maintain adequate porosity is also suggested. The process efficiency of the treatment is depended on many factors, importantly the airflow rate, pressure drop, temperature, pH, packing materials, and the residence time. In addition, the packed material should have good water-holding capacity, pore volume >80% and particle diameter >4 mm and 55% of the total organic content (Oh and Choi 2000). The parameters defining the performance of the biofiltration can be expressed as pollutant loading capacity (L), elimination capacity (EC), and removal efficiency (RE) (Kennes and Veiga 2001).

The biofiltration occurs as two sequential mechanisms, one is the absorption/adsorption of the gas pollutants in biofilm, and other is the biomass growth of the biofilm due to the microbial metabolism due to aeration (Fig. 2).

4.2 Microbiology of Biofiltration

Bacteria, actinomycetes, and fungi are the common microorganisms that are actively involved in the degradation of air contaminants during biofiltration. These indigenous microbial populations are significantly found higher in compost materials, compared to soil, peat, or other biologically inert materials. The organic compounds from the waste gas stream serve as the energy source/carbon for the growth of heterotrophic microorganisms, while the autotrophic populations feed on the oxidizable hydrogen sulfide and ammonia. Heterotrophic organisms are found predominant compared to autotrophs. The degradation of pollutants within a biofiltration unit can be simply explained as Eq. (1),

$$\text{Organic Pollutant} + O_2 \xrightarrow{\text{(microbes)}} CO_2 + \text{salt} + H_2O + \text{heat} + \text{biomass} \quad (1)$$

In addition, inorganic salts and other organic acids are also released depending upon the type of gas stream treated. Recent advancements involve immobilization of microorganisms onto carrier materials, i.e., polypropylene pellets, Ca-alginate, etc., to improve the feasibility of microbial activity. The higher efficiency of treatment depends on the solubility of contaminants and oxygen to the liquid phase, which serves as the medium for the microorganisms. The biodegradation reactions within the biofiltration units are exothermic, and the changes in filter bed temperature are due to the microbial activity. The energy released during the biological reactions can be of maximum 50 kcal h^{-1}, and the temperature gradient within the

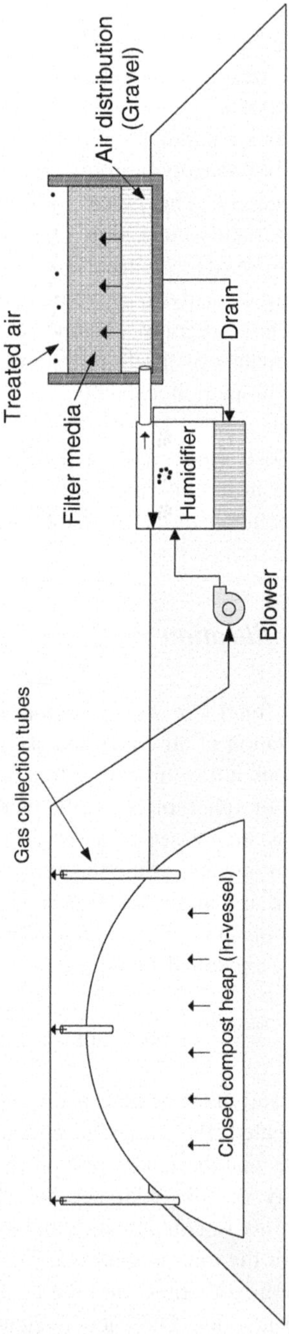

Fig. 2 Schematic diagram of a biofiltration unit treating compost gases

filter can be in the order of 2–10 °C, which varies depending on the inflow of VOC (Hwang et al. 2002; Delhomenie and Heitz 2005). Typically, biofilters contain 10^6–10^{10} CFU of bacteria and actinomycetes per gram of bed and fungi in the range of 10^3–10^6 CFU per gram of bed (Ottengraf 1987).

The greater advantage of the biofiltration is the sustainability of microorganisms that is maintained for a very long period. The choice of microbial consortia could be selected as per the target pollutant; however, indigenous microbial populations could treat a wide range of pollutants. These indigenous microorganisms normally acclimatize themselves to the compounds introduced into the biofilter to metabolize the target pollutants. The acclimatization period for the easily degradable compounds might take 10 days, while for the more resistant compounds suitable inoculum could be introduced initially to reduce the acclimatization period. It could also be noted that the activity of biofilter would sustain for at least two weeks to two months without any VOC feed, while there are enough nutrient supplements and periodic aeration to the biofilter. Biofiltration units can treat gas streams of airflow from 60 to 150,000 m^3/h, with lower VOC concentrations, compared to other physical and chemical treatment methods (Table 4).

4.3 *Key Factors for Design Criteria*

Some of the principal factors to be considered during the operation of biofiltration system are listed below:

- The rate of airflow per surface unit of the biofilter is one of the important parameters to be monitored carefully, i.e., expressed as cubic meters of air per second for each square meter of biofilter surface, should be in the range of 0.3–9.5.
- The pH of the biofilter should be in the range of 6.5–7.5 for higher microbial activity.
- Pressure drop should be taken care regularly. Typical pressure drops range from 1 to 3 in. of water.
- 0.5–2.0 m depth range is recommended for biofilter, with 1 m being the typical depth of a biofilter.
- Optimal moisture content in the range of 20–60% by weight is accepted.
- Temperature range of 10–40 °C is optimal for the growth and metabolism of microorganisms. Temperatures below or higher than the optimal range would lower the microbial an activity or destroy the biomass.
- Off-gas rates are typically around 1000–150,000 m^3/h.

Table 4 VOCs treated at different loading rate and packing media during composting

S. No.	Compound	Packing material and microorganism	Inlet organic load	References
1.	Toluene	Reactor packed with vermiculite or with vermiculite-granular activated carbon as packing material. Fungus *Scedosporium apiospermum* TB1 was inoculated	Inlet toluene concentration of 3–9 g/m^3 that corresponds to toluene load of 220–628 g/m^3 h	Garcia-Pena et al. (2001)
2.	Styrene	The packing material was perlite, of particle size up to 2 mm with a bulk weight of 120 g dm^{-3} mixed microbial cultures of the bacterial and fungal population	The range of inlet styrene concentrations studied was from 1000 to 200 mg m^{-3}. Elimination capacity of 140 g m^{-3} h^{-1} with a removal efficiency above 90%	Paca et al. (2001)
3.	Toluene	The compost balls/pellets were used in biofiltration trials made from municipal waste compost	Operated at a flow rate of 1 m^3 h^{-1} and a toluene concentration of 250–500 ppm	Roy et al. (2003)
4.	Benzene, Toluene and p-xylene (BTX)	Enriched bacterial consortium and *Pseudomonas putida* strain PPO1. peat/perlite-packed columns	Three BTX components was observed at the rate of 40 ± 50 g/h m^3 filter bed	Oh and Bartha (1997)
5.	Toluene and *n*-propanol	Indigenously microbial consortium from a compost-woodchip mixture	The biofilter was fed with toluene vapors at loadings up to 175 g m^{-3} h^{-1} and removal efficiencies of 70–99% were observed	Dixit et al. (2012)
6.	Styrene and acrylonitrile	Indigenously microbial consortium obtained from yard waste compost and activated sludge. Yard waste compost mixed with shredded hard plastics in a 25:75 v/v ratio of plastics: compost as the bed media was inoculated with thickened municipal activated sludge	Under steady-state conditions, maximum elimination capacity of pure styrene was obtained 44 g m^{-3} h^{-1}. Maximum elimination capacity of 103 g m^{-3} h^{-1}, whereas the maximum removal rate of	Dehghanzadeh et al. (2005)

(continued)

Table 4 (continued)

S. No.	Compound	Packing material and microorganism	Inlet organic load	References
			120 g m^{-3} h^{-1} acquired for pure acrylonitrile. Removal of acrylonitrile was not affected by the presence of styrene in the air stream, but styrene was affected in the presence of acrylonitrile	
7.	Toluene and ethyl acetate	Indigenously microbial consortium from a mixture of wood chip and compost	An inhibition effect on toluene removal at high ethyl acetate concentration. The total VOC concentration ranged from 4 to 12.4 g m^{-3}. Removal capacities of about 200 g m^{-3} h^{-1} were achieved by a medium that is porous and contains sufficient moisture and nutrients	Deshusses et al. (1999)
8.	Hexane and methanol	Trickle-bed-air-biofilters (TBABs). Fungi consortium	Hexane loading rates ranged from 0.9 to 13.2 g/m^3 h for both TBABs. The corresponding methanol rates varied from 2.3 to 37.7 g/m^3 h and from 4.6 to 64.5 g/m^3 h for TBABs "A" and "B", respectively	Zehraoui et al. (2013)
9.	Hexane	Biofilters packed with organic filter beds, such as peat moss (PM) and pine sawdust (PS). *Fusarium solani* was used as inoculum	Inlet flow rate of 0.35 L min^{-1} was used to pump hexane-saturated air at the top of the reactor. The biofilters were operated at airflow rate of 0.04 L min^{-1} corresponding to the same EBRT of EB1	Hernandez-Melendez et al. (2008)

(continued)

Table 4 (continued)

S. No.	Compound	Packing material and microorganism	Inlet organic load	References
			biofilter. Minimum 5 days was required to develop EC of 100 g m^{-3} h^{-1} for PS	
10.	Xylene	Biofilter packed with scoria/compost was acclimatized using one liter of the activated sludge as inoculum. The inoculum was enriched with 3 L of nutrient solution and aerated before addition	Under steady-state conditions, 98% removal of inlet xylene concentration of 1.34 g m^{-3} was achieved. While a maximum elimination capacity of 97.5 g m^{-3} h^{-1} was observed for IL of 199.5 g m^{-3} h^{-1}	Amin et al. (2014)

4.4 Key Factors for Restoring Biofilter Performance

The effective performance of the biofilter is majority based on the inlet gas C/N ratio in addition to the principal factors mentioned in section, which directly affects the operation of the microbial system either by nitrogen depletion or excess. Few of the actions needed to restore the biofilter performance include:

- Optimizing the C/N ratio of the system by adding appropriate carbon or nitrogen source.
- Increasing the height of the biofilter by adding more media to compensate settling.
- Replacing the old media by new media, with more active microbial populations and rich nutrient concentration.
- The media could be remixed for the uniform distribution of moisture and porosity.
- The underlying air distribution system could be reconstructed.

5 Economics of Composting

Economics is crucial for any project to be self-sustainable and yielding profits helps in expansion of an industry. There are different costs associated with a compost treatment facility which could be divided into six categories: (1) Site where the composting facility will be built. (2) Feedstock cost, i.e., the cost at which we buy the feedstock. (3) Costing for equipment (4) Labor costs. (5) Overhead costs and (6) Revenue or product sales which determine the overall profit of the process. These six categories have their own factors on which it should be selected: To choose a site, the following needs to be considered including site design, site preparation, accessibility for transportation of feedstock/product, availability of water, water management, surface and subsurface conditions, wind direction and it should be away from buildings to avoid odor issues. The factors within the feedstocks include its characteristics, handling, capacity, and tipping rates. The type of equipment and labor used will affect the product quality which represents the products sales or revenue. The cost of several types of equipment could be obtained in Table 1. The overhead costs include the taxes, manufacturing and nonmanufacturing costs such as fuel, depreciation, site maintenance, utilities (Rynk 1992).

Other essential economic indicators which determine the profitability of the plant include capital costs or CAPEX that refers to the overall investment required for the project and operating costs or OPEX that refers to yearly running expenses of the plant, and there are other profitability metrics such as payback period which refers to the time in years at which the invested money will be recovered back. Return on investment is the rate of return in percentage for the investment over its lifetime. The lifetime of most projects is usually between 15 and 20 years. Generally, the capital cost decreases with the increase in capacity which is due to the economies of scale. The capital cost of the composting facility in Japan was estimated between $75,000 and 125,000/ton/day processing capacity. Similarly, the operating costs ranged between $10 and 100 for every ton of waste processed. The electricity consumption in the plant did not follow the economics of scale; however, with the increase in capacity, the electricity consumption decreased. The electricity consumption ranged between 10 and 200 kWh/wet ton depending on the process and the operational capacity. For a capacity of 10,000 wet tons/year, the electricity consumption was approximately 100 kWh/wet ton. Electricity, chemicals, water, fuel correspond to 10–50% of the operational expenses for most composting facilities (Zhang 2011).

The cost of composting could vary depending on the type of the substrate, process used, etc. For example, composting raw sludge costs €50–80/ton/year. The compost which is used as a fertilizer is sold at a price of €1–13/ton, which is lower grade product which makes composting not economically feasible. The cost of composting varies with the market of supply and demand. In Europe, the soil has abundant carbon and nutrients which make the compost market weaker, and whereas in India for instance, the soil lack nutrients and the producing compost out of organic waste is profitable in India which was sold between $50 and $60/ton (Fersi 2015). The net costs for a compost process post-anaerobic digestion using co-digestion of grass silage and pig manure were €2.8/m^3 (Nolan 2012).

6 Conclusion

Degradation of organic waste through different composting methodologies and the factors influencing the processes have been discussed. This chapter also addressed the diverse types of composters available today, the bioremediation of pollutants during composting, and how to treat VOC emissions during composting. Finally, an insight is given on composting costs which would help the people from various fields including policy-makers and decision-makers about the feasibility of this technology.

References

Acharya CN (1939) Studies on the hot fermentation processes for the composting of town refuse and other waste material. Indian J Agric Sci 9(6):817–833

Amin MM, Rahimi A, Bina B, Heidari M, Moghadam FM (2014) Performance evaluation of a scoria-compostbiofilter treating xylene vapors. J Environ Health Sci Eng 12(1):140

Amir S, Hafidi M, Merlina G, Revel JC (2005) Sequential extraction of heavy metals during composting of sewage sludge. Chemosphere 59:801–810

Atalia KR, Buha DM, Bhavsar KA, Shah NK (2015) A review on composting of municipal solid waste 9(5):20–29

Barriuso E, Houot S, Serra-Wittling C (1997) Influence of compost addition to soil on the behaviour of herbicides. Pestic Sci 49:65–75

Boraste A, Vamsi KK, Jhadav A, Khairnar Y, Gupta N, Trivedi S, Patil P, Gupta G, Gupta M, Mujapara AK, Joshi B (2009) Biofertilizers: a novel tool for agriculture 1(2):23–31

Buyuksonmez F, Rynk R, Hess TF, Bechinski E (1999) Occurrence, degradation and fate of pesticides during composting. Compost Sci& Utilizat 7(4)

Chen L, de Haro Marti M, Moore A, Falen C (2011) The composting process. CIS 1179. University of Idaho—Dairy Manure Compost Production and Use in Idaho

Chiang KY, Huang HJ, Chang CN (2007) Enhancement of heavy metal stabilization by different amendments during sewage sludge composting process. J Environ Econom Manag 17 (4):249–256

Cooperband L (2002) The art and science of composting: a resource for farmers and compost producers. University of Wisconsin, Center for Integrated Agricultural Systems, Madison, Wisconsin

Dehghanzadeh R, Torkian A, Roshani B, Asadi M (2005) Evaluation of biofilter performance in the co-treatment of styrene and acrylonitrile vapors mixture, Processing of the 1st Congress on Biotechniques for Air Pollution Control. UDC-Publisher, La Coruna, Spain, pp 377–384

Delhomenie MC, Heitz M (2005) Biofiltration of air: a review. Crit Rev Biotechnol 25:53–72

Deshusses MA, Johnson CT, Hohenstein GA, Leson G (1999) Biofiltration of high loads of ethyl acetate in the presence of toluene. J Air Waste Manage Assoc 49:973–979

Dixit RM, Deshmukh SC, Gadhe AA, Kannade GS, Lokhande SK, Pandey RA, Vaidya AN, Mudliar SN, Deshusses MA (2012) Treatment of mixtures of toluene and n-propanol vapours in a compost-woodchip-based biofilter. Environ Technol 33(7):751–760

Fang M, Wong JWC (2000) Changes in thermophilic bacteria population and diversity during composting of coal fly ash and sewage sludge. Water Air Soil Pollut 124:333–343

Fersi S, Chtourou N, Jury C, Poncelet F (2015) Economic analysis of renewable heat and electricity production by sewage sludge digestion—a case study. Int J Energy Res 39 (2):234–243

Garcia-Pena EI, Hernández S, Favela-Torres E, Auria R, Revah S (2001) Toluene biofiltration by the fungus *Scedosporium apiospermum* TB1. Biotechnol Bioeng 76:61–69

Hayar S, Munier-Lamy C, Chone T, Schiavon M (1997) Physico-chemical versus microbial release of 14C-atrazine bound residues from a loamy clay soil incubated in laboratory microcosms. Chemosphere 34:2683–2697

Hernandez-Melendez O, Bárzana E, Arriaga S, Hernández-Luna M, Revah S (2008) Fungal removal of gaseous hexane in biofilters packed with poly(ethylene carbonate) pine sawdust or peat composites. Biotechnol Bioeng 100:864–871

Hwang SCJ, Wu SJ, Lee CM (2002) Water transformation in the media of biofilters controlled by Rhodococcus fascians in treating an ethyl acetate-contaminated airstream. J Air Waste Manag Assoc 52:511–520

Juárez MFD, Prahauser B, Walter A, Insam H, Franke-Whittle IH (2015) Co-composting of biowaste and wood ash, influence on a microbially driven-process. Waste Manag 46:155–164

Kastner M, Miltner A (2016) Application of compost for effective bioremediation of organic contaminants and pollutants in soil. Appl Microbiol Biotechnol 100:3433–3449

Katoh M, Kitahara W, Sato T (2016) Role of inorganic and organic fractions in animal manure compost in lead immobilization and microbial activity in soil. Appl Environ Soil Sci, pp 1–9

Kennes C, Veiga MC (2001) Bioreactors for waste gas treatment. Kluwer, Dordrecht

Kupper T, Bucheli TD, Brändli RC, Ortelli D, Edder P (2008) Dissipation of pesticides during composting and anaerobic digestion of source-separated organic waste at full-scale plants. Biores Technol 99(17):7988–7994

Lalander C, Senecal J, Calvo MG, Ahrens L, Josefsson S, Wiberg K, Vinneras B (2016) Fate of pharmaceuticals and pesticides in fly larvae composting. Sci Total Environ 565(15):279–286

Lemmon CR, Pylypiw HM Jr (1992) Degradation of diazinon, chlorpyrifos, isofenphos, and pendimethalin in grass and compost. Bull Environ Contam Toxicol 48(3):409–415

Liu W, Huo R, Xu J, Liang S, Li J, Zhao T, Wang S (2017) Effects of biochar on nitrogen transformation and heavy metals in sludge composting. Biores Technol 235:43–49

Misra RV, Roy RN, Hiraoka H (2003) On-farm composting methods. Food and Agriculture Oragnization of the United Nations, Rome, pp 1–26

Nolan T, Troy SM, Gilkinson S, Frost P, Xie S, Zhan X, Harrington C, Healy MG, Lawlor PG (2012) Economic analyses of pig manure treatment options in Ireland. Biores Technol 105: 15–23

Oh YS, Bartha R (1997) Construction of a bacterial consortium for the biofiltration of benzene, toluene and xylene emissions. World J Microbiol Biotechnol 13:627–632

Oh YS, Choi SC (2000) Selection of suitable packing material for biofiltration of toluene, m- and p-xylene vapors. J Microbiol 38:31–35

Onwwosi CO, Igbokwe VC, Odimba JN, Eke IE, Nwankwoala MO, Iroh IN, Li E (2017) Composting technology in waste stabilization: on the methods, challenges and future prospects. J Environ Manage 190:140–157

Ottengraf SPP (1987) Biological-systems for waste-gas elimination. Trends Biotechnol 5:132–136

Paca JB, Koutsky M, Maryska M, Halecky (2001) Styrene degradation along the bed height of perlite biofilter. J Chem Technol Biotechnol 76:873–878

Porto ALM, Melgar GZ, Kasemodel MC, Nitschke M (2011) Biodegradation of pesticides, pesticides in the modern world. In: Stoytcheva M (ed) Pesticides use and management. ISBN: 978-953-307-459-7

Qian X, Shen G, Wang Z, Guo C, Liu Y, Lei Z, Zhnag Z (2014) Co-composting of livestock manure with rice straw: characterization and establishment of maturity evaluation system. Waste Manag 34:530–535

Roy S, Gendron J, Delhoménie MC, Bibeau L, Heitz M, Brzezinski R (2003) Pseudomonas putida as the dominant toluene-degrading bacterial species during air decontamination by biofiltration. Appl Microbiol Biotechnol 61:366–373

Rynk R (ed) (1992) On-farm composting handbook. In: Natural Resource, agriculture and engineering service, Ithaca, NY. Publication No. NRAES-54
Singh J, Kalamdhad AS (2012) Reduction of heavy metals during composting—a review. Int J Environ Protect 2(9):36–43
Singh J, Kalamdhad AS (2013) Effects of lime on bioavailability and leachability of heavy metals during agitated pile composting of water hyacinth. Biores Technol 138:148–155
Singh J, Kalamdhad AS (2016) Effect of lime on speciation of heavy metals during composting of water hyacinth. Front Environ Sci Eng 10:93–102
Taiwo LB, Oso BA (2004) Influence of composting techniques on microbial succession, temperature and pH in a composting municipal solid waste. Afr J Biotechnol 3:239–243
Varma VS, Kalamdhad AS (2014a) Evolution of chemical and biological characterization during thermophilic composting of vegetable waste using rotary drum composter. Int J Environ Sci Technol 12(6):2015–2024
Varma VS, Kalamdhad AS (2014b) Stability and microbial community analysis during rotary drum composting of vegetable waste. Int J Recycl Organ Waste Agric 3:52
Venkateswaran P, Vellaichamy S, Palanivelu K (2007) Speciation of heavy metals in electroplating industry sludge and wastewater residue using inductively coupled plasma. Int J Environ Sci Tech 4(4):497–504
Wei Y, Li J, Shi D, Liu G, Zhao Y, Shimaoka T (2017) Environmental challenges impeding the composting of biodegradable municipal solid waste: a critical review. Resour Conserv Recycl 122:51–65
Wong JWC, Selvam A (2006) Speciation of heavy metals during co-composting of sewage sludge with lime. Chemosphere 63:980–986
Zehraoui A, Hassan AA, Sorial GA (2013) Biological treatment of n-hexane and methanol in trickle bed air biofilters under acidic conditions. Biochem Eng J 77(15):129–135
Zhang H, Matsuto Y (2011) Comparison of mass balance, energy consumption and cost of composting facilities for different types of organic waste. Waste Manag 31(3):416–422

Author Biographies

Dr. V. Sudharsan Varma earned his PhD degree in Environmental engineering from IIT Guwahati, India, where he worked on the composting of vegetable using different composting methodologies. Currently he is a Post-doctoral fellow at Agricultural Research Organization, Newe Ya'ar Research Center, Institute of Soil, Water and Environmental Sciences, Israel. His current research focusses on the management and composting of livestock manure and biosolids. He has coauthored over 30 publications.

Dr. Shanmugaprakash Muthusamy is currently working as assistant professor in Kumaraguru College of Technology, Coimbatore, India. His current research interests include bioenergy systems, biomaterials, and environmental biotechnology. He has coauthored more than 20 research articles, 3 book chapters and 1 book to his credit. Currently, he is supervising two master students and one PhD scholar.

Dr. Karthik Rajendran is working as a senior postdoctoral researcher at University College Cork, Ireland working on process modelling, techno-economic evaluation, lifecycle assessments and energy systems analysis of bioenergy and bio-based product systems. He has an earned PhD in chemical engineering from UB, Sweden. He had coauthored more than 25 peer-reviewed articles, 10 book chapters, and 2 monographs with an h-index 11 to his credit.

Chapter 8
Role of Bacterial Consortia in Bioremediation of Textile Recalcitrant Compounds

Madhava Anil Kumar, Palanichamy Baskaralingam, Abdur Rawoof Salma Aathika and Subramanian Sivanesan

Abstract The increasing industrial demand for remediating the textile wastewater in an effective way has led to the pervasive acceptance of bioremediation. Bioremediation techniques such as bioaccumulation, biosorption, bioaugmentation, and biodegradation utilize the biological systems to treat the textile effluents containing the recalcitrant dye molecules. Bioremediation is known to be environmentally reliable and is an alternative to the conventional decomposition techniques with the prerequisite to fulfill the efficacy and economic viability. Among the aforementioned bio-remedial measures, biodegradation of the textile dyes is the trustworthy industrial application. Biodegradation of dyes can be achieved using single bacterial strains and co-cultures/consortia. The consortial systems are proven to be advantageous over a single strain as they involve an inductive synergistic mechanism among the co-existing strains. As a result of this co-metabolism, there is a formation of different intermediate metabolites such as toxic aromatic amines which are furthermore mineralized by the other bacterial strains in the consortia.

Keywords Bioremediation · Biodegradation · Bacterial consortia
Textile effluent · Mineralization · Recalcitrant

M. A. Kumar
Analytical and Environmental Science Division & Centralized Instrument Facility, CSIR-Central Salt & Marine Chemicals Research Institute, Bhavanagar 364002, India

M. A. Kumar · A. R. S. Aathika · S. Sivanesan (✉)
Department of Applied Science and Technology, Alagappa College of Technology, Anna University, Chennai 600025, India
e-mail: sivanesh1963@gmail.com

P. Baskaralingam
Department of Chemistry, College of Engineering, Anna University, Chennai 600025, India

S. J. Varjani et al. (eds.), *Waste Bioremediation*, Energy, Environment, and Sustainability, https://doi.org/10.1007/978-981-10-7413-4_8

1 Introduction: Textile Industrial Sector—An Overview

The first inception of dyeing was the wall paintings of Altamira cave, Spain, during 1500–900 BC. The twentieth century evidenced the exponential use of synthetic dyes which replaced the natural ones to a larger scale (Welham 2000). Globally, India stands next to China and Turkey in supplying textiles to the European Union (EU) with a share of approximately 8.1%. Indian textile industry is one of the major export sectors with several diversified units adding to the major country's economy (Goswami 1990). Twenty-seven percentage of the foreign exchange earnings are from the textiles export, contribute about 14% to the industrial production with 3% to the gross domestic product (GDP) of the country and have increased from 3.5 to 4.0% in the last four years. The cumulative global foreign direct investment (FDI) in Indian textile sector from 2000–01 to 2014–15 was approximately 1.5 billion USD, and India's share in global FDI in 2014 was 3%. The size of India's textile market in 2014 was 99.0 billion USD, which is expected to become 226 billion USD by 2023 at a compound annual growth rate (CAGR) of 8.7% which was revealed in the World investment report published in the year 2016 by United Nations, Geneva.

India has several advantages in the textile sector as there is an abundant availability of the concerned raw materials. As the industry requires intensive labor, India is having the advantage of the dense population; the textile industry adds 21% to the total employment either directly or indirectly. India hosts around 2324 textile industries which include 83 composite and 2241 semi-composite industries. Tamil Nadu has 741 industries with 739 belonging to semi-composite processing units and two composite mills. Gujarat stands next to Tamil Nadu with 523 textile industrial units.

The main contribution to wastewater which arises from the textile industries is during dyeing processes, chemical fiber production, and wool scouring. Table 1 shows the water required during the dyeing processes per day. The major environmental issue arises from wet processing, where huge tonnes of dyes are lost in the effluents during dyeing operations.

Table 1 Water required for different textile operations

S. no.	Wet processes	Water required (L/kg)	Solution pH
1	Sizing/slashing	4.35	7.0–9.5
2	Desizing	11.75	6.0–8.0
3	Scouring/ kiering	32.5	10.0–13.0
4	Bleaching	40–48	6.0–6.2
5	Mercerizing	24.5	12.0–3.0
6	Dyeing	105	10.5–11.5

The integrated textile industry is engaged in the production of yarn, fabric, and finished goods from raw fibers and extensively utilizes different variety of dyes (Pandey et al. 2007). It is obvious that the demand for textile dyes prolongs with the increasing population. The pervasive use of dyestuffs is mainly attributed to the simplicity in synthesis, diverse chemical structures, high molar extinction coefficients, and fastness to light and wetness (Zollinger 2003).

Textile industry consumes a large volume of water and chemicals for textile processing. Most of these dyes escape the conventional wastewater treatment, and they show greater stability to light, temperature, and water. The textile effluents are highly colored, and their characteristics mainly depend on the different chemical constituents of the starting materials used in different dyeing steps. The composition of the textile mill effluent (TME) is influenced by the presence of organic and inorganic compounds such as colored materials, surfactants, toxicants, and chlorinated compounds, thus making them persistent environmental pollutants. TME is generally characterized by different water quality parameters such as chemical oxygen demand (COD), biochemical oxygen demand (BOD), pH, intense color, and dissolved solids. Usually, TMEs have BOD/COD ratios around 1:4, describing their non-biodegradable nature (Bilińska et al. 2016).

2 Textile Dyes and Its Toxicity

The textile dyes have phenyl aromatic ring structures that are ionic in nature, and the color intensity of the dye molecules is due to the transition between the distinct molecular orbitals. The chromophores and auxochromes are responsible for intensifying color and they are known to have greater affinities towards each other (Peters and Freeman 1991). The Society of Dyers and Colorists, UK, in association with American Association of Textile Chemists and Colorists (AATCC) published the classification of dyes in the Color Index (C.I.) based on the structure and applications. Table 2 elaborates the degree of fixation onto the support and characteristics of the dyes.

2.1 Dyes and Their Applications

The azo dyes have extensive applications in textile fabrication and have azo with sulfonic (SO_3^{-1}) electron withdrawing groups generating electron deficiency in the molecules, thus making them less susceptible to oxidative catabolism and are hardly biodegraded under aerobic conditions (Sen et al. 2016). The anthraquinone-based reactive dyes are water-soluble dyes used for coloring wool and silk as they have a greater affinity toward these fibers without the aid of auxiliary binding agents. The anthraquinone-based disperse dyes lack water-solubilizing groups and are usually adsorbed by fibers having hydrophobic groups with the aid of mordants. The

Table 2 Characteristics and the estimated fixation of dyes onto the support

S. no.	Dye	Characteristic features	Support	Degree of fixation (%)
1	Acid	Negatively charged and binds with the cationic groups of the fiber	Polyamide	89–95
2	Reactive	Forms a covalent bond with OH, NH, or SH group	Cellulose	50–90
3	Metal complex	Forms strong complexes with metals and dyes	Wool	90–98
4	Direct	Binds to fiber by Van der Waals forces of attraction	Cellulose	70–95
5	Basic	Positively charged and binds with the anionic groups of the fiber	Acrylic	95–100
6	Disperse	Sparingly soluble and penetrates through the fibers	Polyester	90–100
7	Vat	Insoluble and gives soluble leuco derivatives upon reduction and is reoxidized upon exposure to air	Cellulose	80–95
8	Sulfur	Hetero-polyaromatic compounds containing sulfur	Cellulose	60–90

anthraquinone-based vat dyes are known for their brilliant colors and fastness to light and are insoluble in water. The leuco derivative of these dyes (soluble vat) is absorbed by the fiber and is then converted to the insoluble form (Salabert et al. 2015). Sulfonated phthalocyanine dyes are non-volatile, highly water-soluble, and thermally unstable. The sulfonated aromatics are hydrophilic in nature and remain impermeable through the cell membrane, thus making them as recalcitrant compounds. The nitro-substituted sulfonates in the benzene ring are highly persistent than the unsubstituted sulfonates (Lade et al. 2015). Triphenylmethane (TPM) dyes are colorants that find applications in textile, medicine, and laboratory as stains. They have a complex aromatic molecular structure with additional phenyl rings resisting biodegradation. The resonance-stabilized ring structures impart intense color to the fabrics and make them less susceptible to microbial attack (Mohanty et al. 2016).

2.2 *Toxicity of Dyes and Aromatic Amines*

TMEs containing the dyes and their intermediates when discharged as untreated/ partially treated effluents into aqueous ecosystems shall obstruct sunlight penetration, thereby disturbing the photosynthetic activity of the phytoplanktons and deteriorate the dissolved oxygen concentration. It also imparts unpleasant color to the water bodies, leading to the loss of aesthetic values and also adds toxicity to aquatic life systems. The toxicity is mainly due to the presence of trace metals and other auxiliary chemical compounds. The industrial personnel exposed to the dyestuffs are prone to high rates of the development of urinary bladder cancer (Burkinsha 2016). The risk of cancer increases with exposure to certain aromatic amines which are used as precursors in the dye-manufacturing units. Generally, during azo-dye degradation, the cleavage in the azoic group results in the liberation of aromatic amines which are toxic and carcinogenic. The discharge of toxic textile wastewater causes drastic change to the ecosystem, and therefore, it is necessary to treat these TMEs prior disposing them to the environment (Lade et al. 2015).

3 Need for the Best Available Techniques

The textile industries face a great stress to ensure environmental safety and environmental protection activities by inculcating the ways to reduce the generation of waste, treatment of wastewater, and the use of non-polluting dyestuffs. It is mandatory for the textile industries to abide by the rules and regulations of Environment Protection Act (EPA), 1986, the legislation including Water (Prevention and Control of Pollution) Act, 1974, and Water Cess Act, 1978. Larger textile industries generate adequate funds to implement the best available techniques (BAT) for effluent treatment. But it is impossible for small-scale textile units to own an effluent treatment plant because of the exorbitant costs involved in commissioning and setting up the plant. In order to facilitate this, wastewater treatment using common effluent treatment plants (CETPs) is considered as an option. India had only one CETP till 1990 in Hyderabad, and in 1991, Ministry of Environment and Forest (MoEF), Government of India, took an initiative to support the CETPs financially to ensure environmental safety without compensating the economic growth. By 2005, around 88 CETPs were established and the number of CETPs increased to 1191 across the country. Among them, 148 CETPs belong to Tamil Nadu and 187 belong to Maharashtra.

The burgeoning numbers of textile units with the voluminous effluent generation, the issues such as scarcity of space, lack of technical human resource and high cost pose as the serious constraints to abate water pollution, thus resulting in the inappropriate wastewater management (Mishra 2016). Despite the recommendations on the cleaner production technologies and waste reduction techniques, these textile units remain to be highly polluting (Khan and Malik 2014). The different conventional physicochemical treatment techniques and their mechanism for remediating the textile-dyeing effluents are shown in Table 3.

The decolorization efficiency is considerably better, but the aforementioned physicochemical techniques face severe drawbacks such as being economically unfeasible and inability to eliminate the pollutant and their intermediates (Moradeeya et al. 2017; Karthikeyan et al. 2014; Vidhyadevi et al. 2014). These techniques are not very suitable due to their high initial capital and operational costs and they also generate considerable amounts of sludge causing secondary pollution problems, thereby subsequently demanding other treatment options.

Table 3 Conventional physico-chemical treatment techniques

S. no.	Technique(s)	Mechanism
1	Adsorption	The adsorbate gets adsorbed onto the adsorbent surface by physical or chemical forces
2	Membrane filtration	The hydraulic pressure drives the desired separation through the semi-permeable membrane
3	Ion exchange	The removal of ions from aqueous solution by replacing another ionic species using natural and synthetic ion exchangers
4	Coagulation	A coagulant aid is added to the wastewater for destabilizing the colloidal particles, which helps in the agglomeration
5	Flocculation	Flocculation forms larger flocs of the destabilized particles which can be easily removed
6	Ozonation	It is a tertiary treatment for decolorization in which ozone activates H_2O_2
7	Fenton's oxidation	Fe^{2+} salts and H_2O_2 are used to treat the wastewater, and the reaction takes place in the acidic pH, which results in the formation of strong hydroxyl radical and ferric iron. The decolorization is due to the sorption of the dissolved dyes to the formed flocs
8	Photochemical degradation	Degradation is achieved by hydroxyl radical generation during the UV/H_2O_2, UV/TiO_2, UV/O_2, and UV/Fenton's processes, and the molecule is degraded to CO_2 and H_2O
9	Electrolysis	Dyes react on the basis of applied electric current passing through the electrodes. This technique involves electro-coagulation, electro-flotation, and redox electrochemical reactions

4 Bioremediation

Bioremediation techniques utilize the biological systems to treat the pollutants and are environmentally reliable and an alternate to conventional decomposition. These techniques usually involve bioaccumulation, biosorption, bioaugmentation, and biodegradation. Bioaccumulation is defined as the ability of the viable biomass to accumulate the pollutants, which is based on the tolerance and uptake capacities of the biomass. The limitation of this technique is that the microbial growth is inhibited when the pollutant concentrations are too high for bioaccumulation and such microbial cells require metabolic energy (Robinson et al. 2001). Biosorption usually involves the adsorption phenomena, where the pollutants (adsorbate) are adsorbed onto regenerative and eco-friendly adsorbents/biosorbents. The limitation of this process is that it cannot be used for handling voluminous effluents because of the problems associated with the disposal of adsorbed biomass (Kuhad et al. 2004). Bioaugmentation is the process of introducing selected species which may be endogenous or exogenous to a complex ecosystem with pollutants (Joshi et al. 2017; Cronje et al. 2002). The limitation of bioaugmentation technique is that the introduced bacterial strain may fail to grow or survive as they undergo some competitive inhibition with the environmental pollutant (Herrero and Stuckey 2014; El Fantroussi and Agathos 2005). Biodegradation is an economical and effective way of treating wastewater as it is inexpensive, eco-friendly, and environmentally compatible and has less sludge-producing properties (Saratale et al. 2011).

4.1 Phyto-remediation

Phyto-remediation is considered as a green and safe approach to mitigate the problem of pollution and is less disruptive to the environment. The treatment processes involve mechanisms like hyper-accumulation, rhizo-degradation and phyto-stabilization. Several plant species belonging to glycophytes such as *Salsola vermiculata*, *Brassica juncea*, *Typhonium flagelliforme*, *Blumeamal colmii*, *Glandularia pulchella,* and *Portulaca grandiflora* are known to exhibit potential ability to degrade dyes (Kabra et al. 2011). Phyto-remediation is suitable only for sites contaminated with lower concentrations of the pollutant. The limitation of this technique is that the plants require a longer time to remediate and the fate of the intermediates is unknown. Additionally, the plant species employed for remediating the pollutants may find difficulty in growing in water or the soils contaminated with the xenobiotic pollutants.

4.2 Enzyme-Aided Biodegradation

Enzymes are a better option for bio-remediating the environmental contaminants, and the oxidoreductive enzyme systems comprise oxidases and reductases that are involved in the degradation of several environmental pollutants (de Gonzalo et al. 2016). Enzymes like oxidase, reductase, and methylase are known to catalyze bio-transformations of dyes (Kabra et al. 2011). Generally, bacterial cells involve the oxidation and reduction of the chemical bonds by the enzymes mainly belonging to oxidoreductases. These enzymes may be extracellular or intracellular, depending on their type and growth pattern of the bacterial cells. The oxidoreductases help in detoxification of xenobiotic compounds (Mahmood et al. 2016).

Lignin-degrading enzymes such as lignin peroxidase (LiP) and manganese peroxidase (MnP) are known to have a potential application in bioremediation of diverse pollutants. LiP (E.C. 1.11.1.14) is a plant peroxidase containing a heme group and utilizes H_2O_2 as co-factor for oxidizing the phenolic group (Dawkar et al. 2009). MnP (E.C. 1.11.1.13) catalyzes the oxidation of phenolic contaminants, and the reaction is mainly dependent upon the divalent manganese and buffers used (de Gonzalo et al. 2016).

Laccase (EC 1.10.3.2) consists of four histidine copper-binding domains and is the most commonly used enzyme for biodegradation of textile dyes and PAHs (Balaji et al. 2016). It has the ability to degrade the substrates having high redox potential and helps in the removal of hydrogen atom from the hydroxyl group of the substituted mono- and polyphenolic compounds (Akkaya et al. 2016). Tyrosinase (EC1.14.18.1) is a monophenol mono-oxygenase which catalyzes the phenol oxidation with the help of molecular oxygen rather than H_2O_2. The catalysis occurs in two steps such as hydroxylation and oxidation. They are known to have the dual enzyme activities such as cresolase and catecholase (Mahmood et al. 2016).

Azoreductase (E.C. 1.7.1.6) catalyzes the reductive cleavage of azo bonds to release the aromatic amines (Khan and Malik 2016). They are categorized into flavin-dependant and flavin-independent azoreductases on the basis of their function (Rawat et al. 2016). Triphenylmethane reductase is a dinucleotide-binding motif-containing enzyme and is induced during the degradation of TPM dyes (Kim et al. 2008). The veratryl alcohol oxidase (EC 1.1.3.7) is another oxidative enzyme induced by the microbial system during the dye degradation and helps in the cleavage of the phenyl rings (Bourbonnais and Paice 1988).

The limitation of the enzyme-based biodegradation is that these enzymes are prone to denaturation at high temperatures; their activity gets lowered when exposed to higher concentrations of pollutants (Seenuvasan et al. 2017). The production and purification of these enzymes are relatively expensive and require controlled environmental conditions (Seenuvasan et al. 2014). To overcome these limitations and in order to increase the enzyme activity, immobilization of the enzymes onto suitable support matrices are done (Seenuvasan et al. 2013).

4.3 *Microorganisms-Aided Biodegradation*

Microorganisms bear the versatile capacity to remove the pollutants from wastewater by biodegrading the recalcitrant compounds (Mahmood et al. 2016). This is done by screening the microorganisms for the functionality of the target contaminants. Microbial cell-aided biodegradation is more advantageous on the features like occurrence, ability to adapt, cost-effectiveness with the ease of cultivation and manipulation, rapid growth unlike plants, and considerable selectivity toward the environmental pollutants (Komal et al. 2017). Thus, the abovesaid traits make these microbial cells more beneficial when compared with the other bio-remedial tools. The microbial systems utilize enzymes which may be intracellular or extracellular for the degradation of pollutants (Kumar et al. 2016a). The microbial systems can be mutated and genetically engineered for effective bioremediation.

4.3.1 Biodegradation Using Fungi

Ligninolytic fungal systems have the ability to degrade different xenobiotic compounds with the help of oxidases and reductases for the cleavage of hetero-polyaromatic moieties. Certain noteworthy fungal strains such as

Phanerochaete chrysosporium, *Aspergillus niger*, *Aspergillus terricola*, *Pycnoporus sanguineus*, *Trametes versicolor*, *Iperx lacteus,* and *Funalia trogii* are known to degrade distinct dyes (Sen et al. 2016).

4.3.2 Biodegradation Using Yeasts

Yeasts are a class of organisms belonging to fungi and are known to grow faster than the fungal strains with resistance to unfavorable environmental conditions. The yeasts employ putative enzymic mixtures for the biodegradation of dyes. *Saccharomyces cerevisiae*, *Trichosporon beigelii*, *Candida tropicalis*, *Candida zeylanoides*, *Candida lipolytica*, *Kluyveromyces marxianus*, *Galactomyces geotrichum*, *Debaryomyces polymorphus,* and *Issatchenkia occidentalis* are known to degrade different dyes (Jadhav et al. 2007). The use of yeast strains is limited due to their longer hydraulic retention time and other process conditions (Sathya et al. 2017).

4.3.3 Biodegradation Using Algae

Algae are photosynthetic organisms thriving in the marine ecosystems, and the algal biomasses are generally used in stabilization ponds which have an ability to induce azoreductase for the azo-dye degradation. *Chlorella pyrenoidosa*, *C. sorokiniana*, *C. vulgaris*, *Lemnaminus cula*, *Scenedesmus obliquus*, *Oscillateria tenuis*, *Closterium lunula*, *Cosmarium* sp, and *Spirogyra* sp. are the algal strains known to biodegrade different dyes into aromatic amines and other intermediates (Patil et al. 2011).

4.3.4 Biodegradation Using Actinomycetes

Actinomycetes are filamentous bacteria, and dye degradation using actinomycetes is more advantageous than fungal treatment. Reports are available for the biodegradation of distinct classes of dyes like *Streptomyces krainskii* SUK-5, *Nocardia corallina,* and *N. globerula* (Mane et al. 2008). Dye decolorization using actinomycetes is limited because such cells require a longer retention/induction time for decolorizing the dyes and the dye molecules get adsorbed onto the biomass rather than getting degraded (Bagewadi et al. 2011).

4.3.5 Biodegradation Using Bacteria

Bacterial cells are known to decolorize and degrade dyes under aerobic, anaerobic, and anoxic conditions. Dye degradation using bacterial culture is faster, cheaper, and efficient when compared with other the biological systems. The bacterial cultures used for dye degradation are either in their pure form or mixed cultures consisting of identified or unidentified strains (Sakthipriya et al. 2016).

5 Dye Degradation by Pure Bacterial Strains

Pure bacterial cultures or strains belonging to the genus such as *Bacillus*, *Pseudomonas*, *Enterobacter*, *Proteus*, *Aeromonas*, *Klebsiella*, and *Achromobacter* are known to effectively decolorize and degrade the dyes. The use of pure bacterial strains ensures reproducible observations having the merit of being easily interpreted using the biochemical and molecular biological analyses. Pure bacterial strains are known for their ability to degrade dyes and their use is highly advantageous as they give reproducible observations that can be interpreted easily. Dye degradation using a single bacterial strain shows remarkable efficiency in the aerobic, anaerobic, and anoxic conditions (Kumar et al. 2016b).

The contaminated sites harbor a diversified category of microorganisms that are well-adapted to survive in the presence of pollutants and are usually known to develop traits, capable of consuming or assuming the contaminants for their survival (Karthikeyan et al. 2017). These indigenous strains are more effective, advantageous and are termed as extremophiles as they usually adapt themselves to the biotic and abiotic stresses for their survival. The metabolic system of the surviving microorganisms adapts an alternative to routine systems, and this phenomenon is known as acclimation. At lower dye concentrations, the acclimated microbes exhibit negative inhibition, while at higher dye concentrations, the microbial strains get acclimated to a greater extent (Kumar et al. 2015). This is well-perceived by researchers, and some of the salient features of bacterial strain isolation are ease of screening, eco-friendly and economically viable, producing lesser sludge with metabolites of mineralized nature (Kumar et al. 2017b).

6 Dye Degradation by Mixed Bacterial Consortia

The mixed bacterial cultures are composed of identified and unidentified strains indigenous to different environmental pollutants. In a mixed bacterial consortium/ culture, the individual bacterial strains attack the dye molecules at different positions to form the intermediate metabolites usually of low molecular weight fractions. The bacterial consortial systems are advantageous over the individual bacterial cultures as they involve an inductive, synergistic mechanism among the co-existing strains. The intermediate products are usually aromatic amines, and the consortial strains release certain enzymes which act on these amines and decompose them to simpler products (Ayed et al. 2010). Table 4 gives a snapshot of the research work on dye degradation by the mixed bacterial population and bacterial consortium comprising identified strains for the degradation of dyes.

Table 4 Application of bacterial consortia comprising of mixed culture and individual identified strains in dye degradation

S. no.	Bacterial consortia	Name of the dye	References
1	Bacterial consortium RVM11.1 comprising seven different isolates	Reactive violet 5	Moosvi et al. (2005)
2	Bacterial consortium comprising *B. cereus* (BN-7), *P. putida* (BN-4), *P. fluorescence* (BN-5), and *Stenotrophomonas acidaminiphila* (BN-3)	Acid red 88	Khehra et al. (2005)
		Acid red 119	
		Acid red 97	
		Acid blue 113	
		Reactive red 120	
3	Bacterial consortium (NJ38, NJ1, NJLC1, and NJLC3) comprising of gram-negative and gram-positive bacterial isolates	Direct RED 81	Junnarkar et al. (2006)
4	Mixed culture	Reactive black 5	Mohanty et al. (2006)
5	Isolated halophilic and halotolerant bacteria	Remazol black B	Asad et al. (2007)
		Maxilon blue	
		Sulphonyl scarlet BNLE	
		Sulphonyl blue TLE	
		Sulphonyl green BLE	
		Remazol black N	
		Entrazol blue IBC	
5	Consortium of *E. coli* DH5α and *P. luteola*	Reactive red 22	Chen and Chang (2007)
6	Consortium comprising *A. faecalis*, *Sphingomonas sp. EBD*, *B. subtilis*, *B. thuringiensis*, and *Enterobacter cancerogenus*	Direct blue 15	Kumar et al. (2007)
7	Consortium of *Enterobacter* sp., *Serratia* sp., *Yersinia* sp., and *Erwinia* sp.	Reactive red 195	Jirasripongpun et al. (2007)
8	Bacterial consortium JW-2 comprising *Paenibacillus polymyxa*, *Micrococcus luteus*, and *Micrococcus* sp.	Reactive violet 5R	Moosvi et al. (2007)
9	Consortium comprising *Pseudomonas aeruginosa* and *Bacillus circulans* and NAD1 and NAD6 isolates	Reactive black 5	Dafale et al. (2008)

(continued)

Table 4 (continued)

S. no.	Bacterial consortia	Name of the dye	References
10	Bacterial consortium TJ-1 comprising *Aeromonas caviae*, *Proteus mirabilis,* and *Rhodococcus globerulus*	Acid orange 7	Joshi et al. (2008)
11	Bacterial consortium DMC comprising *Pseudomonas aeroginosa* PAO1, *Stenotrophomonas maltophila,* and *Proteus mirabilis*	Direct black 22	Mohana et al. (2008)
12	Consortium SKB-II comprising *B. vallismortis* and *B. megaterium*	Congo red	Tony et al. (2009a, b)
		Brodeaux	
		Ranocid fast blue	
		Blue BCC	
13	Consortium comprising *B. vallismortis*, *B. pumilus*, *B. cereus*, *B. subtilis,* and *B. megaterium*	Direct red 28	Tony et al. (2009a, b)
14	Consortium-GR comprising *Proteus vulgaris* and *Miccrococcus glutamicus*	Scarlet R	Saratale et al. (2009)
		Green HE4BD	Saratale et al. (2010)
15	Consortium DAS comprising three *Pseudomonas* sp.	Reactive orange 16	Jadhav et al. (2010)
16	Consortium-IV comprising *Sphingobacterium* sp. ATM, *B. odysseyi* SUK3, and *P. desmolyticum* NCIM 2112	Orange 3R	Tamboli et al. (2010)
17	Consortium comprising *S. paucimobilis*, *B. cereus* ATCC14579, and *B. cereus* ATCC11778	Methyl orange	Ayed et al. (2010)
18	Mixed bacterial strains from CETPs	Mixed dyes	Rajeswari et al. (2011)
19	Consortium comprising *Providencia* sp. SDS and *Pseudomonas aeuroginosa* strain BCH	Red HE3B	Phugare et al. (2011)
20	Mixed culture SB4	Reactive violet 5R	Jain et al. (2012)
21	Consortium GG-BL comprising *Galactomyces geotrichum* MTCC 1360 and *Brevibacillus laterosporus* MTCC 2298	Rubine GFL	Waghmode et al. (2012)
22	Consortium-AVS comprising *Kocuria rosea* MTCC 1532, *Pseudomonas desmolyticum* NCIM 2112, and *Micrococcus glutamicus* NCIM 2168	Methylene blue	Kumar et al. (2012)
		Acid Blue 15	
23	Consortium comprising *P. putida*, *B. subtillis,* and *P. aeruginosa*	Acid Blue	Shah (2014)

(continued)

Table 4 (continued)

S. no.	Bacterial consortia	Name of the dye	References
24	Acclimated mixed culture	Reactive black HEBL	Kumar et al. (2014)
25	Acclimated mixed bacterial culture	Acid red 88	Kumar et al. (2015)
26	Consortium BMP1/SDSC/01 comprising *Bacillus* sp., *B subtilis*, *B.* cereus, *B. mycoides*, *Pseudomonas* sp., and *Micrococcus* sp.	Azo dyes	Mahmmod et al. (2015)
27	Consortium comprising *Proteus* spp., *Pseudomonas* spp., and *Acinetobacter* spp.	Methyl red Carbol fuchsin	Joshi et al. (2015)
28	Consortium ETL-A comprising *B. subtilis*, *Stenotrophomonas* sp., *P. stutzeri*, and *P. aeruginosa*	Reactive orange M_2R	Shah (2016)
29	Consortium comprising *Bacillus* sp. L10, *B. flexus* strain NBN2, *B. cereus* strain AGP-03, *B. cytotoxicus* NVH 391-98	Direct Blue 151	Lalnunhlimi and Krishnaswamy (2016)
		Direct Red 31	
30	Microbial Community	Actual textile wastewater	Forss et al. (2017)
31	Concocted bacterial consortium comprising *Achromobacter xylosoxidans* strain APZ, *K. pneumoniae* strain AHM, and *B. mannanilyticus* strain AVS	Textile mill effluent	Kumar et al. (2017c)

The consortial systems are proven to be advantageous over a single bacterial strain as they involve an inductive synergistic mechanism among the co-existing strains. Azo-dye degradation usually involves the cleavage of the azo bond that occurs to release the aromatic amines, and these amines are oxidized to reactive electrophilic species that covalently bind to the deoxyribonucleic acid (DNA), leading to severe toxicity and mutagenicity (Pandey et al. 2007). However, the advantageous bacterial consortia are favorable for the mineralization of the liberated aromatic amines (Lade et al. 2015; Kumar et al. 2017a).

7 Futuristic Scope of Bacterial Consortia-Aided Bioremediation

The bacterial consortium catalyzing the bio-transformation of dyes and effluents would lead to the detoxification and mineralization that would provide an insight into a strategic textile industrial wastewater treatment. The non-toxic and mineralized products of dyes and effluents would eliminate the threat of secondary pollution problems when they are disposed to the environment. The demand for the best available technologies for remediating the problems of textile colorants and effluents can be justified by employing the bacterial consortium for bio-transforming the dyes and mineralizing the liberated aromatic amines which are expected to result in the economic viability. Furthermore, the bio-transformation yielding the products of detoxified and mineralized nature can be tailored to selectively downstream these products. By implementing the effective concoction of the bacterial consortium, the scope and marketability of the processes in the wastewater treatment would be ameliorated. As for the pursuit of bacterial consortia-aided bioremediation, this beneficial phenomenon would be extended by immobilizing the bacterial consortium onto the suitable supports that would ease the separation of bacterial cells from the treated effluents which could be a potential lead for designing the pilot plant assembly.

8 Conclusions

The biodegradation of textile dyes is a user-friendly bio-remedial approach to mitigate the impacts and toxicities of the intermediate metabolites. The biodegradation of textile dyes using bacterial strains and their consortial systems provides an efficient and trustworthy way to mineralize and detoxify the dyes. These systems utilize inducible consortial oxidoreductases for the detoxification and mineralization into least/reduced toxic metabolites. Presently, the design of such consortial systems is gaining pervasive importance as they work under adverse conditions. Further, tailoring of process parameters is required for utilizing the bacterial consortia for making the process viable and serving the necessity of environmental safety and health.

References

Akkaya A, Ozseker EE, Akdogan HA (2016) Degradation of dyes by laccase. Anal Lett 49 (6):790–798

Asad S, Amoozegar MA, Pourbabaee AA (2007) Decolorization of textile azo dyes by newly isolated halophilic and halotolerant bacteria. Biores Technol 98(11):2082–2088

Ayed L, Khelifi E, Jannet HB (2010) Response surface methodology for decolorization of azo dye methyl orange by bacterial consortium: produced enzymes and metabolites characterization. Chem Eng J 165(1):200–208

Bagewadi ZK, Vernekar AG, Patil AY et al (2011) Biodegradation of industrially important textile dyes by actinomycetes isolated from activated sludge. Biotechnol, Bioinfor Bioeng 1(3):351–360

Balaji N, Kumar KS, Vinodhini G et al (2016) Immobilization of laccase onto micro-emulsified magnetic nanoparticles for enhanced degradation of a textile recalcitrant. J Environ Biol 37 (6):1489–1496

Bilińska L, Gmurek M, Ledakowicz S (2016) Comparison between industrial and simulated textile wastewater treatment by AOPS—Biodegradability, toxicity and cost assessment. Chem Eng J 306:550–559

Bourbonnais R, Paice MG (1988) Veratrol oxidases from the lignin-degrading basidiomycete *Pleurotus sajor-caju*. Biochem J 255(2):445–450

Burkinsha SM (2016) Physico-chemical aspects of textile coloration, society of dyers and colorists. Wiley, United Kingdom

Chen BY, Chang JS (2007) Assessment upon species evolution of mixed consortia for azo dye decolorization. J Chin Inst Chem Eng, 38(3–4):259–266

Cronje GL, Beeharry AO, Wentzel MC et al (2002) Active biomass in activated sludge mixed liquor. Water Res 36(2):439–444

Dafale N, Rao NN, Meshram SU et al (2008) Decolorization of azo dyes and simulated dye bath wastewater using acclimatized microbial consortium-biostimulation and halo tolerance. Biores Technol 99(7):2552–2558

Dawkar VV, Jadhav UU, Ghodake GS et al (2009) Effect of inducers on the decolorization and bio-degradation of textile azo dye Navy blue 2GL by *Bacillus* sp. VUS. Biodegradation 20 (6):777–787

de Gonzalo G, Colpa DI, Habib MHM et al (2016) Bacterial enzymes involved in lignin degradation. J Biotechnol 236:110–119

El Fantroussi S, Agathos SN (2005) Is bioaugmentation a feasible strategy for pollutant removal and site remediation? Curr Opin Microbiol 8(3):268–275

Forss J, Lindh MV, Pinhassi J et al (2017) Microbial biotreatment of actual textile wastewater in a continuous sequential rice husk biofilter and the microbial community involved. PLoS ONE 12 (1):e0170562

Goswami O (1990) Sickness and growth of India's textile industry: Analysis and policy options. Econ Polit Wkly 25(45):2496–2506

Herrero M, Stuckey DC (2014) Bioaugmentation and its application in wastewater treatment: a review. Chemosphere 140:119–128

Jadhav JP, Parshetti GK, Kalme SD et al (2007) Decolorization of azo dye methyl red by *Saccharomyces cerevisiae* MTCC 463. Chemosphere 68(2):394–400

Jadhav JP, Kalyani DC, Telke AA et al (2010) Evaluation of the efficacy of a bacterial consortium for the removal of color, reduction of heavy metals, and toxicity from textile dye effluent. Biores Technol 101(1):165–173

Jain K, Shah V, Chapla D et al (2012) Decolorization and degradation of azo dye-reactive violet 5R by an acclimatized indigenous bacterial mixed cultures-SB4 isolated from anthropogenic dye contaminated soil. J Hazard Mater 213–214:378–386

Jirasripongpun K, Nasanit R, Niruntasook J et al (2007) Decolorization and degradation of C.I. Reactive Red 195 by *Enterobacter* sp. Thammasat International. J Sci Technol 12(4):6–11

Joshi T, Iyengar L, Singh K et al (2008) Isolation, identification and application of novel bacterial consortium TJ-1 for the decolourization of structurally different azo dyes. Biores Technol 99 (15):7115–7121

Joshi PA, Jaybhaye S, Mhatre K (2015) Biodegradation of dyes using consortium of bacterial strains isolated from textile effluent. Eur J Exp Biol 5(7):36–40

Joshi N, Dholakiya RN, Kumar MA et al (2017) Recycling of starch processing industrial wastewater as a sole nutrient source for the bioflocculant production. Environ Progr Sustain Energy. doi:https://doi.org/10.1002/ep.12608

Junnarkar N, Murty DS, Bhatt NS et al (2006) Decolorization of diazo dye direct red 81 by a novel bacterial consortium. World J Microbiol Biotechnol 22(2):163–168

Kabra AN, Khandare RV, Kurade MB et al (2011) Phytoremediation of a sulphonated azo dye Green HE4B by *Glandularia pulchella* (Sweet) Tronc. (Moss Verbena). Environ Sci Pollut Res 18(8):1360–1373

Karthikeyan S, Kumar MA, Maharaja P et al (2014) Process optimization for the treatment of pharmaceutical wastewater catalyzed by poly sulpha sponge. J Taiwan Inst Chem Eng 45 (4):1739–1747

Karthikeyan V, Kumar MA, Mohanapriya P et al (2017) Biodegradation of remazol brilliant blue R using isolated bacterial culture (*Staphylococcus* sp. K2204). Environ Technol. doi:https://doi.org/10.1080/09593330.2017.1369579

Khan S, Malik A (2014) Environmental and health effects of textile industry wastewater. Environmental Deterioration and Human Health, Springer, Netherlands

Khan S, Malik A (2016) Degradation of reactive black 5 dye by a newly isolated bacterium *Pseudomonas entomophila* BS1. Can J Microbiol 62(3):220–232

Khehra MS, Saini HS, Sharma DK et al (2005) Comparative studies on potential of consortium and constituent pure bacterial isolates to decolorize azo dyes. Water Res 39(20):5135–5141

Kim MH, Kim Y, Park HJ et al (2008) Structural insight into bioremediation of triphenylmethane dyes by *Citrobacter* sp. triphenylmethane reductase. J Biol Chem 283(46):31981–31990

Komal J, Kumar MA, Thiruvengadaravi KV et al (2017) Indigenously acclimatized bacterial consortium for anthracene biotransformation. Energy Sources, Part A: Recovery, Utilization, Environ Eff 39(5):528–537

Kuhad RC, Sood N, Tripathi KK et al (2004) Developments in microbial methods for the treatment of dye effluents. Adv Appl Microbiol 56:185–213

Kumar K, Devi SS, Krishnamurthi K et al (2007) Decolorization and detoxification of direct blue-15 by a bacterial consortium. Biores Technol 98(16):3168–3171

Kumar MA, Kumar VV, Premkumar MP (2012) Chemometric formulation of bacterial consortium-AVS for improved decolorization of resonance-stabilized and hetero-polyaromatic dyes. Biores Technol 123:344–351

Kumar MA, Vijayalakshmi A, Lincy EAR et al (2014) Biotransformation of Reactive Black HEBL into 3-nitroso-3-azabicyclo (3.2.2) nonane by an acclimated mixed culture. International Journal of ChemTech Research 6(9): 4172–4179

Kumar MA, Kumar VV, Ponnusamy R et al (2015) Concomitant mineralization and detoxification of acid red 88 by an indigenous acclimated mixed culture". Environ Progr Sustain Energy 34 (5):1455–1466

Kumar MA, Harthy DK, Kumar VV et al (2016a) Detoxification of a triphenylmethane textile colorant using acclimated cells of *Bacillus mannanilyticus* strain AVS. Environ Progr Sustain Energy 36(2):394–403

Kumar MA, Priyadarshini R, Seenuvasan M et al (2016b) Biotransformation and detoxification of a greater tinctorial textile colorant using an isolated bacterial strain. J Environ Biol 37(6):1497–1506

Kumar MA, Poonam S, Kumar VV et al (2017a) Mineralization of aromatic amines liberated during the degradation of a sulfonated textile colorant using *Klebsiella pneumoniae* strain AHM. Process Biochem 57:181–189

Kumar MA, Zamana PA, Kumar VV et al (2017b) *Achromobacter xylosoxidans* strain APZ for phthalocyanine dye degradation: Chemo-metric optimization and canonical correlation analyses. Journal of Water Process Engineering 18:73–82

Kumar MA, Vigneshwaran ME, Priya M (2017c) Concocted bacterial consortium for the detoxification and mineralization of azoic-cum-sulfonic textile mill effluent. J Water Process Eng 16C:199–205

Lade H, Kadam A, Paul D et al (2015) Biodegradation and detoxification of textile azo dyes by bacterial consortium under sequential microaerophilic/ aerobic processes. EXCLI J 14:158–174

Lalnunhlimi S, Krishnaswamy V (2016) Decolorization of azo dyes (Direct Blue 151 and Direct Red 31) by moderately alkaliphilic bacterial consortium. Braz J Microbiol 47(1):39–46

Mahmood R, Sharif F, Ali S et al (2015) Enhancing the decolorizing and degradation ability of bacterial consortium isolated from textile effluent affected area and its application on seed germination. Sci World J 2015:628195

Mahmood S, Khalid A, Arshad M et al (2016) Detoxification of azo dyes by bacterial oxidoreductase enzymes. Crit Rev Biotechnol 36(4):639–651

Mane UV, Gurav PN, Deshmukh AM et al (2008) Degradation of textile dye reactive navy-blue Rx (Reactive blue-59) by an isolated actinomycete *Streptomyces krainskii* SUK-5. Malays J Microbiol 4(2):1–5

Mishra AK (2016) Smart materials for waste water applications. Scrivener Publishing, Wiley, New Jersey

Mohana S, Shrivastava S, Divecha J et al (2008) Response surface methodology for optimization of medium for decolorization of textile dye Direct Black 22 by a novel bacterial consortium. Biores Technol 99(3):562–569

Mohanty S, Dafale N, Rao NN (2006) Microbial decolorization of reactive black-5 in a two-stage anaerobic-aerobic reactor using acclimatized activated textile sludge. Biodegradation 17 (5):403–413

Mohanty J, Shinde MN, Barooah N et al (2016) Reversible insulin hexamer assembly promoted by ethyl violet: pH-controlled uptake and release. J Phys Chem Lett 7(19):3978–3983

Moosvi S, Keharia H, Madamwar D (2005) Decolourization of textile dye reactive violet 5 by a newly isolated bacterial consortium RVM 11.1. World J Microbiol Biotechnol 21(5):667–672

Moosvi S, Kher X, Madamwar D (2007) Isolation, characterization and decolorization of textile dyes by a mixed bacterial consortium JW-2. Dyes Pigm 74(3):723–729

Moradeeya P, Kumar MA, Thorat RB et al (2017) Screening of nanocellulose for biosorption of chlorpyrifos from water: chemometric optimization, kinetics and equilibrium. Cellulose 24:1319–1332

Pandey A, Singh P, Iyengar L (2007) Bacterial decolorization and degradation of azo dyes. Int Biodeterior Biodegradation 59(2):73–84

Patil PD, Gude VG, Mannarswamy A et al (2011) Optimization of direct conversion of wet algae to biodiesel under supercritical methanol conditions. Biores Technol 102(1):118–122

Peters AT, Freeman HS (1991) Colour chemistry: the design and synthesis of organic dyes and pigments. Elsevier Applied Science. Elsevier Science Pub, New York

Phugare SS, Kalyani DC, Patil AV et al (2011) Textile dye degradation by bacterial consortium and subsequent toxicological analysis of dye and dye metabolites using cytotoxicity, genotoxicity and oxidative stress studies. J Hazard Mater 186(1):713–723

Rajeswari K, Subashkumar R, Vijayaraman K (2011) Biodegradation of mixed textile dyes by bacterial strains isolated from dyewaste effluent. Res J Environ Toxicol 5:97–107

Rawat D, Mishra V, Sharma RS (2016) Detoxification of azo dyes in the context of environmental processes. Chemosphere 155:591–605

Robinson T, McMullan G, Marchant R et al (2001) Remediation of dyes in textile effluent: a critical review on current treatment technologies with a proposed alternative. Biores Technol 77(3):247–255

Sakthipriya N, Doble M, Sangwai JS (2016) Efficacy of *Bacillus subtilis* for the biodegradation and viscosity reduction of waxy crude oil for enhanced oil recovery from mature reservoirs. Energy Sources, Part A: Recovery, Utilization Environ Eff 38(16):2327–2335

Salabert J, Sebastián RM, Vallribera A (2015) Anthraquinone dyes for superhydrophobic cotton. Chem Commun 51(75):14251–14254

Saratale RG, Saratale GD, Kalyani DC et al (2009) Enhanced decolorization and biodegradation of textile azo dye Scarlet R by using developed microbial consortium-GR. Biores Technol 100 (9):2493–2500

Saratale RG, Saratale GD, Chang JS et al (2010) Decolorization and biodegradation of reactive dyes and dye wastewater by a developed bacterial consortium. Biodegradation 21(6):999–1015

Saratale RG, Saratale GD, Chang JS et al (2011) Bacterial decolorization and degradation of azo dyes: a review. J Taiwan Inst Chem Eng 42(1):138–157

Sathya DJH, Turakhia AM, Kumar MA et al (2017) Bioethanol from saccharificed lignocellulosic rich *Aloe vera* rinds using *Saccharomyces cerevisiae* MTCC 4779. Energy Sources, Part A: Recovery, Utilization, Environ Eff. doi:https://doi.org/10.1080/15567036.2017.1328004

Seenuvasan M, Malar GCG, Preethi S et al (2013) Fabrication, characterization and application of pectin degrading Fe_3O_4–SiO_2 nanobiocatalyst. Mater Sci Eng, C 33:2273–2279

Seenuvasan M, Kumar KS, Malar GCG et al (2014) Characterization, analysis, and application of fabricated Fe_3O_4-chitosan-pectinase nanobiocatalyst. Appl Biochem Biotechnol 172(5):2706–2719

Seenuvasan M, Sanjayini SJ, Kumar MA et al (2017) Cellulase mediated saccharification of lignocellulosic rich pseudostem of *Musa cavendish* for bio-ethanol production by *Saccharomyces cerevisiae* MTCC 4779. Energy Sources, Part A: Recovery, Utilization, Environ Eff 39(6):570–575

Sen SK, Raut S, Bandyopadhyay P et al (2016) Fungal decolouration and degradation of azo dyes: a review. Fungal Biol Rev 30(3):112–133

Shah MP (2014) Microbial degradation of acid blue dye by mixed consortium. Int J Environ Bioremediat Biodegradation 2(3):125–132

Shah M (2016) Microbial degradation of Reactive Orange M2R dye by bacterial consortium ETL-A. J Microb Biochem Technol 8(6):483–487

Tamboli DP, Gomare SS, Kalme SS et al (2010) Degradation of Orange 3R, mixture of dyes and textile effluent and production of polyhydroxyalkanoates from biomass obtained after degradation. Int Biodeterior Biodegradation 64(8):755–763

Tony BD, Goyal D, Khanna S (2009a) Decolorization of textile azo dyes by aerobic bacterial consortium. Int Biodeterior Biodegradation 63(4):462–469

Tony BD, Goyal D, Khanna S (2009b) Decolorization of Direct Red 28 by mixed bacterial culture in an up-flow immobilized bioreactor. J Ind Microbiol Biotechnol 36(7):955–960

Vidhyadevi T, Murugesan A, Kalaivani SS et al (2014) Optimization of the process parameters for the removal of reactive yellow dye by the low cost *Setaria verticillata* carbon using response surface methodology: Thermodynamic, kinetic and equilibrium studies. Environ Progr Sustain Energy 33(3):855–865

Waghmode TR, Kurade MB, Lade HS et al (2012) Decolorization and biodegradation of Rubine GFL by microbial consortium GG-BL in sequential aerobic/microaerophilic process. Appl Biochem Biotechnol 167(6):1578–1594

Welham A (2000) The theory of dying (and the secret of life). J Soc Dyers Colour 116(5):140–143

Zollinger H (2003) Color chemistry: syntheses, properties, and applications of organic dyes and pigments. Wiley, Zurich

Author Biographies

Dr. M. Anil Kumar is a Scientist at CSIR-Central Salt & Marine Chemicals Research Institute, Bhavnagar, India. He graduated with M. Tech., Environmental Science and Technology (2012) and Ph.D. degree in Technology (2017). His areas of expertise are Environmental Biotechnology, Chemo-metric Optimization and Nano-biotechnology. To credit, he has got more than 35 publications in National and International journals with awards for his oral presentations.

Dr. P. Baskaralingam is currently working as Assistant Professor in the Department of Chemistry at Anna University, Chennai, India. He has published more than 25 research articles including national and international journals in various fields. His research expertise is in adsorption, wastewater treatment, Environmental science, bio-fuels and Catalysis. Currently, he is leading research projects in the field of wastewater treatment using membrane technology and scheming heterogeneous catalysis for bio-fuel applications. He is a life member of the Indian Society for Technical Education (ISTE).

Ms. A.R. Salma Aathika is a Research Scholar pursuing Ph.D. under exchange availing Fulbright fellowship at Massachusetts Institute of Technology, United States. She is graduated with M. Tech., Environmental Science and Technology (2015) and her areas of interest are Bioenergy from food waste and Environmental Management. She has got two publications in well reputed International Journals.

Prof. S. Sivanesan graduated with M.Sc. degree in Environmental Chemistry and Ph.D. degree in Atmospheric Science and Technology. He has post-doctoral experiences at Italy, Germany, Japan and United States. He is presently serving as Dean at Alagappa College of Technology, Anna University, Chennai. His areas of expertise include Atmospheric Science, Environmental Biotechnology, Bio-energy and Environmental Management. He has several scientific contributions that include 150 publications in leading peer-reviewed journals, authored 7 books and monographs. He has supervised 22 Ph.D. theses, 03 M.S (By Research) theses and 77 Dissertation Projects. He serves as an invited reviewer and editorial member in peer-reviewed journals. Apart from being a highly productive Environmental Scientist, he was also an effective administrator with outstanding achievements. He was honored by several awards at various occasions and few of them are-Best Environmental Engineer Award 2015 by Institute of Chemical Engineers & Engineers without borders, and Tamil Nadu Scientists (TANSA 2013) Award in Environmental Science by Government of Tamil Nadu.

Chapter 9
Polycyclic Aromatic Hydrocarbons from Petroleum Oil Industry Activities: Effect on Human Health and Their Biodegradation

Sunita J. Varjani, Rutu R. Joshi, P. Senthil Kumar, Vijay Kumar Srivastava, Vipin Kumar, Chiranjib Banerjee and R. Praveen Kumar

Abstract Nowadays pollution control and abatement are critical issues faced by environmental scientists due to rapid industrialization. Petroleum industry is one of the major industries which release hydrocarbon pollutants in environment. Polycyclic aromatic hydrocarbons (PAHs) are the priority pollutants which are released into the environment by exploration activities of petroleum industries. The indiscriminate accumulation of petroleum hydrocarbon pollutants can be hazardous to the human life and aquatic biota. Due to toxicity of these pollutants, establishing efficient and environment-friendly method to degrade and detoxify these pollutants is an important research challenge. Various physiochemical methods are applied all over the world to remediate of petroleum hydrocarbon pollutants. Bioremediation technique has been developed for treatment of crude oil pollutants using biological

S. J. Varjani (✉)
Gujarat Pollution Control Board, Sector-10A, Gandhinagar 382010
Gujarat, India
e-mail: drsvs18@gmail.com

R. R. Joshi
School of Biological Sciences and Biotechnology, University and Institute
of Advanced Research, Gandhinagar 382007, Gujarat, India

P. Senthil Kumar
Department of Chemical Engineering, SSN College of Engineering,
Chennai 603110, India

V. K. Srivastava
GSFC University, Fertilizernagar, Vadodara 391750, Gujarat, India

V. Kumar · C. Banerjee
Department of Environmental Science and Engineering, Indian Institute
of Technology (Indian School of Mines), Dhanbad 826004, India

R. Praveen Kumar
Department of Biotechnology, Anna Bio Research Foundation,
Arunai Engineering College, Tiruvannamalai 606603, Tamil Nadu, India

S. J. Varjani et al. (eds.), *Waste Bioremediation*, Energy, Environment, and Sustainability, https://doi.org/10.1007/978-981-10-7413-4_9

agents like bacteria, fungi, algae, and plants. Applications of certain microorganisms have gained importance in the field of applied environmental microbiology. The application of microbes to degrade pollutants is getting attention due to its environmental and economic benefits. They can be used to change bioavailability and toxicity of petroleum hydrocarbons present in polluted soil and aqueous environment. This paper explores hydrocarbons present in petroleum crude. The effect of petroleum hydrocarbon pollutants on human health and environment is also discussed. This chapter also explains microbial degradation of these pollutants.

Keywords Biodegradation · Petroleum hydrocarbons · Environment pollution
Polyaromatic hydrocarbons (PAHs)

1 Introduction

Polyaromatic hydrocarbons (PAHs) are chemical compounds that contain more than one fused benzene ring (Bisht et al. 2010; Nikitha et al. 2017) which consists of numerous carbon atoms joined together to form multiple rings. More than 10,000 different PAH compounds exist in environment (Rubailo and Oberenko 2008; Abdel-Shafy and Mansour 2016). They are commonly found in petroleum fuels, coal products, and tar (Rubailo and Oberenko 2008; Varjani and Upasani 2016c). Different types of anthropogenic activities are the sources of PAHs present in environment (Samanta et al. 2002; Varjani and Upasani 2017c) such as fuel combustion, pyrolytic processes, spillage of petroleum products, waste incinerators, and domestic heaters (Bamforth and Singleton 2005). Mostly, PAHs are formed from the incomplete combustion of plant or animal matter, or carbon fuels, such as coal or petroleum (Varjani and Upasani 2017b). PAHs such as naphthalene, anthracene, phenanthrene, fluoranthene, and pyrene having four or more fused ring structure are typically recalcitrant to biodegradation (Haritash and Kaushik 2009; Varjani 2017a).

They are also called polynuclear aromatic hydrocarbons (Stogiannidis and Laane 2015). Other activities that release PAHs include driving, agricultural burning, roofing or working with coal tar products, sound- and water-proofing, coating pipes, steel making, and paving with asphalt (Choi et al. 2010). PAHs are persistent organic pollutants (POPs) which are toxic, carcinogenic, and mutagenic so their presence in environment is the harmful effect on human health (Haritash and Kaushik 2009; Gupte et al. 2016; Varjani and Upasani 2017a). Chemical, physical, and biological processes are used in remediation of PAHs (Ukiwe et al. 2013; Varjani 2017b). Among all these techniques, biological processes by application of microorganisms for bioremediation of pollutants are preferred as they are economically viable and environment-friendly (Varjani and Upasani 2016c). In this chapter sources, applications, types, characteristics and properties, and toxicity of PAHs are discussed. It also includes the recent literature available for degradation of PAHs by bacteria and fungi.

2 Source and Uses of PAHs

Different industrial, agricultural, and medical activities are involved in releasing large quantity of hazardous hydrocarbon substances in the environment and create pollution problems (Kim et al. 2013; Sarria-Villa et al. 2015; Varjani and Upasani 2016b). PAHs are the product of incomplete combustion of organic compounds. PAHs are formed as a result of incomplete combustion of organic compound at high temperature (500–800 °C) to low temperature (100–300 °C) for long duration (Stogiannidis and Laane 2015; Sarria-Villa et al. 2015). They are also known as polyarenes or polynuclear aromatic hydrocarbons (Haritash and Kaushik 2009; Ganesh et al. 2014). They are organic compound which is toxic and persistent for the environment (Varjani et al. 2015; Ghosal et al. 2016) which are released from different natural and anthropogenic activities.

Natural sources include forest and grass fires, oil seeps, volcanoes, chlorophyllous plants, fungi, and bacteria. Anthropogenic sources of PAHs include: (a) Petroleum, combustion of fossil fuel—including motor vehicle emission and power generation, (b) Effluent from industries and wastewater treatment plant, (c) Electric power generation, (d) Refuse incineration, (e) Home heating, (f) Production of coke, carbon black, coal tar, asphalt roads, roofing tar, (g) Internal combustion engines, (h) Smoked foods (meat, fish, etc), (i) Cigarette and tobacco smoke, (j) Wood burning (Freeman and Cattell 1990), and (k) Hazardous waste sites, coal gasification sites, chimneys of industries as well as aluminum production industrial units (Rubailo and Oberenko 2008; Abdel-Shafy and Mansour 2016; Nikitha et al. 2017).

2.1 Characteristics and Properties of PAHs

Many PAHs have toxic, mutagenic, and carcinogenic properties. Polycyclic aromatic hydrocarbons are lipophilic, nonpolar molecules (Lundstedt et al. 2003; Bojes and Pope 2007). PAHs are not very soluble in water hence they persist in the environment. PAHs are highly lipid soluble and thus readily absorbed from gastrointestinal tract of mammals (Choi et al. 2010; Bisht et al. 2010). Polycyclic aromatic hydrocarbons (PAHs) are organic compounds that are mostly colorless, white, or pale-yellow solids (Lundstedt et al. 2003; Bojes and Pope 2007). In PAHs structure mainly carbon and hydrogen are present. But it also contains N, S, and O atoms which are heteroatoms. It is also considered to be known as PAH (Clemente et al. 2014; Varjani et al. 2015). Generally, their molecular weight ranges from 166 to 328 (Chiou et al. 1998). It generally depends on number of carbon it possesses as well as the position of fused rings and elements present in its structure. They are group of several hundred chemically related compounds (Haritash and Kaushik 2009; Choi et al. 2010). They have toxic effects on organisms through various ways (Arulazhagan and Vasudevan 2011; Varjani and Upasani 2016c). PAHs enter in the environment through various routes and are usually found as a mixture containing two or more of PAHs. According to genotoxicity and abundance, US Environmental Protection Agency (USEPA) have listed 16 PAH compounds as priority pollutants (Lundstedt et al. 2003; Bojes and Pope 2007).

2.2 *Types of PAHs*

Chemically PAHs are comprised of two or more benzene rings. Polycyclic aromatic hydrocarbons have two or more single or fused aromatic rings with a pair of carbon atoms shared between rings in their molecules. According to their arrangements, they are defined as linear, cluster, or angular (Rubailo and Oberenko 2008; Sarria-Villa et al. 2015). Types of PAH according to ring arrangement are summarized in Table 1. Commonly there are two types of categories small PAHs having six or less than six fused aromatic rings and large PAHs having more than six aromatic rings in their structure (Mohsen et al. 2009; Ghosal et al. 2016).

2.3 *Uses of PAHs*

Apart from industrial emissions or effluents, PAHs can be released in the environment through incomplete combustion of organic material, *viz.*, coal, oil, and wood. PAHs are used as intermediates in thermosetting plastics, lubricating materials, and other chemical industries (Rubailo and Oberenko 2008; Gupte et al. 2016). PAHs are present in asphalt, which is used for roads construction (Nikitha et al. 2017). Precise PAHs and specific refined products are also used in the field of electronics, functional plastics, and liquid crystals. Table 2 represents industrial application(s) of some polyaromatic hydrocarbons (PAHs).

3 Toxicity of PAHs

Different types of petroleum components are directly and indirectly more or less harmful for living organisms and surrounding environment (USEPA 2012; Varjani 2017a) as they are reported as potential carcinogens. Some are mutagenic and teratogenetic. The most dangerous PAHs are benzo[a]pyrene, benz[a]anthracene, and dibenz[a,h]anthracene (ATSDR 1995; USEPA 2012; Lammel et al. 2015; Varjani and Upasani 2017a). These compounds also appear to be less bioavailable (Varjani and Upasani 2017b). There solubility rate in water is very low (Samanta et al. 2002; Nikitha et al. 2017). In a study, benzo[a]pyrene was estimated to be responsible for 48–52% of the added risk of sediment and for 44–54% sediment in another. Benzo[a]pyrene is responsible for DNA binding, sister chromatid

Table 1 Classification of selected PAHs according to benzene ring arrangement

No.	Structure of PAHs		Name
1.	Linear	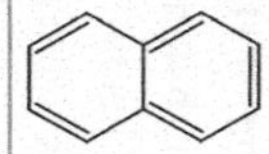 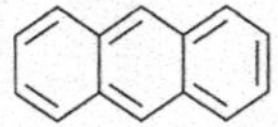 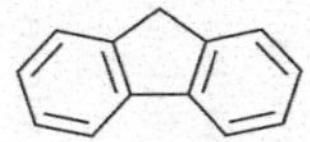	Naphthalene, Anthracene, Fluorene
2.	Angular	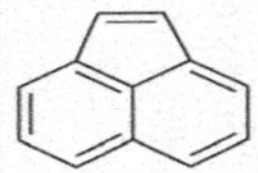 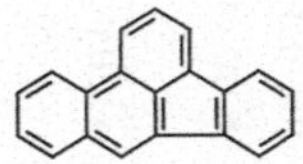 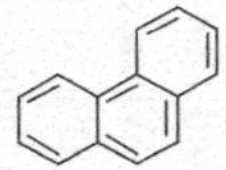	Acenaphthylene, Benzo[k]fluoranthene, Phenanthrene
		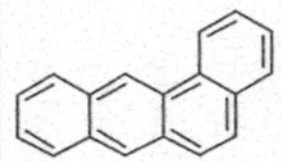 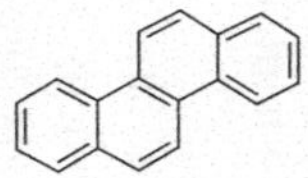 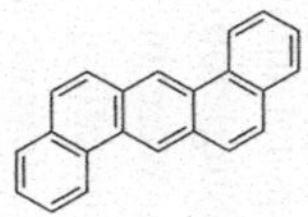	Benzo[a]pyrene, Chrysene. Dibenz[a,h]anthracene
		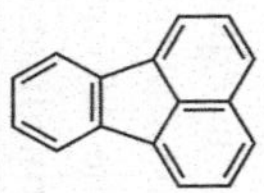 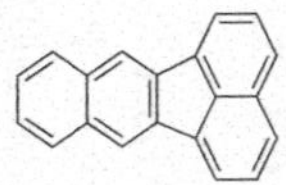 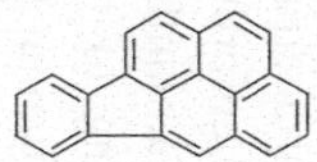	Fluoranthene, Benzo(k)fluoranthene, Indeno[1,2,3]pyrene
3.	Clustered	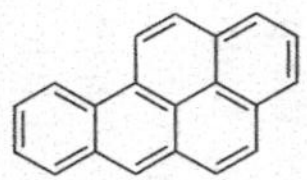 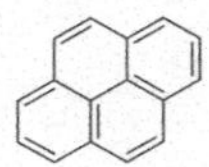 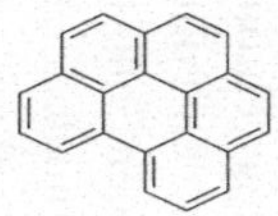	Benzo[a]pyrene Pyrene, Benzo[g,h]anthracene

Source Abdel-Shafy and Mansour (2016), Kim et al. (2013), Bojes and Pope (2007)

Table 2 General uses of some polyaromatic hydrocarbons (PAHs)

No.	Name of PAHs	Application	References
1.	Acenaphthene	Manufacture of dyes, plastics, pigments, pharmaceuticals, and pesticides	Abdel-Shafy and Mansour (2016)
2.	Anthracene	Manufacture of dyes and pigments; diluents for wood preservatives	Nikitha et al. (2017)
3.	Fluoranthene	Manufacture of dyes, pharmaceuticals, and agrochemicals	ATSDR (1995), Abdel-Shafy and Mansour (2016)
4.	Fluorene	Manufacture of dyes, pigments, pesticides, thermoset plastic, and pharmaceuticals	Abdel-Shafy and Mansour (2016), Nikitha et al. (2017)
5.	Phenanthrene	Manufacture of pesticides and resins	ATSDR (1995), Nikitha et al. (2017)
6.	Pyrene	Manufacture of pigments	ATSDR (1995), Abdel-Shafy and Mansour (2016), Nikitha et al. (2017)

exchange, chromosomal aberrations, point mutation, and transformations in mammalian cell cultures (Bamforth and Singleton 2005). When concentrated benz[a]anthracene and dibenz[a,h]anthracene were included 90% of the added risk could be estimated. According to this estimation, these compounds are very easily combined with environmental factors. Several attempts have been made to classify PAHs, and one of this is called as toxicological equivalent factor (TEF). Benzo[a]pyrene is often assumed to be the most harmful PAH (Kim et al. 2013; Sarria-Villa et al. 2015).

Generally, that PAHs compounds have more toxicity which contain more number of benzene rings. PAHs toxicity is measured by using LD50 values (Bamforth and Singleton 2005). In toxicological studies, it is also important to consider the possible toxic metabolites of PAHs. The first step in degradation of aromatic hydrocarbons in eukaryotes is formation of trans-dihydrodiol (Das and Chandran 2011; Gupte et al. 2016). These trans-dihydrodiols are in balance with trans-diol epoxides. The dihydrodiols are converted to catechols. Catechol is then converted into to intermediates of tricarboxylic acid cycle (ATSDR 1995). Toxic metabolites affect DNA which leads to DNA damage. These effects are shown in animals very slowly. The reactive species formed, quinones and trans-diol epoxides, can also cause DNA scission (Bamforth and Singleton 2005; Varjani and Upasani 2017a).

4 Effect of PAHs on Human Health and Environment

Petroleum hydrocarbon pollutants are mostly known as carcinogens (Nikitha et al. 2017). They can also damage other body parts of human beings. They are reported to be harmful for either short term or long term (Varjani et al. 2015; Varjani 2017a). They can lead to genetic and immunotoxic effects.

Short-term effects of petroleum hydrocarbon pollutants are symptoms such as eye irritation, nausea, skin irritation, vomiting, diarrhea, and inflammation (Kim et al. 2013; Yan et al. 2015; NHHS 2017; Varjani and Upasani 2017a). Anthracene, benzo(a)pyrene, and naphthalene are known as direct skin irritants. Anthracene and benzo(a)pyrene are reported as skin sensitizers means they can cause an allergic skin response in animals and humans (Oluwaseun et al. 2017).

Long-term effects persist for long time period and slowly damage body parts. The symptoms are categorized as chronic effect such as decreased immune function, cataracts, kidney and liver damage (e.g., jaundice), breathing problems, asthma-like symptoms, and lung function abnormalities. Repeated contact of PAHs with skin may induce redness and skin inflammation (ATSDR 1995; Ghosal et al. 2016). Naphthalene is responsible for red blood cells' breakdown if inhaled or ingested in large amounts (Bisht et al. 2010; Olajire and Essien 2014).

Genotoxic effect causes genetic damage in living beings. Mainly PAHs are not genotoxic by themselves. But they need to be metabolized to diol epoxides. Diol epoxides react with DNA, thus inducing genotoxic damage (Bamforth and Singleton 2005).

Mostly PAHs are organic compound which are playing role as carcinogenic compounds. Epoxides and dihydrodiols of some PAHs bind with cellular proteins and DNA. Because of this activity cell is damaged. It causes mutations, developmental malformations, tumors, and cancer (NAHH 2017). It also has increased risk of mostly skin, lung, bladder, gastrointestinal cancers, and stomach cancer, etc. PAHs are directly or indirectly connected in food, skin contacts, and respiration, etc. Hence it causes lung cancer by inhalation, stomach cancer by ingesting PAHs in food, and skin cancer by skin contact (Haritash and Kaushik 2009).

Different sources of PAHs and their effect(s) on human health are shown in Fig. 1. PAHs cause harmful effect on male or female fertility organs. Embryo-toxic effects occur due to benzo(a)anthracene, benzo(a)pyrene, and naphthalene. High amount of benzo(a)pyrene during pregnancy is responsible for birth defects and decreased body weight in the offspring in mice (ATSDR 1995; Digg et al. 2012). It is not known that this type of effect particularly occurs in humans or not. The effects of PAHs in living organisms are low birth weight, premature delivery and heart malformations, lower intelligence quotient at age three, increased behavioral problems at age six and eight, and childhood asthma (ATSDR 1995; Kim et al. 2013; Yan et al. 2015). Presence of PAHs is suppressing immune reaction in rodents. The defined mechanisms of PAH-induced immunotoxicity are still not clear; however, it appears that immunosuppressant which may be involved in mechanism by which PAH induces cancer (Kim et al. 2013).

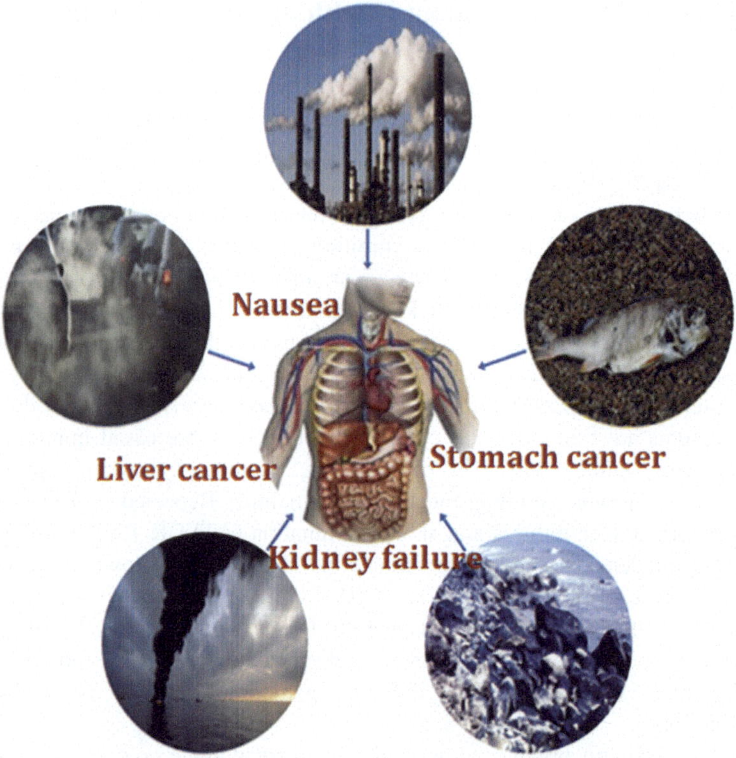

Fig. 1 Different sources of PAHs and their effect(s) on human health (Dudhagara et al. 2016)

5 Remediation of PAHs or Degradation of PAHs

Crude oil is a mixture of hydrocarbons composed of mainly heteroatomic and non-heteroatomic hydrocarbons (Arulazhagan and Vasudevan 2011; Kostka et al. 2011; Varjani and Upasani 2017a). PAHs typically enter in the environment through various natural and anthropogenic activities. Petroleum industry is one of them (Das and Chandran 2011; Lammel et al. 2015; Varjani and Upasani 2016a, b). PAHs are found in soil and river sediment near industrial sites (Wilson and Jones 1993; Yan et al. 2015). Oil spills, creosote, coal mining dust, and other fossil fuel sources also include pollution of PAHs in the environment (Okparanma et al. 2011 Varjani 2017b). Two- and three-ring PAHs can spread widely while dissolved in water or as gases in the atmosphere (Leahy et al. 1990; Sakari 2012). PAHs have higher molecular weight as compared to alkanes hence they can disperse locally or regionally, adhere to particulate matter that is suspended in air or water until the particles land or settle out of the water column (Bamforth and Singleton 2005; Haritash and Kaushik 2009). PAHs have a strong attraction for organic carbon, and thus highly organic sediments in rivers, lakes, and ocean can be a substantial sink for PAHs (Yan et al. 2015; Azab et al. 2016; Varjani 2017a).

Chemical, physical, and biological processes are used in degradation of PAHs (Chaillan et al. 2004; Ukiwe et al. 2013). All techniques are involved to decrease the level of hydrocarbon pollutants in the environment (Varjani and Upasani 2017a). Biological processes mainly apply living organisms for remediation of pollutants (Varjani and Upasani 2016c). In this process, plants are used which is known as phytoremediation (Ukiwe et al. 2013; Yan et al. 2015; Varjani and Upasani 2017b).

Biodegradation is one of the best suitable techniques for treating soil, water, or sediments contaminated with PAHs, which is used to degrade or detoxify environmental pollutants. It can be aerobic and/or anaerobic (Chaillan et al. 2004; Varjani 2017a). Biodegradation is considered as a clean up method that presents possibility to eliminate organic contaminants with help of natural biological activity available in substrate (Qi et al. 2017; Varjani et al. 2015). In this process, indigenous microorganisms are used to remove pollutants from the polluted site (Okparanma et al. 2011; Varjani and Upasani 2016c). Biodegradation leads complete mineralization of pollutants which is affected by different factors such as soil type, temperature, soil pH, type of pollutant(s), oxygen level of soil, nutrient content of soil (Zohair et al. 2006; Yan et al. 2015; Varjani 2017a). These factors are capable to degrade pollutants and show effect on microbial growth (Arulazhagan and Vasudevan 2011; Lammel et al. 2015; Varjani and Upasani 2017a).

There are number of bacteria living in environment which play role in degradation of PAHs. Some bacteria can utilize low molecular weight (LMW) PAHs as their carbon source. The main mechanism requires the presence of molecular oxygen which initiates enzymatic attack of PAHs ring (Butler and Mason 1997). In the starting phase, oxidation of arenes and catalyzed dioxygenase takes place in aerobic bacterial system. In this system, cis-dihydrodiols are the primary by-products by a multicomponent enzyme system (Shuttleworth and Cerniglia 1995). These dihydrodiols are replaced by intradiol or extradiol ring cleaving dioxygenases through either a meta-cleavage or an ortho-cleavage pathway. It plays an important role as center intermediates such as protocatechuate and catechols. These are converted into tricarboxylic acid (TCA) intermediates for further process (Shuttleworth and Cerniglia 1995; Gibson and Parales 2000). The bacterial naphthalene dioxygenase system is useful in oxidizing bi- and tri-cyclic PAH substrates, i.e., LMW PAHs such as naphthalene, anthracene, and phenanthrene. (Gibson and Parales 2000). High molecular weight (HMW) PAHs exhibit more persistence in contaminated environment and genotoxicity due to increased molecular weight. HMW PAHs contain more benzene ring, so these PAHs take more time period for degradation as compared with LMW PAHs, e.g., half-life of 3-ring molecule phenanthrene in soil and sediment range is 16–126 days, while half-life of 5-ring molecule benzo[a]pyrene range is 229–1400 days (Shuttleworth and Cerniglia 1995; Vasconcelos et al. 2011).

Schematic degradation pathway of PAHs by bacteria and fungi is shown in Fig. 2. The principal mechanism for aerobic bacterial metabolism of PAHs is the first oxidation of benzene ring by the action of dioxygenase enzymes and made *cis*-dihydrodiols (Rubailo and Oberenko 2008). Dihydrodiols are metabolized by series of enzymes converted to catechol to carbon dioxide and water. In the fungal degradation, it is categorized as two types of remediation (a) Lignolytic and (b) non-lignolytic fungi. Lignolytic fungi (white-rot fungi) produce lignin peroxidase enzymes. PAHs are converted into PAH-QUINONES by ring fission and release the CO_2 in the environment (Fig. 2). Non-lignolytic fungi (*Chrysosporium pannorum, Cunninghamella elegans*) produce endo-oxygenase enzymes which take part in oxidation of PAH compounds and covert into ARENE OXIDE. It is then converted into phenol and trans-dihydrodiol (Fig. 2). Organisms from the genus *Pseudomonas, Sphingomonas, Acinetobacter, and Rhodococcus* can oxidize naphthalene and other PAHs using dioxygenase enzymes (Okparanma et al. 2011; Varjani 2017a). Addition of straw, wood chips, and other lignin-rich substrates help in enhanced degradation of these pollutants by fungi. *Methanotrophs* also have ability to degrade PAHs via action of methane monooxygenase gene (Okparanmaet al. 2011; Lammel et al. 2015). Species of genera *Arthrobacter, Thiobacter*, *Bacillus, Stenotrophomonas, Escherichia,* and *Alcaligenes* are reported as anthracene and phenanthrene degraders (Gupte et al. 2016; Varjani 2017a). Some microorganisms from genera *Pseudomonas* and *Mycobacterium* are capable to transforming and degrading PAHs under aerobic conditions. It is also noted that

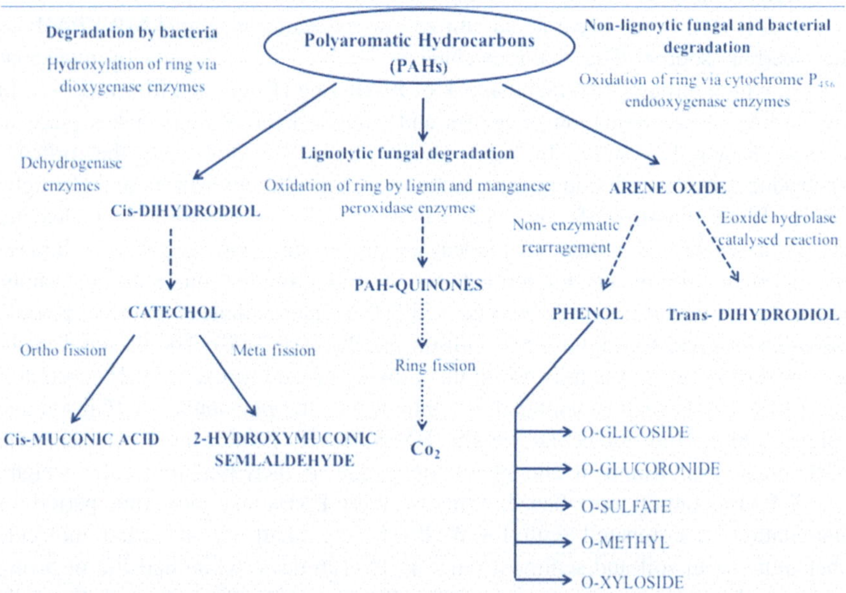

Fig. 2 Degradation pathway of PAHs by bacteria and fungi (*source* Bamforth and Singleton 2005; Gupte et al. 2016)

anthracene could be completely mineralized by *Sphingomonas, Nocardia, Beijerinckia, Paracoccus,* and *Rhodococcus* with dihydrodiol as initial oxygenated intermediate (Chaillan et al. 2004; Ukiwe et al. 2013).

The main purpose of remediation is not only to remove of pollutant(s) but also to maintain a quality of the environment. Bioavailability is an important factor for degradation process (Chaillan et al. 2004; Kim et al. 2013; Varjani and Upasani 2017a). Bioavailability of PAHs depends on biotic and abiotic factors interaction. Other living organisms are also used to remove and reduce the level of this type of pollutant in environment. Algae and some invertebrates such as *protozoans*, *mollusks,* and many *polychaetes* have limited ability to metabolize PAHs and bio-accumulate unequal concentrations of PAHs in their tissues. However, PAH metabolism can vary substantially across invertebrate species (Wilson and Jones 1993; Yan et al. 2015). Most vertebrates metabolize and excrete PAHs relatively quickly.

PAHs transform slowly to a wide range of degradation products. Biological degradation by microbes is a dominant form of PAH transformation in the environment. Soil-consuming invertebrates such as earthworms speedup PAH degradation, either through direct metabolism or by improving conditions for microbial transformations (Lundstedt et al. 2003; Stankovic et al. 2014). Abiotic degradation in atmosphere and top layers of surface waters can produce nitrogenated, halogenated, hydroxylated, and oxygenated PAHs; some of these compounds can be more toxic, water-soluble, and mobile than their parent PAHs (Sarria-Villa et al. 2015; Stankovic et al. 2014).

6 Conclusion

Polycyclic aromatic hydrocarbons (PAHs) are a group of compounds known as polycyclic organic matters (POMs) possessing potential health hazards. They play a vital role in environmental pollution and related problems. Besides other approaches, de-aromatization of these pollutants might be a good option to target and curb PAH pollution. It is a difficult task to remove these pollutants totally from the environment, but it is possible to reduce their quantity in the environment. Using living organisms to degrade PAHs and mineralize or transfer them into CO_2 and H_2O is nowadays considered as the safe option for their remediation. Voluminous research groups have been evolved to work for different bioremediation tools in the form of efficient bacteria and fungi as potential degraders. It is a big challenge for scientists working in the field of bioremediation to achieve complete bioremediation of PAH polluted site(s). More detailed research is necessary to determine exactly what is going on with PAHs in polluted environment. There are still various aspects of bioremediation of these persistent pollutants that remain unknown or otherwise have insufficient information, which requires future attention. The development and newer approaches focus to target specific PAHs. Development of precise, effective, and composite technology to treat complex mixtures is still a thrust area of research.

References

Abdel-Shafy HI, Mansour MSM (2016) A review on polycyclic aromatic hydrocarbons: source, environmental impact, effect on human health and remediation. Egypt J Pet 25:107–123

Agency for Toxic Substances and Disease Registry (ATSDR) (1995) Toxicological profile for polycyclic aromatic hydrocarbons (PAHs): U.S. Department of Health and Human Services, Public Health Service, Atlanta, GA. URL: https://www.atsdr.cdc.gov/toxprofiles/tp69.pdf. Last accessed: 18 Aug 2017

Arulazhagan P, Vasudevan N (2011) Biodegradation of polycyclic aromatic hydrocarbons by a halo tolerant bacterial strain *Ochrobactrum* sp. VA1. Mar Pollut Bull 62:388–394

Azab AM, Shaban WM, Zaki MS, Authman MMN, Zaher MFA (2016) Monitoring of petroleum hydrocarbons in sediment and gastropods from Suez Gulf. Red Sea Life Sci J 13(7):46–59

Bamforth SM, Singleton I (2005) A Review: Bioremediation of polycyclic aromatic hydrocarbons: current knowledge and future directions. J Chem Technol Biotechnol 80:723–736

Bisht S, Pandey P, Sood A, Sharma S (2010) Biodegradation of naphthalene and anthracene by chemo-tactically active rhizobacteria of populus deltoids. Braz J Microbiol 41:922–930

Bojes HK, Pope PG (2007) Characterization of EPAs 16 priority pollutant polycyclic aromatic hydrocarbons (PAHs) in tank bottom solids and associated contaminated soils at oil exploration and production sites in Texas. Regul Toxicol Pharmacol 47(3):288–295

Butler CS, Mason JR (1997) Structure-function analysis of the bacterial aromatic ring-hydroxylating dioxygenases. Adv Microb Physiol 38:47–84

Chaillan F, Flèche LA, Bury E, Phantavong YH, Grimont P, Saliot A, Oudot J (2004) Identification and biodegradation potential of tropical aerobic hydrocarbon-degrading microorganisms. Res Microbiol 155(7):587–595

Chiou TC, McGroddy ES, Kile ED (1998) Partition characteristics of polycyclic aromatic hydrocarbons on soils and sediments. Environ Sci Technol 32(2):264–269

Choi H, Harrison R, Komulainen H, Saborit MJ (2010) Polycyclic aromatic hydrocarbons WHO guidelines for indoor air quality: selected pollutants (https://www.ncbi.nlm.nih.gov/books/NBK138709/). Last accessed: 18 Aug 2017

Clemente RA, Torres-Palma RA, Peñuela AG (2014) Removal of polycyclic aromatic hydrocarbons in aqueous environment by chemical treatments: a review. Sci Total Environ 478:201–225

Das N, Chandran P (2011) Microbial degradation of petroleum hydrocarbon contaminants: an overview. In: Biotechnology research international 2010:13 (doi:https://doi.org/10.4061/2011/941810)

Detoxifying Heterocyclic Aromatic Amines (HCAs) and polycyclic aromatic hydrocarbons (PAHs) with Chlorella vulgaris, A natural approach to human health (NAHH), Biofoundation (https://biofoundations.org/detoxifying-heterocyclic-aromatic-amines-hcas-and-polycyclic-aromatic-hydrocarbons-pahs-with-chlorella-vulgaris/) (Last assessed: 18.08.2017)

Digg DL, Harris KL, Rekhadevi PV, Ramesh A (2012) Tumor microsomal metabolism of the food toxicant, benzo(a)pyrene, in ApcMin mouse model of colon cancer. Tumor Biol 33(4):1255–1260

Dudhagara DR, Rajpara RK, Bhatt JK, Gosai HB, Sachaniya BK, Dave BP (2016) Distribution, sources and ecological risk assessment of PAHs in historically contaminated surface sediments at Bhavnagar coast, Gujarat, India. Environ Pollut 213:338–346

Freeman Cattell D J (1990) Wood burning as a source of atmospheric polycyclic aromatic hydrocarbons. Environ Sci Technol 24:1581–1585

Ganesh KA, Vijayakumar L, Joshi G, Magesh Peter D, Dharani G, Kirubagaran R (2014) Biodegradation of complex hydrocarbons in spent engine oil by novel bacterial consortium isolated from deep sea sediment. Biores Technol 170:556–564

Ghosal D, Ghosh S, Dutta T K, Ahn Y (2016) Current state of knowledge in microbial degradation of polycyclic aromatic hydrocarbons (PAHs): a review. Front Microbiol 7:1–27 (Article 1369) (https://doi.org/10.3389/fmicb.2016.01369)

Gibson DT, Parales RE (2000) Aromatic hydrocarbon dioxygenases in environmental biotechnology. Curr Opin Biotechnol 11(3):236–243

Gupte A, Tripathi A, Patel H, Rudakiya D Gupte S (2016) Bioremediation of polycyclic aromatic hydrocarbon (PAHs): a perspective. Open Biotechnol J 10:363–378

Haritash AK, Kaushik CP (2009) Biodegradation aspects of polycyclic aromatic hydrocarbons (PAHs): a review. J Hazard Mater 169:1–15. http://biomonitoring.ca.gov/downloads/polycyclic-aromatic-hydrocarbons-pahs-fact-sheet. Polycyclic Aromatic Hydrocarbons (PAHs) Fact Sheet, biomonitoring, California. Last accessed: 18 Aug 2017

Kim K, Jahan SA, Kabir E, Brown RJC (2013) A review of airborne polycyclic aromatic hydrocarbons (PAHs) and their human health effects. Environ Int 60:71–80

Kostka JE, Prakash O, Overholt WA, Green SJ, Freyer G, Canion A, Delgardio J, Norton N, Hazen TC, Huettel M (2011) Hydrocarbon-degrading bacteria and the bacterial community response in gulf of Mexico beach sands impacted by the deepwater horizon oil spill. Appl Environ Microbiol 77:7962–7974

Lammel G, Dvorska A, Klanova J, Kohoutek J, Kukuka P, Prokes R, Sehili MA (2015) Long-range atmospheric transport of polycyclic aromatic hydrocarbons is worldwide problem —results from measurements at remote sites and modelling. Acta Chim Slov 62:729–735

Leahy JG, Colwell RR (1990) Microbial degradation of hydrocarbons in the environment. Microbiol Rev 54:305–315

Lundstedt S, Haglund P, Oberg L (2003) Degradation and formation of polycyclic aromatic compounds during bioslurry treatment of an aged gasworks soil. Environ Toxicol Chem 22 (7):1413–1420

Mohsen A, Simin N, Chimezie A (2009) Biodegradation of polycyclic aromatic hydrocarbons (PAHs) in petroleum contaminated soils. Iran J Chem Eng 28(3):53–59

Nikitha T, Satyaprakash M, Satya Vani S, Sadhana B, Padal SB (2017) A review on polycyclic aromatic hydrocarbons: their transport, fate and biodegradation in the environment. Int J Curr Microbiol Appl Sci 6(4):1627–1639

Okparanma RN, Ayatamuno MJ, Davis DD, Allagoa M (2011) Mycoremediation of polycyclic aromatic hydrocarbons (PAH) contaminated oil-based drill-cuttings. Afr J Biotech 10 (26):5149–5156

Olajire A, Essien JP (2014) Aerobic degradation of petroleum components by microbial consortia. J Pet Environ Biotechnol 5. doi:https://doi.org/10.4172/2157-7463.1000195

Oluwaseun O, Alegbeleye A, Opeolub OB, Jackson V (2017) Bioremediation of polycyclic aromatic hydrocarbon (PAH) compounds: (acenaphthene and fluorene) in water using indigenous bacterial species isolated from the Diep and Plankenburg rivers, Western Cape, South Africa. Braz J Microbiol 48:314–325

Qi Y, Wang C, Lv C, Lun Z, Zheng C (2017) Removal capacities of polycyclic aromatic hydrocarbons (PAHs) by a newly isolated strain from oil field produced water. Int J Environ Res Public Health 14:215

Rubailo AI, Oberenko AV (2008) Polycyclic aromatic hydrocarbons as priority pollutants. J Sib Fed Univ Chem 4:344–354

Sakari M (2012) depositional history of polycyclic aromatic hydrocarbons: reconstruction of petroleum pollution record in peninsular Malaysia. Water Research Unit & School of Science and Technology, Malaysia CHE-6 (http://cdn.intechopen.com/pdfs/29372/InTechDepositional_history_of_polycyclic_aromatic_hydrocarbons_reconstruction_of_petroleum_pollution_record_in_peninsular_malaysia.pdf) (Last assessed: 09/06/2017)

Samanta SK, Singh OV, Jain RK (2002) Polycyclic aromatic hydrocarbons: environmental pollution and bioremediation. Trends Biotechnol 20(6):243–248

Sarria-Villa R, Ocampo-Duque W, Páez M, Schuhmacher M (2015) Presence of PAHs in water and sediments of the Colombian Cauca river during heavy rain episodes, and implications for risk assessment. Sci Total Environ 540:455–465

Shuttleworth KL, Cerniglia CE (1995) Environmental aspects of PAH biodegradation. Appl Biochem Biotechnol 54(1–3):291–302

Stankovic D, Krstic B, Nikolic N (2014) Effect of traffic on the soil contamination with polycyclic aromatic hydrocarbons PAHs. Biotechnol Equip 22(2):736–741

Stogiannidis E, Laane R (2015) Source characterization of polycyclic aromatic hydrocarbons by using their molecular indices: an overview of possibilities. In: Whitacre DM (ed) Reviews of environmental contamination and toxicology. Springer International Publishing Switzerland (https://doi.org/10.1007/978-3-319-10638-0_2)

Ukiwe LN, Egereonu UU, Njoku CP, Nwoko IAC, Allinor IJ (2013) Polycyclic aromatic hydrocarbons degradation techniques: a review. Int J Chem 5(4):43–55

U.S. Environmental Protection Agency (2012) 2012 edition of the drinking water standards and health advisories: Office of Water, U.S. Environmental Protection Agency, Washington, D.C. EPA 822-S-12-001. URL: https://www.epa.gov/sites/production/files/2015-09/documents/dwstandards2012.pdf. Last accessed: 18 Aug 2017

Varjani SJ (2017a) Review on microbial degradation of petroleum hydrocarbon. Biores Technol 223:277–286

Varjani SJ (2017b) Biodegradation of petroleum hydrocarbon pollutants in marine environments. In: Prasad R, Kumar N (eds) Microbes and sustainable agriculture. IK International, New Delhi, pp 89–99

Varjani SJ, Upasani VN (2016a) Core Flood study for enhanced oil recovery through ex-situ bioaugmentation with thermo- and halo-tolerant rhamnolipid produced by *Pseudomonas aeruginosa* NCIM 5514. Biores Technol 220:175–182

Varjani SJ, Upasani VN (2016b) Carbon spectrum utilization by an indigenous strain of *Pseudomonas aeruginosa* NCIM 5514: production, characterization and surface active properties of biosurfactant. Biores Technol 221:510–516

Varjani SJ, Upasani VN (2016c) Biodegradation of petroleum hydrocarbons by oleophilic strain of *Pseudomonas aeruginosa* NCIM 5514. Biores Technol 222:1950201

Varjani SJ, Upasani VN (2017a) Review on A new look on factors affecting microbial degradation of petroleum hydrocarbon pollutants. Int Biodeterior Biodegrad 120:71–83

Varjani SJ, Upasani VN (2017b) Critical review on biosurfactant analysis, purification and characterization using rhamnolipid as a model biosurfactant. Biores Technol 232:389–397

Varjani SJ, Upasani VN (2017c) Crude oil degradation by *P. aeruginosa* NCIM5514: influence of process parameters. Indian J Exp Biol 55:493–497

Varjani SJ, Rana DP, Jain AK, Bateja S, Upasani VN (2015) Synergistic ex-situ biodegradation of crude oil by halotolerant bacterial consortium of indigenous strains isolated from on shore sites of Gujarat, India. Int Biodeterior Biodegrad 103:116–124

Vasconcelos U, Franca FP, Oliveira FJS (2011) Removal of high-molecular weight polycyclic aromatic hydrocarbons. Quim Nova 34(2):218–221

Wilson SC, Jones KC (1993) Bioremediation of soil contaminated with polynuclear aromatic hydrocarbons (PAHs): a review. Environ Pollut 81(3):229–249

Yan Z, Jiang J, Cai H, Zhou Y, Lee Y, Krumholz LR (2015) Complex interactions between macrophyte *Acorus calamus* and microbial fuel cells during pyrene and benzo(a)pyrene degradation in sediments. Sci Rep (https://doi.org/10.1038/srep10709)

Zohair A, Salim A, Adeola A, Beck JA (2006) Residues of polycyclic aromatic hydrocarbons (PAHs), polychlorinated biphenyls (PCBs) and organochlorine pesticides in organically farmed vegetables. Chemosphere 63:541–553

Author Biographies

Dr. Sunita Varjani is Scientific Officer at Gujarat Pollution Control Board, Gandhinagar, Gujarat, India. Her major areas of research are Industrial and Environmental Microbiology/Biotechnology and Molecular biology. She has delivered talks as invited speaker in many international conferences including the countries India and China. She is recipient of visiting scientist fellowship from EPFL, Lausanne, Switzerland (2017).

Rutu R. Joshi is presently a trainee at Solid and Hazardous Waste Management Division of CSIR-NEERI, Nagpur, Maharashtra, India. She has earned M.Sc. in Environmental Science from School of Biological Sciences and Biotechnology, University and Institute of Advanced Research, Gandhinagar-382007, Gujarat, India.

Dr. P. Senthil Kumar is currently working as Associate Professor in the Department of Chemical Engineering, Chennai, Tamilnadu, India. His research area is Waste Water Treatment and Alternative Fuels. He recently received "IEI Young Engineers Award 2017-2018" in Environmental Engineering Discipline and received "Hiyoshi Think of Ecology Award 2017".

Prof. Vijay Kumar Srivastava is a Professor and Dean at School of Technology, GSFC University, Vadodara, Gujarat, India. His current research expertise is in solid waste management, waste water treatment and second generation biofuels. He has published 43 research papers, 2 books, number of chapters and technical articles. He is Fellow and Life member of 15 professional bodies.

Dr. Vipin Kumar is an Assistant Professor in the Department of Environmental Science and Engineering at Indian Institute of Technology (Indian School of Mines), Dhanbad, India. Dr. Vipin Kumar's research interest includes microbial remediation of pollutants and biological treatment of solid waste.

Dr. Chiranjib Banerjee is an INSPIRE Faculty in the Department of Environmental Science and Engineering at Indian Institute of Technology (Indian School of Mines), Dhanbad, India. Dr. Banerjee's research interest includes algal biotechnology and microbial remediation of pollutants.

Dr. R. Praveen Kumar is working as Head, Department of Biotechnology, Arunai Engineering College, Tiruvannamalai, Tamil Nadu, India. His area of research includes Renewable Energy from Biomass and Municipal Waste. He has done provisional registration of four patents and published more than 52 papers in peer-reviewed journals. He was awarded with Outstanding Young Investigator.

Chapter 10
Bioreduction of Hexavalent Chromium Using Moderate Thermophilic and Thermophilic Microorganisms

Ana B. Segretin and Edgardo R. Donati

Abstract Hexavalent chromium is one of the toxic heavy metals considered to be potentially harmful for life of a wide range of organisms, including human life, due to its high mobility in soil and water media. Wastewaters containing chromium(VI) are produced from different sources including many industries such as electroplating, leather tanning procedures, and chromite ore processing. When these practices are not conveniently regulated, contamination problems associated with irresponsible waste disposal of effluents appear in the surrounding environments, such as water bodies and sediments. Conventional treatment of chromium(VI) implies its reduction to chromium(III)—a less toxic species—and its later precipitation. This process can be carried out through widely known physicochemical techniques although it is expensive mainly when low concentrations of the pollutant need to be treated. Microorganism-based techniques are considered very cost-effective alternatives, especially for the first step of the treatment, i.e., chromium(VI) reduction. Isolation and identification of chromium(VI)-resistant and chromium(VI)-reducing strains are fundamentally significant for this purpose. Microorganisms capable of living when exposed to high temperatures seem to be great candidates for bioremediation applications due to the increase of the speed of the process. The objective of the present chapter is to study thermophilic chromate-resistant and chromate-reducing bacteria toward Cr(VI) bioreduction. Cr (VI) removal by immobilized bacterial cells and their advantages are also summarized for a better understanding on how to transfer this technology from laboratory to a large-scale application for wastewater treatments. An example of bioreduction of chromium(VI) by immobilized bacteria isolated from the geothermal area of Domuyo (Neuquén-Argentina) is presented as case study.

Keywords Bioremediation · Chromium reducing bacteria · Thermophilic microorganisms · Hexavalent chromium · Immobilization · Wastewater

---

A. B. Segretin · E. R. Donati (✉)
Centro de Investigación y Desarrollo en Fermentaciones Industriales
(CCT La Plata-CONICET; UNLP), Universidad Nacional de La Plata,
Facultad de Ciencias Exactas, La Plata, Argentina
e-mail: donati@quimica.unlp.edu.ar

S. J. Varjani et al. (eds.), *Waste Bioremediation*, Energy, Environment,
and Sustainability, https://doi.org/10.1007/978-981-10-7413-4_10

1 Introduction

Heavy metal contamination of water has become a serious problem to the environment. The anthropogenic activities such as mining, processing, and application metals on diverse industrial activities have greatly increased during the past few years and have lead to several contamination problems. Over the last few decades, the unregulated discharge of both liquid and solid waste products having high levels of heavy metals that come from some industrial processes have become evident and consequently, legislation focus toward the protection of the environment has become more rigid. Even if control measures exist, it is still difficult to manage metal contamination; hence, their removal and remediation have become a priority concern.

Chromium is a heavy metal widely used in industrial processes. According to Dhal et al. (2013), 90% from the total chrome ore is used in diverse metallurgical industries such as steel, alloy, and nonferrous alloy production. Refractory and chemical industries only represent 5% of the total chrome consumption. Between these last applications, the most notable are iron and steel production, cement, glass, ceramics, leather tanning, plating, wood preservation, and pigment industries. The particular properties of chromium such as its resistance to ordinary corrosive agents make it a good material in electroplating to develop protective coatings and in alloys such as stainless steel that results in a corrosion and oxidation resistant material. Between the chemicals produced based on chromium, the largest consumption is for the production of paint and ink pigments. Other minor applications are leather tanning, drilling mud, catalysts, wood, and water treatment (Dhal et al. 2013).

Chromium is the 24th element on the periodic chart, and it is the 21st most abundant element in the Earth's crust at about 100 ppm (Barnhart 1997). The two most common oxidation states of Cr present in the environment are Cr(III) and Cr (VI). They differ in charge, chemical and biochemical reactivity and physico-chemical properties. Cr(III) is essential for certain processes of living organisms as, for example, the control of glucose and lipid metabolism in mammals (Losi et al. 1994). Conversely, Cr(VI) is usually toxic on biological systems. It was found that exposure to hexavalent chromium for long periods of time leads to a variety of health harms. Chromate represents a toxic element because it can freely diffuse across cell membranes and it has a strong oxidative potential causing severe cell damage. The formation of free radicals during the reduction of Cr(VI)–Cr(III) that may occur inside the cell as a mechanism of tolerance is also detrimental to cell functioning processes. Furthermore, when a significant concentration of Cr(III) is formed inside the cell, additional adverse effects may take place because of its high capability to coordinate various organic compounds resulting in inhibition of some metalloenzyme systems (Reynolds et al. 2009). As regards the effects on the environment, microbial communities present in contaminated soils can be modified by affecting their growth and activity (Turpeinen et al. 2004). Hence, it may be of priority concern to develop remediation strategies for the removal of chromium contaminants, especially the toxic Cr(VI) form.

1.1 Remediation of Cr Contaminants: Physicochemical Versus Biological Methods

Removal of chromate by conventional techniques involves a chemical reduction first step followed by alkali precipitation or the use of an ion exchange or adsorption matrix. These procedures result in large amounts of sludge and, in the particular case of the use of specific capture materials, it is a very expensive strategy and presents diverse interferences in the presence of other heavy metals. Since none of these alternatives are cost-effective when thinking about remediation strategies, special attention has been put on biological-based methods.

Biotransformation and biosorption are the most commonly used technologies. As regards biotransformation, the reduction of hexavalent chromium to a less toxic reduced form (trivalent chromium) is considered the best alternative. In the case of biosorption, chromate is accumulated inside or on the surface of the biomass used which has to be disposed off. This represents a disadvantage over the biotransformation in which the contaminant is converted to an innocuous or less toxic form. Chromate reduction takes place by different mechanisms. Under aerobic conditions, microbial chromate reduction is usually carried out by soluble reductases. When the reduction takes place in the absence of oxygen, both soluble and membrane-bound reductases have been found to be implicated. A large array of compounds have been identified to function as electron donors during the reduction process (i.e., carbohydrates, proteins, fats, hydrogen, and NAD(P)H) (Kanmani et al. 2012).

Since the discovery of the first bacteria, *Pseudomonas dechromaticans*, with chromate-reducing activity, in the 1970s (Romanenko and Korenkov 1977), research for finding new microorganisms which present similar mechanisms potentially useful on bioremediation strategies has continued. Much information has been gathered on the topic during the last years of investigation, and Cr(VI)-reducing bacteria belonging to diversified genera have been isolated from a variety of environments, encouraging the development of new technologies for remediation of chromium-contaminated sites (Kanmani et al. 2012).

1.2 Thermophilic Cr(VI) Reducer Microorganisms

Temperature is a key parameter that must be taken into account while dealing with microorganism-based remediation technologies. Cell growth may be significantly affected beyond the microorganism optimum temperature affecting the viability of the cells. On one hand, low temperatures may significantly decrease the fluidity of the cell membrane which is detrimental for the correct functioning of the substrate transport systems. This effect usually results in drastic reduction of cell growth rates. On the other hand, the exposure of microorganisms to temperatures higher to the one of optimum growth affects proteins by causing thermal denaturation, which is in most cases irreversible. The loss of protein structure causes enzymes to fail on their specific functions, including chromate reductase activity is usually lost. Other detrimental effects of high-temperature exposure are alteration of membrane arrangement and inactivation of the synthesis of protein due to alterations on ribosome conformation. As Cr(VI) reduction is usually enzyme mediated, the effects of temperature on the elements involved in this mechanism must be studied. The optimum temperature for Cr(VI) reduction depends mainly on the nature of the species and their capability of adaptation that to deal with temperature changes.

There are many microorganisms which can sustain high-temperature conditions. They have specific membrane composition and enzymes capable of function at high temperatures without suffering drastic denaturation processes. These thermophilic microorganisms have been found to be useful in many bioremediation processes, mainly in those where expensive cooling systems are needed. If microorganisms can develop on high-temperature conditions, these systems can be avoided, which is very convenient to reduce remediation costs and increase the process efficiency.

Some thermophilic bacteria have demonstrated to be successful on reducing chromate at high-temperature conditions. Many of them show a wide range of temperature tolerance which makes them interesting for bioremediation, as they do not require a strict temperature control of the process.

Bhowmick et al. (2009) found that some *Thermoanaerobacter* strains could reduce Cr(VI) successfully under thermophile conditions being its optimum temperature of 60 °C and optimum pH at 6.5. When doing the study of fractionation of the cell-free extracts on these bacteria, it was determined that chromate reduction activity was present in both the cytoplasm and the cell membrane fractions (Bhowmick et al. 2009). *Thermus scotoductus* is another example of bacteria isolated from a gold mine in South Africa that has shown to be able to reduce a variety of metals, including Cr(VI). These microorganisms are able to reduce Cr(VI) aerobically using a complex organic medium amended with chromate up to concentrations of 0.5 mM. Suspension of *T. scotoductus* cells has been also found to reduce Cr(VI) aerobically under nongrowth conditions at optimum temperature and pH of 80 °C and 7, respectively. Additionally, it was determined that the Cr(VI) reduction was performed by a cytoplasmic, constitutively expressed enzyme. Another thermophilic microorganisms found with chromate-reducing activity is *Methanothermobacter thermautotrophicus*, an obligate thermophilic methanogen; Experiments with several chromate concentrations (0.2, 0.4, 1, 3, and 5 mM) in the form of potassium dichromate ($K_2Cr_2O_7$) have been done achieving complete reduction at 0.2 and 0.4 mM Cr(VI) concentrations. However, reduction decreases to an extent of 43.6, 13.0, and 3.7% at higher chromate concentrations of 1, 3, and 5 mM, respectively, probably caused by toxic effect of aqueous chromate to cells at these concentrations (Singh et al. 2015). A Cr(VI)-tolerant, *Bacillus dabaoshanensis* sp. nov. is a rod-shaped, endospore-forming, and facultative anaerobic bacteria. It was first isolated from a heavy-metal-contaminated soil. It presents high tolerance to chromate with a minimum inhibitory concentration of 600 mg L^{-1}, and it showed to successful on reducing Cr(VI) under both aerobic and anaerobic conditions at a pH range of 5.5–10.0 (optimum pH 7.0) and a very wide range of temperatures, from 15 to 50 °C, being its optimum condition between 30 and 37 °C. Although it is not a strictly thermophilic bacteria, its chromate-reducing activity works at moderate thermophilic conditions (Cui et al. 2015). Another member of the *Bacillus* genus, *Bacillus thermoamylovorans* SKC1, first isolated from a municipal sewage, was reported have chromate reduction activity using L-arabinose as electron donor under an optimum temperature of 50 °C and neutral pH. Experiments showed that this microorganisms were able to reduce Cr(VI) having initial concentrations of chromate up to 150 mg L^{-1}. In addition to chromium, this strain was found capable of reducing selenite and tellurite, as well as soluble forms of Fe(III) (Slobodkina et al. 2007).

Reduction of Cr(VI) to the less toxic oxidation state has been found to be an additional chromosome-encoded chromate resistance mechanism in most of the cases studied, in addition to the plasmid-encoded metal tolerance (Viti et al. 2003). The tolerance/resistance parameter is not absolute, but it depends on the medium used (Mergeay 1995). Numerous chromium-resistant bacteria are isolated, and each species varies in their degree of resistance. This difference is due to the capability of a particular strain to resist the toxicity. Generally, toxicity and reduction experiments are performed using nutrient-rich media because it encourages bacterial growth (Rehman et al. 2008; Shakoori et al. 2000). Unfortunately, in such media interferences can take place due to complexation of Cr(VI) with organic components, thus masking or underestimating the real toxicity of Cr(VI) (Megharaj et al. 2003).

Natural scenarios having chromate contamination problems are usually complex in cation composition. This makes it important to test bacteria under different metal concentrations to infer how they may behave under the exposure of real contaminated effluxes. Bacteria being able to deal with the presence of other contaminants are advantageous because they can perform the desired reductive activity while withstanding the presence of these additional metallic ions (Srinath et al. 2001).

2 Case Study: Bioreduction of Cr(VI) Using an Immobilized *Paenibacillus Dendritiformis* Strain

2.1 Introduction

The bioremediation of Cr(VI)-contaminated wastewater can present several difficulties because of Cr(VI) toxicity to microorganisms, which may cause cell damage and loss of activity. Whole cell immobilization has some advantages over free cells such as being more stable and ease of regeneration because the depletion of nutrient source is minimal by having a continuous feeding system. Furthermore, high biomass loading, possibility of reuse, and elimination of cell separation steps are advantageous when thinking about large-scale optimization processes (Katiyar and Katiyar 1997).

Although the many advantages mentioned above, immobilization like systems can present new challenges when dealing with living microorganisms. The poor diffusion of oxygen and nutrients inside the matrix and across the column support may cause metabolism depletion affecting the chromate reduction activity. Benazir et al. (2010) found that Cr(VI) reduction by immobilized *Bacillus* cells was similar to that of free cells but the immobilized cells of *Pseudomonas aeruginosa* exhibited little lesser reduction efficiency than the free cells. Tucker et al. (1998) also reported that Cr(VI) was removed by immobilized cells at considerably slower rates than by suspended cells, suggesting that transport limitations may be controlling the rate of removal of these metals in immobilized cell systems.

Nevertheless, Farag and Zaki (2010) who worked with immobilized cells of *Pseudomonas* S4, found that immobilized cells could be reused three times without losing their chromium reduction activity, with each cycle spanning for three days achieving 80% of reduction efficiency of Cr(VI). Hence, they concluded that the use of immobilized cells is preferable due to its capability for using several times without losing efficiency, which makes it more economical.

The choice of the immobilization support is a key factor when using microorganisms for environmental bioremediation on columns or bioreactors. Support materials can be distinguished in two large groups, natural gels such as agar, carrageenan, alginate, and synthetic matrices like polyurethane, polyethylene glycol, and polyacrylamide. Natural gel matrices are biodegradable and subject to abrasion, whereas synthetic gels are not biodegradable and have better mechanical properties. On this work, calcium alginate has been chosen as the immobilization matrix.

Immobilized cell systems are a relevant strategy for industrial bioremediation processes based on their prolonged efficiency and for being more economical and environmentally friendly than the classic remediation techniques. Research on the transfer limitations in these systems and methods to overcome these limitations by employing better supports and better reactors are of prime necessity.

The present study focuses on a chromate reduction system using immobilized moderate thermophilic strain of *Paenibacillus dendritiformis*. This strain was isolated from a hot spring belonging to the geothermal area of Domuyo in the province of Neuquen, Argentina.

Domuyo geothermal area is emplaced in the southern slope of the Domuyo hill, the highest peak in Patagonia with an elevation of 4.709 m.a.s.l. Domuyo is not a stratovolcano; however, there are magmatic chambers near the base of the hill that control the geothermal activity of the area (Pesce 2010). There the surface manifestations that include fumaroles, hot springs, and geysers are mostly of neutral pH and high temperatures. So far, none microbial biodiversity studies have been done in this area. This work focuses on the application of a thermophilic *P. dendritiformis* strain isolated on this environment for the reduction of Cr(VI) using an immobilized cell system on a packed column (Fig. 1).

2.2 Materials and Methods

2.2.1 Sample Collection and Enrichment Conditions

The Cr(VI)-resistant and reducing bacterial strain of *P. dendritiformis* was isolated from a hot spring sample a geothermal area of Domuyo in the province of Neuquén in Argentina. Water samples were collected from the manifestation of Los Tachos (abbreviated "LT"). Los Tachos is located on the valley of the Cavunco stream; it is a wide, high temperature hot spring of water–vapor mixed type with sodium chloride, as its major component (Mas 2010). Figure 2 shows images of the sampling location.

Samples were collected in sterile plastic jars and kept at room temperature. Temperature and pH at the sampling sites were measured in situ with a Hanna HI

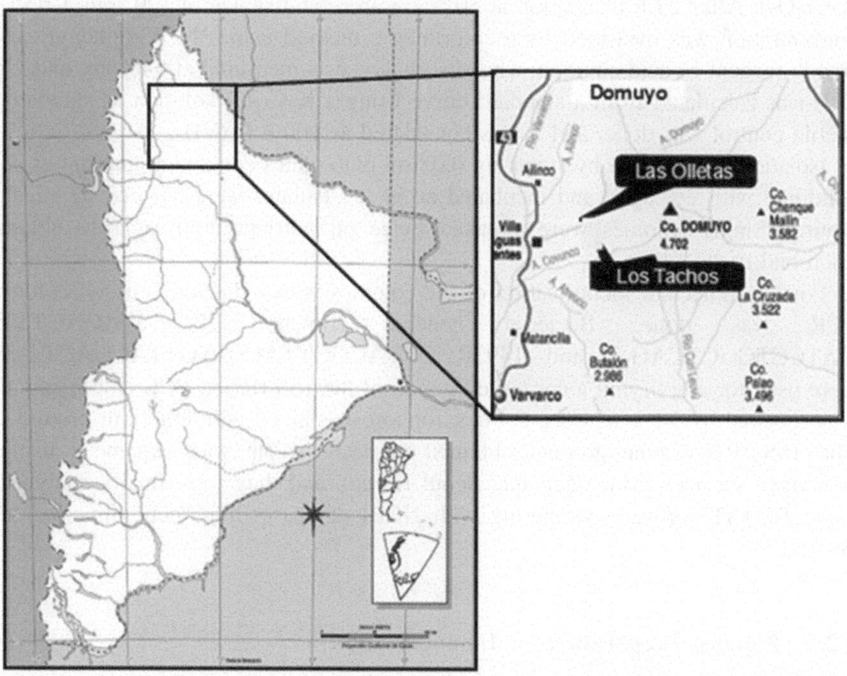

Fig. 1 Location of the geothermal area of Domuyo and the sampling places in a map of Neuquén province and Argentina

Fig. 2 Isolation of the Cr(VI) resistant-reducing *Paenibacillus dendritiformis* from a geothermal area of Domuyo in the province of Neuquén in Argentina

8424 portable instrument properly calibrated against calibration standards. The pH value of the sample collected was 7.0, and the temperature was 57 °C. Enrichment cultures on nutritive broth (Biokar) were made using the water sample from Los Tachos at 50 °C and pH7.

2.2.2 Evaluation of Chromate Reduction Activity, Isolation of Strains and Molecular Identification

Chromium reduction activity was tested on 20 ppm Cr(VI) nutritive broth (as $K_2Cr_2O_7$). After 24 h incubation at 50 °C in an orbital shaker at 120 rpm, Cr(VI) concentration was measured by a colorimetric method using 'N, N-diphenyl carbazide reagent in acid solution. The absorbance was measured at 540 nm, and Cr (VI) was calculated from a standard curve using a $K_2Cr_2O_7$ solution as standard. Sterile control was done, and it was considered as 100% Cr(VI).

Isolations were made by streaking 0.05 ml onto plates containing nutrient broth solidified with 2% agar, and incubated at 50 °C. Isolates were purified by streak plating. Single colonies were streaked twice on nutrient broth agar to obtain microbial monoculture.

For the molecular identification of the colonies with reduction activity, colony PCR was done. Bacterial general primers 27F: 5′-AGAGTTT GATCCTGGCTCAG-3′ and 1492R: 5′-TACGGTTACCTTGTTACGACTT-3′ were used for amplifying a conserved region of the 16S rRNA. PCR amplification was checked by 1.5% agarose gel electrophoresis stained with ethidium bromide. The 16S rRNA gene product obtained of each isolate was sequenced using Macrogen services (Macrogen Inc, Seoul Korea), and they were finally analyzed using BLAST software comparing with NCBI database (http://ncbi.nlm.nih.gov/BLAST).

2.2.3 Biomass Preparation for Immobilization

For the immobilization of cells, 3.5 L of *P. dendritiformis* culture was performed on Erlenmeyer flasks containing nutrient broth amended with 50 ppm Cr(VI) as

$K_2Cr_2O_7$ at 50 °C under shaking conditions (120 rpm) till mid-log phase (18 h). Cells were harvested under sterile conditions by centrifugation (8000 rpm) at 4 °C for 15 min, washed carefully with sterile distillated water and re-suspended in 19 mL of nutrient broth before being used for immobilization.

2.2.4 Cell Immobilization Process

Concentrated biomass was added into sterilized 1.5% sodium alginate solution under sterile conditions. To prepare the gel beads, the alginate–cell mixture was dripped into sterile 2% $CaCl_2$ solution. The beads were then hardened by resuspending them into a fresh 2% $CaCl_2$ sterile solution overnight at 4 °C. Bead mechanical resistance assessment was performed by exposing fifty alginate beads in a shaking flask with 50 mL of nutrient broth and kept them at 50 °C under shaking at 150 rpm for 24 h. Then, the content was filtered, and counting of the number of gel beads was performed. Mechanical resistance of alginate beads was calculated and expressed as a "fracture frequency" parameter, as it is showed in the equation below.

$$\text{Fracture frequency } \% = \frac{N}{Nt} \times 100$$

where *N* represents the number of fractured beads and *Nt* is the initial number of gel beads put on the Erlenmeyer flask.

2.2.5 Preparation of Packed Column System

The continuous reactor used in this study was constructed from 18.5 cm long and 3.5 cm internal diameter Pyrex glass column. It was packed with 88 g of *Paenibacillus dendritiformis* immobilized alginate beads mixed with 59 g of glass beads of 1 mm diameter. The total volume of the column was 160 mL, and the bed volume was 52 mL. An up-flow mode was used during the assay, and the flow rates used were 3.8 and 9.8 mL h^{-1}, which were regulated by peristaltic pump corresponded to retention times of 13 and 5 h, respectively. The column was aerated by an air pump, and the medium at the reservoir was vigorously mixed with the air using a magnetic stirrer to favor oxygen dissolution. Effluent was taken out through an outlet port at the top of the reactor. The column reactor was operated under 50 ° C inside an incubator machine. The feed Cr(VI) concentration placed at the reservoir was 50 ppm The reactor was operated with nutrient broth amended with Cr(VI) at desired concentration at pH 7. Samples collected were centrifuged at 6000 rpm for 5 min, and the supernatant was used to determine Cr(VI) concentration. Figure 3 describes the reactor designed for the Cr(VI) reduction assays.

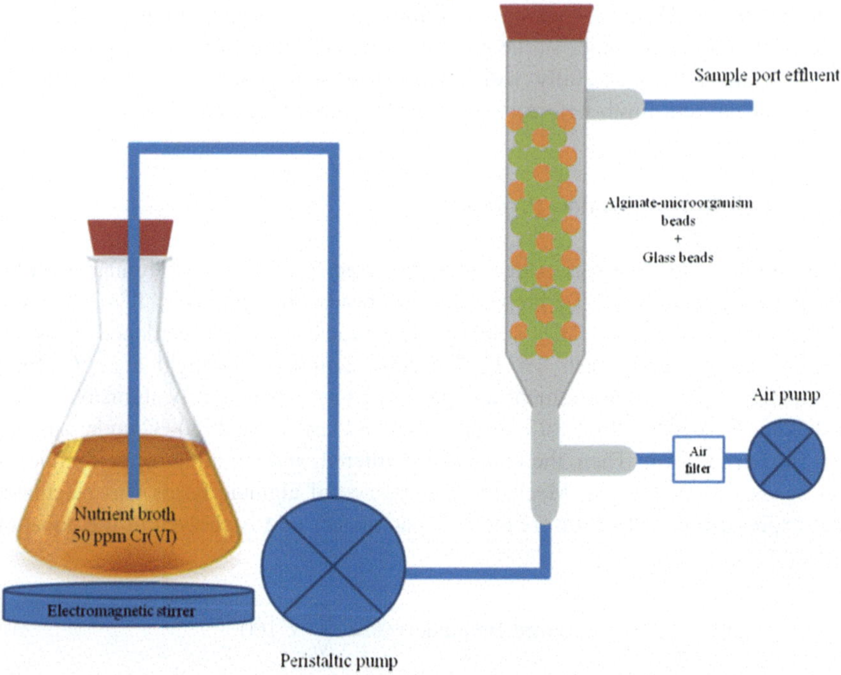

Fig. 3 Design of reactor system for Cr(VI) reduction by immobilized *P. dendritiformis*

2.2.6 Analytical Methods

Chromium reduction activity was estimated using hexavalent chromium specific colorimetric reagent 1,5-diphenylcarbazide (DPC) in acid solution (Bartlett and James 1996). Spectrophotometric measurements were made immediately at 540 nm. Cr(VI) concentration in the sample was calculated from a standard curve using a $K_2Cr_2O_7$ solution as standard. Total chromium concentration was measured using atomic absorption spectroscopy. All measurements were done in duplicates.

2.3 *Results*

2.3.1 Evaluation of Chromate Reduction Activity on Isolates

LT strain isolated from an enrichment culture from the geothermal area of Los Tachos was found to reduce chromate efficiently after 24 h of incubation. It showed a 99.72% of chromate reduction, considering 100% as the initial value of Cr(VI) concentration on the culture. Sterile control presented a chromate reduction of 33%. Total chromium remained constant (Fig. 4).

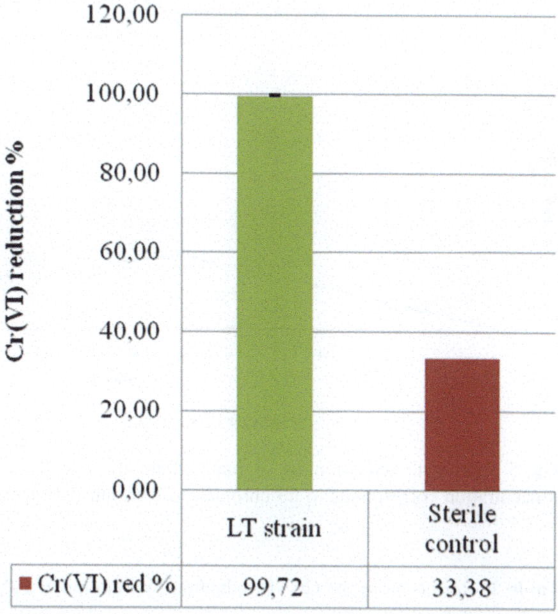

Fig. 4 Chromate reduction percentage from LT strain isolation. 100% was considered as the initial value of Cr(VI) concentration for each condition

2.3.2 Identification of the Isolate

Isolate was identified by sequencing of the almost complete 16S rRNA gene. Table 1 shows the results of the comparison of the sequence with the database of NCBI, the closest BLAST hit, its accession number, and the identity percentage.

Even if chromate reduction is well characterized among the *Bacillus* genera (Garbisu et al. 1998; Camargo et al. 2003; Pal and Paul 2004), no *P. dendritiformis* reports with this activity have been found to actual date.

Cr(VI) reduction by immobilized P. dendritiformis cells

Packed bed columns with immobilized cells of *P. dendritiformis* were tested under two different flow rates: 9.8 and 3.8 mL h^{-1}. Samples taken during the first 17 h were analyzed to see Cr(VI) reduction activity. Figure 5 shows the chromate concentration for the two conditions tested. In both cases, chromate has been reduced, but for flow rate 9.8 mL h^{-1} depletion at the efflux is over the permitted limit of Cr(VI) on water (0.5 ppm). In the case of the slowest rate tested, concentration of chromate drops under the detection limit of the analytic technique (<0.18 ppm). This means that the system reached an allowed Cr(VI) concentration,

Table 1 Result obtained from sequence analysis with BLAST database

Isolate	BLAST hit	Accession N°	Identity %
LT3	*Paenibacillus dendritiformis*	HQ625389.1	99

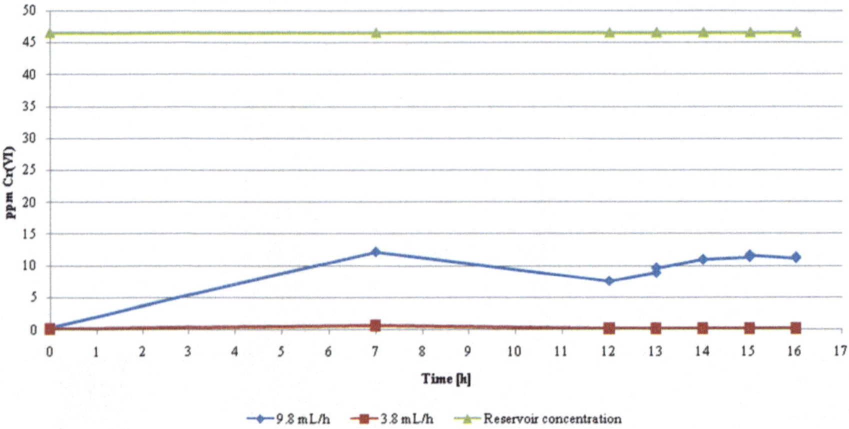

Fig. 5 Chromate reduction by alginate immobilized cells at different flow rates. Initial time concentration corresponds to no chromate added nutrient broth added to stabilize the column

Table 2 Results from the Cr(VI) reduction by immobilized *P. dendritiformis* during a 17 h

Flow rate mL/h	Cr(VI) efflux concentration (17 h)	% Cr(VI) reduction	Total Cr(VI) reduced (17 h)	Fracture frequency %
9.8	11.34 ppm	75.6	114.102	0
3.8	<0.18 ppm	99.6	52.25	0

and thus, the efflux is safe from undesirable contaminant on its oxidized form. This last system was monitored during 120 more hours, and it showed to keep Cr(VI) concentration below the detection limit during all this time (data not shown), making the conditions tested promising for the design of a potential remediation strategy for Cr(VI) contaminated effluxes. Table 2 summarizes all the results previously discussed from the study.

2.4 Conclusions

The Ca-alginate immobilized cells of *P. dendritiformis* were used up to five days on continuous system obtaining Cr(VI) reduction with values below the limit concentration allowed on water. This indicates that this strain could be used on a promising system for the development of a continuous bioprocess in the treatment of Cr(VI) contaminated effluents. In order to implement these techniques on a larger scale, further investigation on factors affecting both microbial growth and chromate reduction processes is necessary. Therefore, for the better understanding of all these factors, a lot of research and development of efficient strategies are needed at all scales: laboratory, pilot, and in situ studies. Understanding of the reduction pathways, the possible formation of toxic intermediates, and if more stable products are implicated in the process, is also necessary.

References

Barnhart J (1997) Chromium chemistry and implications for environmental fate and toxicity. J Soil Contam 6:561–568

Bartlett RJ, James BR (1996) Chromium. In: Sparks DL (ed) Methods of Soil Analysis, vol 3. Chemical Methods. ASA and SSSA, Madison, WI, pp 683–701

Benazir FJ, Suganthi R, Rajvel D, Padmini Pooja M, Mathithumilan B (2010) Bioremediation of chromium in tannery effluent by microbial consortia. African J Biotechnol 9:3140–3143

Bhowmick DC, Bal B, Chatterjee NS, Ghosh AN, Pal S (2009) A low-GC Gram-positive Thermoanaerobacter-like bacterium isolated from an Indian hot spring contains Cr (VI) reduction activity both in the membrane and cytoplasm. J Appl Microbiol 106:2006–2016

Camargo FAO, Okeke BC, Bento FM, Frankenberger WT (2003) In vitro reduction of hexavalent chromium by a cell-free extract of *Bacillus* sp. ES 29 stimulated by Cu2+. Appl Microbiol Biotechnol 62:569–573

Cui X, Wang Y, Liu J, Chang M, Zhao Y, Zhou S, Zhuang L (2015) *Bacillus dabaoshanensis* sp. nov., a Cr (VI)-tolerant bacterium isolated from heavy-metal-contaminated soil. Arch Microbiol 197:513–520

Dhal B, Thatoi HN, Das NN, Pandey BD (2013) Chemical and microbial remediation of hexavalent chromium from contaminated soil and mining/metallurgical solid waste: a review. J Hazard Mater 250:272–291

Farag S, Zaki S (2010) Identification of bacterial strains from tannery effluent and reduction of hexavalent chromium. J Environ Biol 31:877–882

Garbisu C, Alkorta I, Llama MJ, Serra JL (1998) Aerobic chromate reduction by *Bacillus subtilis*. Biodegradation 9:133–141

Kanmani P, Aravind J, Preston D (2012) Remediation of chromium contaminants using bacteria. Int J Environ Sci Technol 9:183–193

Katiyar SK, Katiyar R (1997) Microbes in control of heavy metal pollution. Adv Microbiol Biotechnol 19:330–344

Losi ME, Amrhein C, Frankenberger WT Jr (1994) Environmental biochemistry of chromium. Reviews of Environmental Contamination and Toxicology. Springer, New York, pp 91–121

Mas LC (2010) History and present status of the Neuquén geothermal project. Proc World Geotherm Congr 2010:7

Megharaj M, Avudainayagam S, Naidu R (2003) Toxicity of hexavalent chromium and its reduction by bacteria isolated from soil contaminated with tannery waste. Curr Microbiol 47:51–54

Mergeay M (1995) Heavy metal resistance in microbial ecosystems. Mol Microbiol Ecol Man 6:7–17

Pal A, Paul AK (2004) Aerobic chromate reduction by chromium-resistant bacteria isolated from serpentine soil. Microbiol Res 159:347–354

Pesce AH (2010) The Domuyo Geothermal Area, Neuquén, Argentina. Geotherm Resour Counc Trans 37:309–314

Rehman A, Zahoor A, Muneer B, Hasnain S (2008) Chromium tolerance and reduction potential of a *Bacillus* sp.ev3 isolated from metal contaminated wastewater. B Environ Contam Toxicol 81:25–29

Reynolds MF, Peterson Roth EC, Bespalov IA, Johnston T, Gurel VM, Menard HL, Zhitkovich A (2009) Rapid DNA double-strand breaks resulting from processing of Cr-DNA cross-links by both MutS dimers. Cancer Res 69:1071–1079

Romanenko VI, Korenkov VN (1977) A pure culture of bacterial cells assimilating chromates and bichromates as hydrogen acceptors when grown under anaerobic conditions. Mikrobiologiya 46:414–417

Shakoori AR, Makhdoom M, Haq RU (2000) Hexavalent chromium reduction by a dichromate-resistant gram-positive bacterium isolated from effluents of tanneries. Appl Microbiol Biotechnol 53:348–351

Singh R, Dong H, Liu D, Zhao L, Marts AR, Farquhar E, Briggs BR (2015) Reduction of hexavalent chromium by the thermophilic methanogen *Methanothermobacter thermautotrophicus*. Geochim Cosmochim Ac 148:442–456

Slobodkina GB, Bonch-Osmolovskaya EA, Slobodkin AI (2007) Reduction of chromate, selenite, tellurite, and iron (III) by the moderately thermophilic bacterium *Bacillus thermoamylovorans* SKC1. Microbiology 76:530–534

Srinath T, Khare S, Ramteke PW (2001) Isolation of hexavalent chromium reducing Cr-tolerant facultative anaerobes from tannery effluent. J Gen Appl Microbiol 47:307–312

Tucker MD, Barton LL, Thomson BM (1998) Reduction of Cr, Mo, Se and U by *Desulfovibrio desulfuricans* immobilized in polyacrylamide gels. J Ind Microbiol Biotechnol 20:13–19

Turpeinen R, Kairesalo T, Häggblom MM (2004) Microbial community structure and activity in arsenic-, chromium-and copper-contaminated soils. FEMS Microbiol Ecol 47:39–50

Viti C, Pace A, Giovannetti L (2003) Characterization of Cr(VI)-resistant bacteria isolated from chromium-contaminated soil by tannery activity. Curr Microbiol 46:1–5

Author Biographies

Dr. Edgardo R. Donati is Full Professor at Faculty of Sciences (University of La Plata, Argentina), Superior Researcher for CONICET (National Council of Science and Technology, Argentina), and the Head of CINDEFI (Biotechnology Institute). He has leading numerous national and international projects and published more than 100 papers, several books and book chapters in his major areas of interest which are environmental microbiology and biotechnology, biomining, bioremediation, and extremophiles. He is member of the editorial board of World Journal Microbiology and Biotechnology and Mine Water and The Environment among other journals.

Biot. Ana B. Segretin has a degree in biotechnology and she is currently a pHD student under Dr. Donati' supervision. She is working on chromium bioreduction and she has published some papers in the topic.

Chapter 11
Mechanism and Action of *Aureobasidium pullulans* on Biosorption of Metals

Ekambaram Nakkeeran, Ravichandran Rathna and Ravichandran Viveka

Abstract In the present scenario, management of heavy metals discharged from industrial effluents has become one of the consequential environmental challenges. Conventional effluent treatment methods are very expensive and generate hazardous sludge. Biosorption using microbial sources is a promising eco-friendly alternative for heavy metals removal. In the past few decades, filamentous fungi, *Aureobasidium pullulans,* have gained attention due to its high biomass generation, easy upscaling, and its tolerance toward heavy metals than other fungal species. *Aureobasidium pullulans* is nonpathogenic microorganism used in food, cosmetics, and pharmaceutical industries. This chapter focuses on morphology, mode of sequestration, and accumulation of various heavy metals by *A. pullulans*.

Keywords *Aureobasidium pullulans* · Biosorption · Kinetics model Isotherm · Mechanism of action

1 Introduction

The global industrialization and urbanization led to the development of environmental pollution associated with public health and ecological balance. In developed and developing countries, major environmental problem is the discharge of heavy metals like cadmium (Cd), chromium (Cr), copper (Cu), nickel (Ni), arsenic (As), lead (Pb), manganese (Mn), and zinc (Zn), directly or indirectly into the surroundings mainly due to anthropogenic activity. The industrial effluents generated

E. Nakkeeran (✉) · R. Rathna · R. Viveka
Department of Biotechnology, Sri Venkateswara College of Engineering (Autonomous), Sriperumbudur Tk 602 117, TamilNadu, India
e-mail: nakkeeran@svce.ac.in

R. Rathna
e-mail: rathnaravi.v@gmail.com

R. Viveka
e-mail: aarthyravi94@gmail.com

S. J. Varjani et al. (eds.), *Waste Bioremediation*, Energy, Environment, and Sustainability, https://doi.org/10.1007/978-981-10-7413-4_11

from various industries such as mining and smelting process, domestic and agricultural waste, photography, pharmaceutical, and other metal-based industries are reported as the sources for heavy metals contamination. These heavy metals are known for its toxicity even at low concentration, bioaccumulation, non-degradable nature, and high solubility in aquatic environments. Heavy metals mainly affect the water bodies and associated species; it enters the aquatic living system which is consumed by the secondary and tertiary consumers, thereby, enters into the food chain and gets biomagnified. High concentration of heavy metals in the ecosystem is the serious threat to the bio-life.

The sediments of river Ganga and its tributaries were reported to be polluted with hazardous heavy metals due to heavy discharge of industrial effluent and domestic sewage (Paul 2017). The Ria of Huelva estuary, in southwestern Spain, is known throughout the world for distribution of heavy metal likes Cu, Zn, Mn, Cr, Ni, Cd, and As in its sediments (Morillo et al. 2004). The benthic and pelagic fish species from the Egyptian Red Sea accumulated heavy metals like Cu, Zn, Pb, Cd, iron (Fe), and Mn in the liver, gills,and muscles. It also reported that essential heavy metals found accumulated in the liver (metabolized), while others in the gills (El-Moselhy et al. 2014). The water quality index (WQI) of Nile River water quality was extended from 55 (marginal) to 27 (poor) due to the discharge of agricultural, industrial, and municipal wastes with a significant level of heavy metal contaminants (Abdel-Satar et al. 2017). In 2012, prawn and shrimp collected from local markets of Malad suburban areas of Mumbai city, India found contaminated by ten heavy metals, namely Cu, Zn, Mn, Fe, Co, Cr, Ni, Pb, Cd, and Hg at permissible limits (Zodape 2014). In 2014, freshwater species, namely *Hampala microlepidota*, *Barbonymus schwanenfeldii*, *Mystacoleucus marginatus*, *Hemibagrus nemurus*, *Cyclocheilicthys apogon,* and *Oreochromis niloticus* from Galas River and Beranang mining pool were detected with contamination of heavy metals, namely Cu, Zn, Pb, Ni, Mn, and Cd (Baharom and Ishak 2015). According to United States Agency for Toxic Substances and Disease Registry, As, Pb, and Hg were placed in first, second, and third in the priority list of hazardous substances and Pb, Cd, Ni, Zn, Cu, Mn, etc., were also present in the list (Kamunda et al. 2016; Baharom and Ishak 2015).

In recent years, the impact of heavy metal contamination is quite a serious issue globally. Researchers mainly focused on developing more efficient and economical method/process for the treatment of heavy metal contaminants. The convention methods used for treatment of heavy metal contaminants are chemical precipitation, flotation, absorption, electrochemical deposition, chemical oxidation/reduction, ion exchange, coagulation/flocculation, membrane technology, reverse osmosis, electrodialysis, etc., but these methods have hitherto been largely restricted to problems such as cost, energy requirement, toxic sludge, and non-eco-friendly. Ion exchange process proved to be the effective method for heavy metal removal at low cost but could not be suitable for high concentration of heavy metals (Dizge et al. 2009). In the past decades, membrane technology has been extensively used for heavy metal removal; it was effective in removing the heavy metal but process generated toxic sludges and expensive as well. For the past few years, adsorption using activated carbon showed to be a significant method for heavy metal removal. However, it is difficult to scale-up and expensive (Irazusta and de Figueroa 2014). The technological and economical limitations of the conventional methods led to the development of novel technology for the treatment of heavy metal contaminants (Tunali et al. 2006).

Biosorption emerged as a suitable alternative method for the removal of heavy metals and found to be effective, eco-friendly, and economically feasible. Biosorption primarily uses microbes like bacteria, fungi, yeast, and algae having heavy metal sequestering activity through adaptation (intrinsic) or genetically engineered (induced). These microbes have the ability to biotransform heavy metal from parts per trillion to parts per billion concentrations (Wang and Chen 2006). Microbes in the polluted environment suffer a great deal of energy requirement for its survival. In order to meet its basic energy requirement, it metabolizes the pollutant as its sole energy or carbon source through evolution and adaptation process, thereby reduces the toxicity level (Tunali et al. 2006). Recently, researchers are mainly focused on employing these microbes as an effective biosorbent, study the metabolic pathway for biosorption and translational research to industrial scale (Wang and Chen 2006). In general, the mechanism involved in heavy metal treatment is surface adsorption onto the cell, intracellular partitioning and extracellular precipitation (Suh et al. 1997). Fungal and yeast species are frequently encountered in metal-contaminated habitats with the ability to resist and adapt to the environment polluted with toxic compounds (Gadd 1993). The fungal species isolated from the heavy metal-contaminated site found to have modifications in its biochemical mechanisms and genetic material (Gadd 1993).

Identification of potential biosorbent with the ability to treat a high volume of heavy metal-contaminated effluents is the main goal for environmental researchers. Some of the microbes used as a biosorbent are bacteria (*Pseudomonas* sp.—Uranium (D'Souza et al. 2006), *Citrobacter* sp.—lead and cadmium (Macaskie et al. 1987), *Bacillus* sp.—lead and copper (Tunali et al. 2006)), fungi (*Rhizopus arrhizus*—mercury (Tobin et al. 1984), *Cladosporium cladosporioides*—cyanides (Patil and Paknikar 1999), *Mucor hiemalis*—mercury (Hoque and Fritscher 2016), *Aspergillus niger*—cadmium), yeast (*Saccharomyces cerevisiae* (Wang and Chen 2006)), and algae (*Chlorella vulgaris*—cadmium (Mehta and Gaur 2005), *Ecklonia radiata*—chromium (Mehta and Gaur 2005), *Spirulina* sp.—zinc, lead (Mehta and Gaur 2005)). However, recent studies are widely focused on *Aureobasidium pullulans* for heavy metal treatment than other filamentous fungi due to its yeast-like nature, amenable manner, physiological homogeneity, metal tolerance ability, and ubiquitous nature (Gadd and Griffiths 1980b). *Aureobasidium pullulans* is extensively used in food, cosmetics, and pharmaceutical industries. Figure 1 represents the applications of *A. pullulans* in various sectors. This chapter mainly focuses on the morphology, mode of sequestration, and accumulation of various heavy metals by *A. pullulans*.

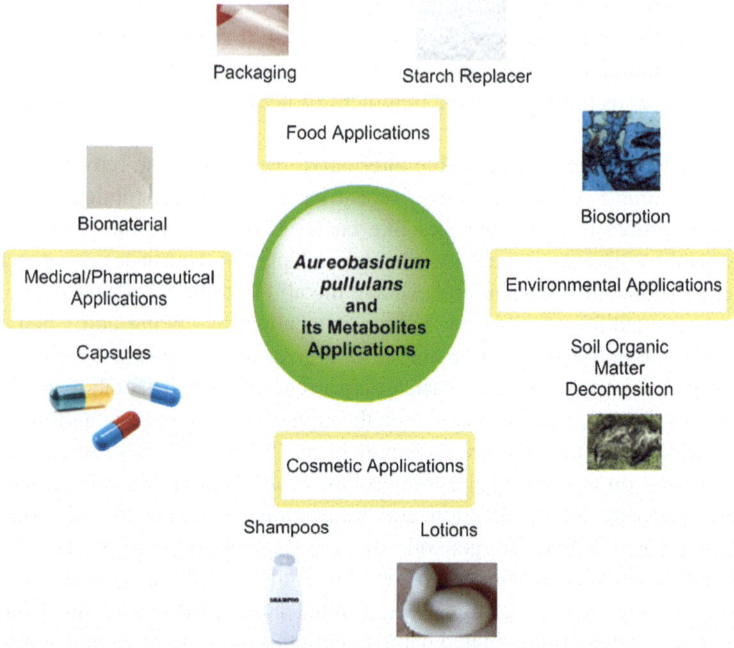

Fig. 1 Applications of *A. pullulans* in various sectors

2 *A. Pullulans* as Biosorbent

Aureobasidium pullulans is a polymorphic yeast-like fungus with mycelium and melanin-pigmented chlamydospores (Gadd and Mowll 1985). It generally referred as black yeast due to the formation of melanin pigment on its surface over a period of time (Gadd and Griffiths 1980a). *Aureobasidium pullulans* is extensively used as a model system for synthesis of biopolymers, and its genomic and molecular aspects have been explored extensively. *Aureobasidium pullulans* has been widely used for the production of biopolymers, enzymes, antibiotics, biosurfactants, etc., due to its safe and eco-friendly nature. Therefore, it is used in food, cosmetics, biomedical, and pharmaceutical industries. These industries generate a large amount of stable biomass residual waste. Substantial quantities of yeast-like fungi biomass could be easily obtained either by submerged or solid-state fermentation. Thereby, modest fermentation process with economical growth media is sufficient for high biomass yield. The cellular growth, metabolism, and pigmentation of *A. pullulans* depend on the essential metal ions like Cu (Gadd and Griffiths 1980a), Zn (Reeslev et al. 1993). Extracellular polysaccharide substances present in the yeast-like cell influence the accumulation of heavy metals (Suh et al. 1999). The cell intactness, detoxification mechanism, and morphogenetic traits of *A. pullulans* make it an ideal candidate for the heavy metal treatment (Ghaedi et al. 2014; Gadd and Mowll 1985). Table 1

Table 1 Heavy metals removal by *A. pullulans* as a biosorbent

Metals class	Metal	Form	Metal uptake rate	References
Toxic metals	Pb	Cell suspension	215.6 mg/g cell	Suh et al. (1999)
	Cu	Free cells	35 mg/dry weight	Gadd and Rome (1988)
	Cu	Encapsulated	197 mg/dry weight	Gadd and Rome (1988)
	Cu	Laboratory cultivated and maintained at 25 °C	0.1 mg/dry weight	Gadd and Rome (1988)
	Cu	Yeast-like cells, laboratory cultivated and washed with distilled water	15 nmol/10^7 cells	Gadd and Rome (1988)
	Zn	Yeast-like cells, laboratory cultivated and washed with distilled water	25 nmol/10^7 cells	Gadd et al. (1987)
	Cd	Yeast-like cells, laboratory cultivated and washed with distilled water	10 nmol/10^7 cells	Gadd et al. (1987)
	Co	Yeast-like cells, laboratory cultivated and washed with distilled water	20.9 nmol/10^7 cells	Gadd et al. (1987)

represents few attempts made on *A. pullulans* as a biosorbent. Mutational and metabolic engineering investigation on this microbe is not complicated. Using recombinant DNA technology, researchers might manipulate the genetic material of black yeasts for the environment bioremediation purposes.

Biosorption utilizes typical, mutated, engineered, free, immobilized, flocculent as well as pretreated cell types. The use of living and dead cell is still in controversy based the capacity to uptake metal ions and growth media requirements. In general, there are two modes of metal ion uptake which are considered biphasic, namely passive and active modes (Suh et al. 1997). Former is biosorption, while latter is bioaccumulation. Biosorption could be performed by both living and dead cells which are non-metabolic dependent, whereas bioaccumulation could be performed using living cells alone which is metabolic dependent (Viraraghavan and Srinivasan 2011).

2.1 **A. Pullulans** ***Morphogenetic Nature***

The polymorphic *A. pullulans* has the ability to grow as budding yeast as well as polykaryotic hyphal extensions (Guar et al. 2010) with high morphogenetic diversity (Andrews et al. 1994). The genetic diversity of *A. pullulans* was studied using multilocus molecular analysis by rDNA technology and described the new potent varieties within the species and redefined *A. pullulans* into four different species, namely *A. pullulans* (de Bary) G. Arnaud var. *pullulans*, *A. pullulans*

Table 2 Morphological display of *A. pullulans*

Morphology	Elucidation	Pictorial representation
Conidia	Ellipsoidal amerospores cells	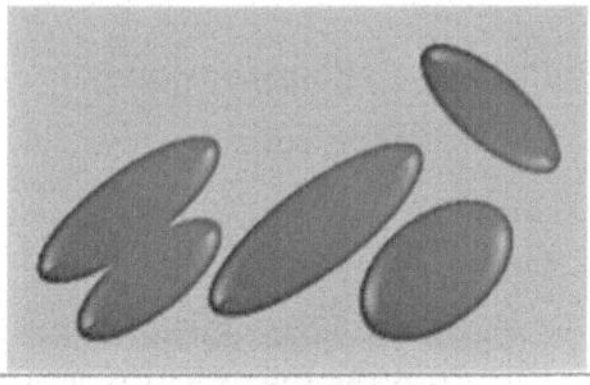
Yeast-like cells	Elongated singled cells	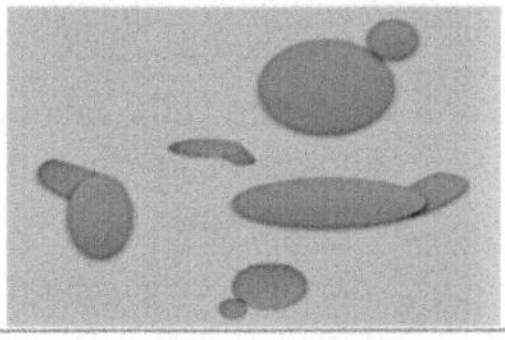
Blastospores	Budding in yeast	
Chlamydospores	Cells formed with hyphae	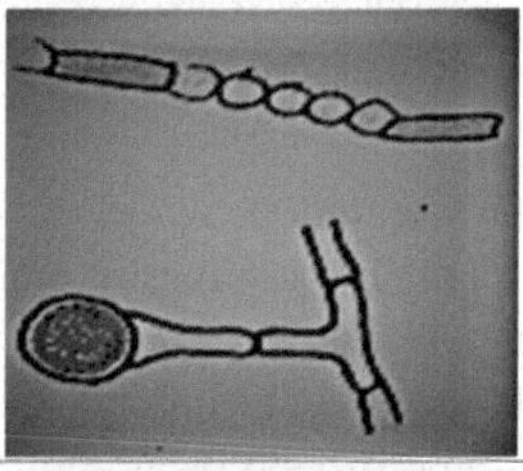
Mycelium	Aseptate polykaryotic hyphal extensions	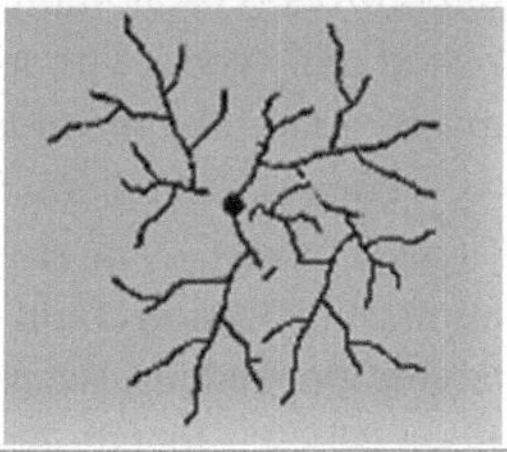

(de Bary) G. Arnaud var. *melanogenum*, *A. pullulans* (de Bary) G. Arnaud var. *subglaciale,* and *A. pullulans* (de Bary) Arnaud var. *namibiae* (Zalar et al. 2008). In general, *A. pullulans* has five distinct morphological forms, namely conidia, yeast-like cells, blastospores, chlamydospores, and mycelium (Shabtai and Mukmenev 1995). The morphological display of *A. pulluans* is described in Table 2. The main morphogenetic transition in *A. pullulans* is yeast-like cells, mycelium and melanin-pigmented chlamydospores (Dominguez et al. 1978).

The melanin-pigmented chlamydospores were found to accumulate large amount of metal ions than hyaline cell (Gadd et al. 1987; Gadd and Rome 1988). There is no report stating the significance and mechanism of accumulation of metal ions by melanin in biosorption process. Researchers reported that high proportion of melanin-producing microbes have the ability to survive in environmentally stressed biospheres (Bell and Wheeler 1986). In addition, melanin is the most stable and resistant bioactive material that confers protection against extreme conditions (Jacobson 2000; Bell and Wheeler 1986). Heavy metal uptake by the yeast-like protoplast was found to be higher than intact cell, and protoplast was found to be more susceptible to toxicity than intact cells (Gadd et al. 1987). The metal uptake of encapsulated *A. pullulans* biosorbent is lower than the free *A. pullulans* biosorbent due to the penetration of microbes on the capsule membrane (Park et al. 2001). Investigation on lead accumulation by live and dead cells of *A. pullulans* revealed the higher accumulation of lead in live cells than the death cells due to the presences of extracellular polymeric substances, which was found contradictory to the results obtained by *Saccharomyces cerevisiae* (Suh et al. 1999). Hence, heavy metal accumulation by *A. pullulans* depends mainly on its morphogenetic nature.

3 Mechanism of Action

The mechanism of metal biosorption by fungi generally involves four distinct processes, namely (i) intracellular accumulation, (ii) extracellular accumulation, (iii) cell surface precipitation, and (iv) extracellular polymeric substance interaction with metal ions. In general, metals first interact with the reactive functional group present in the cell wall of the fungi and deposition of inorganic forms of metal. The pleomorphic nature of *A. pullulans* that as yeast-like cells, mycelium and melanin pigment chlamydospores, and other transitory forms helps in the uptake of heavy metals (Gadd and Mowll 1985). The former two cell types showed biphasic cation uptake characteristics (metabolism-independent binding of cadmium to cell surfaces followed by metabolism-dependent uptake of cadmium into the cytoplasm) while the latter cell type uptakes metal ions through surface precipitation (Gadd and Mowll 1985). Based on the cell metabolism, biosorption takes place either by metabolism-dependent or metabolism-independent processes. In general, when metal ions come in contact with cell membrane of the microbe, it either transported across the membrane via cellular metabolism which is possible in viable cells alone or interacts with the functional groups available on the cell membrane (cell surface adsorption) by forming complexation, exchange of ions, or physical adsorption which is metabolism-independent processes (Veglio and Beolchini 1997). In *A. pullulans*, initially heavy metals bind to the charged sites on the cell surface through the rapid metabolism-independent process and diffuse into the cell cytosol by the slower metabolism-dependent process (Suh et al. 1997). The mechanism of action of *A. pullulans* is illustrated in Fig. 2.

The cell wall of *A. pullulans* has discrete functional sites such as carboxylate, amine, phosphate, imidazole, hydroxyl, which has the capacity to interact with

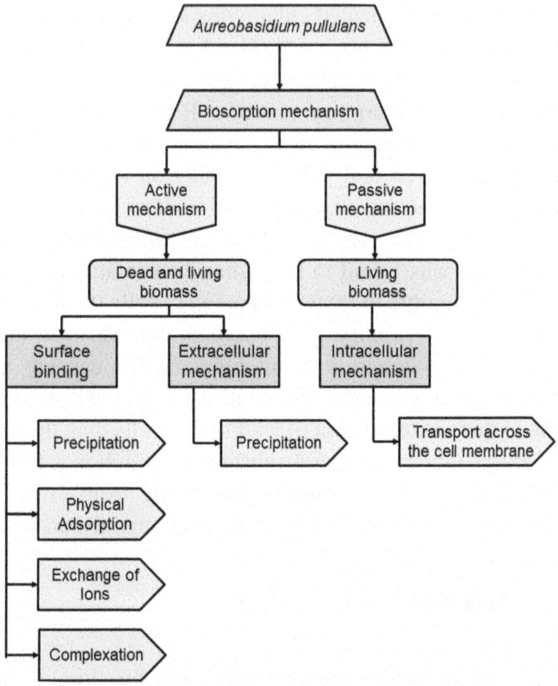

Fig. 2 Biosorption mechanism of *A. pullulans*

metal ions. These functional groups form complexation with metals ions (Ghaedi et al. 2014). The presence of melanin pigment on *A. pullulans* acts as a barrier for intracellular accumulation of metals and helps in surface binding (Gadd and Mowll 1985). The melanin pigment contains free carboxyl and hydroxyl groups, which are the main binding sites for metal ions; thereby the metal binding capacity of melanin pigment helps in surface sorption mechanism (Gadd and Rome 1988). The presence of the extracellular polymeric substance in *A. pullulans* acts as a barrier for the penetration of metal ions into the cell and gets accumulated on the surface of the cells (Suh et al. 1999). Therefore, extracellular polymeric substance and melanin pigment on *A. pullulans* may be solely responsible for the transition of heavy metals from toxic to non-toxic or less toxic.

In *A. pullulans*, intracellular accumulation occurs through apparent stoichiometric exchange of ions across the cell membrane. In the case of yeast-like cells and mycelium, uptake of metal ions is due to the apparent stoichiometric exchange of ions. The ions released in exchange of metal ions from the surroundings significantly influence the heavy metal uptake (Gadd and Mowll 1985; Mowll and Gadd 1984). The free flow ions from the cell were catalyzed by nitrilotriacetate (NTA) or glucose indicating that the heavy metals are in ionic state or bound to low molecular compounds which could actively expelled from the cell (Mowll and Gadd 1984). Intracellular accumulation of metal takes place in living and dead cells of *A. pullulans*; however, living cells found to adsorb significant quantities of metal

(Suh et al. 1999). Researchers are mainly focused on identifying the ideal microbial source for biosorption and paying less attention to mode of sequestration of metal uptake by microbes.

4 Factors Influencing Biosorption Capacity

For effective utilization of *A. pullulans* as a biosorbent, factors influencing the biosorption capacity must be studied. The important factors influencing the capacity of biosorption are given in Fig. 3.

Aureobasidium pullulans is a common environmental epiphyte found in the nature. *Aureobasidium pullulans* commonly include yeast-like cells with actin-like filament protruding outwards, blastospores (conidia) arising from hyphae and chlamydospores with thick cell wall, vacuoles, and mycelial nuclei (Oller 2005). Chlamydospores and hyphal cells have the tendency to secrete extracellular polysaccharide substance while yeast-like cells show the presence of polysaccharide capsule (Oller 2005).

Addition of copper sulfate to 3–4 days old *A. pullulans* solid plate culture showed the formation of blackening of the cells by inhibiting growth. Thereby, copper ions found to influence the morphology of cell by inducing the formation of melanin and mucilaginous material (Gadd and Griffiths 1980a). In the case of liquid culture, copper ions found to stimulate the formation melanin pigmentation, septa, and mycelial structure, whereas mucilage was absent (Gadd and Griffiths 1980a). *Aureobasidium pullulans* was able to survive in the presence of copper due to the stimulation of copper-dependent enzyme (tyrosine oxidase) melanin pathway (Gadd and Griffiths 1980a).

pH is the most important factor that regulates the transition of swollen cell to chlamydospore in *A. pullulans*. pH 6 is considered suitable for transition of yeast

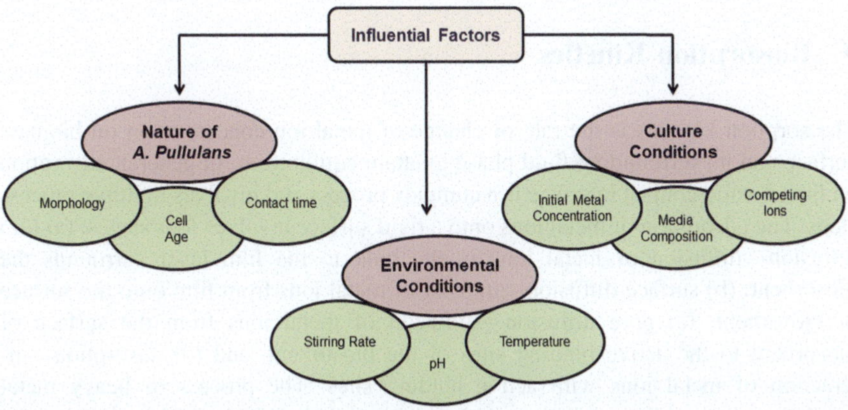

Fig. 3 Factors influencing the biosorption capacity

cells while the transition was absent at pH below 3 (Bermejo et al. 1981). Regfilez et al. (1980) reported that permeation of fungal membranes is reduced at lower pH. Yeast-like cell of *A. pullulans* found to have reduced biosorption capacity than other *Cladosporium* sp. due to the absence of chlamydospores and melanin pigment (Fomina and Gadd 2003). Park et al. (2001) reported the optimum pH and temperature for maximum uptake of cadmium ions were 6 and 27 °C, respectively. In the case of lead, effect of pH on biosorption rate showed Lineweaver–Burk plot and sorption rate was varied at various pH. This was in accordance with competitive inhibition of enzymes (Suh et al. 1997). The growth rate of *A. pullulans* was affected at low pH due to the destruction of cell wall caused by protonation of the functional groups on the surface of the cell (Ghaedi et al. 2014). At high temperature, cells and enzymes present in the *A. pullulans* was destroyed, thereby reduces the growth and metal ion uptake capacity (Ghaedi et al. 2014).

Increase in stirring speed and agitation enhances the removal rate of adsorptive pollutants (Wang and Chen 2006). However, high stirring disrupts the cell surface and reduces the growth rate. Contact time maximizes the availability of functional groups present on the cell surface to the heavy metals (Ghaedi et al. 2014). Competing ions present in the media affect the biosorption of metal ions by *A. pullulans* through competitive inhibition. Mowill and Gadd (1984) reported that calcium inhibited the cadmium uptake of yeast-like cells and mycelium while divalent cation ionophore like zinc, cobalt, mercury could not accelerate the cadmium uptake rate. Biosorption capacity of cells at its exponential phase was higher than stationary phase. Bioaccumulation of lead by live *A. pullulans* was higher than the cell at dead phase (Suh et al. 1999). At stationary and death phase, existence of extracellular polymeric substance was low than the lag phase. Copper binding capacity of mixed *Cladosporium cladosporioides* and *A. pullulans* biomass was higher compared to the individual fungal biomass (Fomina and Gadd 2003). Table 3 represents the various factors and its operating conditions influencing the biosorption capacity.

5 Biosorption Kinetics

The sorption kinetics is the rate of change of metal ion concentration on biomass surface and its surrounding fluid phase to attain equilibrium. In general, adsorption on liquid–solid contact interface is a complex process and involves multiple process steps. The adsorption of metal ions onto a rigid surface involves four stages: (a) film diffusion—diffusion of metal ions in the bulk to the film layer surrounds the biosorbent, (b) surface diffusion—transfer of metal ions from film onto the surface of biosorbent, (c) pore diffusion—diffusion of metal ions from the surface of biosorbent to the active binding sites of the biosorbent, and (d) adsorption—interaction of metal ions with active binding sites. The process of heavy metal

Table 3 Factors influencing the biosorption capacity of *A. pullulans*

Metal	Biosorbent capacity (mg/g)	Operating conditions						References
		pH	Biomass (g/L)	Stirring rate (rpm)	Temperature (°C)	Cell age (h)	Contact time (h)	
Cd	0.023	6	160	200	27	24	50	Park et al. (2001)
Cd	0.13	6	0.05	200	26	48	160	Radulović et al. (2008)
Cu	64	5	1	180	25	24	24	Ghaedi et al. (2014)
Cu	1.04	6	2.58	200	26	48	160	Radulović et al. (2008)
Fe	372.66	6	924.21	200	26	48	160	Radulović et al. (2008)
Zn	17.88	6	1.502	200	26	48	160	Radulović et al. (2008)
Mn	73.27	6	181.70	200	26	48	160	Radulović et al. (2008)
Pb	17.68	6	0.58	200	26	48	160	Radulović et al. (2008)
Pb	25	5	1	180	25	24	24	Ghaedi et al. (2014)
Ni	1	6	2.47	200	26	48	160	Radulović et al. (2008)
Ni	45	5	1	180	25	24	24	Ghaedi et al. (2014)
Cr	0.77	6	1.92	200	26	48	160	Radulović et al. (2008)
Co	43	5	1	180	25	24	24	Ghaedi et al. (2014)

sorption by *A. pullulans* could be studied using different kinetic models. The most commonly used kinetic models are pseudo-first-order and pseudo-second-order models. The kinetic equations for metal adsorption are as follows (Shi et al. 2013).

$$\frac{dC_p}{dt} = -k_d C_p + k_a C_m \tag{1}$$

where k_a is absorption coefficient rate (Lg^{-1} min^{-1}), k_d is desorption coefficient rate (min^{-1}), C_p is metal ion concentration at contaminated site ($\mu g\ g^{-1}$), and C_m is the metal ion concentration.

Lagergren pseudo-first-order rate expression is summarized as follows (Ho and McKay 1998).

$$\frac{dq}{dt} = k_{a1}(q_e - q_t) \tag{2}$$

where q_e and q_t are the amount of metal ion adsorbed on biosorbent at equilibrium and at any time t, respectively ($mg\ g^{-1}$), and k_{a1} is the equilibrium rate constant of the first-order kinetics (min^{-1}).

On integration Eq. 2 with boundary limits, $t = 0$ to $t = t$ and $q_t = 0$ to $q_t = q_t$

$$\ln\left(\frac{q_e}{q_e - q_t}\right) = k_{a1}t \tag{3}$$

The linear form of Eq. 3 is represented as

$$\log(q_e - q_t) = \log(q_e) - \frac{k_{a1}}{2.303}t \tag{4}$$

The most popular form of representation is

$$q_t = q_e(1 - e^{-k_{a1}t}) \tag{5}$$

The pseudo-second-order rate kinetics is summarized as follows (Ho and McKay 1998).

$$\frac{dq}{dt} = k_{a2}(q_e - q_t)^2 \tag{6}$$

where k_{a2} is the equilibrium rate constant of the second-order kinetics (g mg^{-1} min^{-1}).

Rearrange to group like terms

$$\frac{dq}{(q_e - q_t)^2} = k_{a2}dt \tag{7}$$

Integrating Eq. 7 for the boundary conditions $t = 0$ to $t = t$ and $q_t = 0$ to $q_t = q_t$

$$\frac{1}{(q_e - q_t)} = \frac{1}{q_e} + k_{a2}t \tag{8}$$

The most popular form of representation is

$$q_t = q_e\left(1 - \frac{1}{\frac{1}{q_e e^{-k_{a2}t}}}\right) \tag{9}$$

Lead and cobalt sorption by *A. pullulans* biomass found to obey first-order kinetics, whereas for copper and nickel, it obeys second-order kinetics (Ghaedi et al. 2014).

5.1 Biosorption Equilibrium Isotherm Model

Equilibrium isotherm models for the assessment of sorption performance are classified into two types, namely empirical equations and mechanistic models. Mechanistic models are used to analyze the experimental behavior and predict the biosorption process. There are several empirical models are available to describe the equilibrium sorption distribution ranging from single compartmental model to complex compartmental model. The most widely and commonly used isotherm for the single solute system is Langmuir and Freundlich isotherm model.

Langmuir isotherm: The Langmuir isotherm is known as semiempirical isotherm based on four assumptions: (i) the surface of the adsorbent should be homogeneous, (ii) adsorbed molecules do not interact, (iii) adsorption occurs through the same mechanism, and (iv) monolayer adsorption on the homogeneous surface. This is applicable at constant pH, and it obeys Henry's law by the following equation.

$$q_e = \frac{q_{max} k_a C_e}{1 + k_e C_e} \tag{10}$$

where q_e is the equilibrium metal ions uptake (mg of adsorbate/g of dried adsorbent), q_{max} is the maximum monolayer uptake capacity (mg g^{-1}), k_a is the Langmuir adsorption coefficient (L mg^{-1}) for binding of adsorbate on the biosorbent sites, and C_e is the equilibrium concentration (mg L^{-1}).

Freundlich isotherm: Freundlich isotherm shows the relationship between the concentration of solute on the heterogeneous surfaces of the adsorbent and concentration of the solute in the aqueous medium. This isotherm could be applied only at constant pH and solutions with low concentration. The simplest known sorption equation can be expressed as follows.

$$q_e = k_e C_e^{\frac{1}{n}} \tag{11}$$

where q_e is the equilibrium uptake capacity (mg g^{-1}), k_e (L mg^{-1}) and n are Freundlich constants related to biosorption capacity and sorption efficiency, respectively, and C_e is the equilibrium concentration.

The experimental results for lead biosorption by *A. pullulans* were in accordance with Freundlich isotherm model while the rate of biosorption followed Michaelis–Menten kinetics (Suh et al. 1997). Yeast-like cells, hyphae, and chlamydospores of *A. pullulans*, nature of surface binding of copper found to obey Freundlich isotherm model, copper uptake followed Michealis–Menten kinetics, Lineweaver–Burk kinetics, and copper efflux kinetics found to exhibit two compartmental systems (Gadd and Mowll 1985). Biosorption equilibrium data for metals such as lead, copper, nickel, and cobalt revealed that Freundlich isotherm was more suitable for interpretation of adsorption than Langmuir isotherm (Ghaedi et al. 2014). Cadmium uptake by *A. pullulans* followed Michaelis–Menten kinetics (Mowll and Gadd 1984). Zinc uptake kinetics of yeast-like cell protoplast was lower while chlamydospore protoplast showed no intracellular uptake but showed significant kinetics due to the surface binding uptake (Gadd et al. 1987). Copper absorption by the melanin and pigment cells of *A. pullulans* revealed that Freundlich isotherm agrees with Langmuir isotherm equation over moderate solute concentration range (Gadd and Rome 1988).

6 Future Perspectives

Aureobasidium pullulans, yeast-like fungi found to be a promising source as a biosorbent for heavy metals removal. Biosorption using microbes is still in pilot scale due to the shortcomings of the process. Biosorption is a beneficial and attractive field of research in the present era. Development of this process is limited due to inadequate knowledge on the biosorption mechanism and kinetic studies. Researchers must focus on the process parameters, mechanism of action, mode of sequestration, thermodynamic and kinetic studies by microbe associated with biosorption. Real-time field studies are necessary for commercialization of biosorption. The advent of genetic engineering and advances in molecular biology could help in developing novel microorganisms that have the ability to treat simultaneously more than two heavy metals at a time. In near future, biosorption technology would combat heavy metals using multidisciplinary field of biotechnological, non-biotechnological, and nanotechnological processes.

7 Conclusions

The management of heavy metals released from industrial effluents is a major consequential environmental challenge. The advancements in the conventional effluent treatment found to treat heavy metal-contaminated effluents but resulted in the generation of toxic sludge and uneconomical. Biosorption using microbes, especially by fungal species found to be a promising eco-friendly alternative for heavy metals removal. *Aureobasidium pullulans* is a distinct and excellent biological system for heavy metals biosorption. Its unique features are high biomass generation, morphology, easy upscaling, and its tolerance toward heavy metals. Further, studies on *A. pullulans* could be focused to elucidate the biosorption mechanism and mode of sequestration for effective utilization in biosorption technology.

References

Abdel-Satar AM, Ali MH, Goher ME (2017) Indices of water quality and metal pollution of Nile River, Egypt. Egypt J Aquat Res 43(1):21–29

Andrews JH, Harris RF, Spear RN, Lau GW, Nordheim EV (1994) Morphogenesis and adhesion of *Aureobasidium pullulans*. Can J Microbiol 40(1):6–17

Baharom ZS, Ishak MY (2015) Determination of heavy metal accumulation in fish species in Galas River, Kelantan and Beranang mining pool, Selangor. Proc Environ Sci 30:320–325

Bell AA, Wheeler MH (1986) Biosynthesis and functions of fungal melanins. Annu Rev Phytopathol 24(1):411–451

Bermejo JM, Dominguez JB, Goni FM, Uruburu F (1981) Influence of pH on the transition from yeast-like cells to chlamydospores in *Aureobasidium pullulans*. Antonie Van Leeuwenhoek 47 (5):385–392

Dizge N, Keskinler B, Barlas H (2009) Sorption of Ni (II) ions from aqueous solution by Lewatit cation-exchange resin. J Hazard Mater 167(1):915–926

Dominguez JB, Goni FM, Uruburu F (1978) The transition from yeast-like to chlamydospore cells in Pullularia pullulans. Microbiol 108(1):111–117

D'Souza SF, Sar P, Kazy SK, Kubal BS (2006) Uranium sorption by *Pseudomonas* biomass immobilized in radiation polymerized polyacrylamide bio-beads. J Environ Sci Health A 41 (3):487–500

El-Moselhy KM, Othman AI, El-Azem HA, El-Metwally MEA (2014) Bioaccumulation of heavy metals in some tissues of fish in the Red Sea, Egypt. Egypt J Basic Appl Sci 1(2):97–105

Fomina M, Gadd GM (2003) Metal sorption by biomass of melanin-producing fungi grown in clay-containing medium. J Chem Technol Biotechnol 78(1):23–34

Gadd GM (1993) Interactions of fungi with toxic metals. New Phytol 124(1):25–60

Gadd GM, Griffiths AJ (1980a) Effect of copper on morphology of *Aureobasidium pullulans*. Trans Br Mycol Soc 74(2):387–392

Gadd GM, Griffiths AJ (1980b) Influence of pH on toxicity and uptake of copper in *Aureobasidium pullulans*. Trans Br Mycol Soc 75(1):91–96

Gadd GM, Mowll JL (1985) Copper uptake by yeast-like cells, hyphae, and chlamydospores of *Aureobasidium pullulans*. Exp Mycol 9(3):0–40

Gadd GM, Rome L (1988) Biosorption of copper by fungal melanin. Appl Microbiol Biotechnol 29(6):610–617

Gadd GM, White C, Mowll JL (1987) Heavy metal uptake by intact cells and protoplasts of *Aureobasidium pullulans*. FEMS Microbiol Ecol 3(5):261–267

Gaur R, Singh R, Gupta M, Gaur MK (2010) *Aureobasidium pullulans*, an economically important polymorphic yeast with special reference to pullulan. Afr J Biotech 9(47):7989–7997

Ghaedi M, Brazesh B, Karimi F, Ghezelbash GR (2014) Equilibrium, thermodynamic, and kinetic studies on some metal ions biosorption using black yeast *Aureobasidium pullulans* biomass. Environ Prog Sustain Energy 33(3):769–776

Ho YS, McKay G (1998) A comparison of chemisorption kinetic models applied to pollutant removal on various sorbents. Process Saf Environ Prot 76(4):332–340

Hoque E, Fritscher J (2016) A new mercury-accumulating *Mucor hiemalis* strain EH8 from cold sulfidic spring water biofilms. Microbiology 5(5):763–781

Irazusta V, de Figueroa LI (2014) Copper resistance and oxidative stress response in *Rhodotorula mucilaginosa* RCL-11 yeast isolated from contaminated environments in Tucumán, Argentina. In: Bioremediation in Latin America, Springer International Publishing, pp 241–253

Jacobson ES (2000) Pathogenic roles for fungal melanins. Clin Microbiol Rev 13(4):708–717

Kamunda C, Mathuthu M, Madhuku M (2016) Health risk assessment of heavy metals in soils from witwatersrand gold mining basin, South Africa. Int J Environ Res Public Health 13(7):663

Macaskie LE, Dean AC, Cheetham AK, Jakeman RJ, Skarnulis AJ (1987) Cadmium accumulation by a *Citrobacter* sp.: the chemical nature of the accumulated metal precipitate and its location on the bacterial cells. Microbiology 133(3):539–544

Mehta SK, Gaur JP (2005) Use of algae for removing heavy metal ions from wastewater: progress and prospects. Crit Rev Biotechnol 25(3):113–152
Morillo J, Usero J, Gracia I (2004) Heavy metal distribution in marine sediments from the southwest coast of Spain. Chemosphere 55(3):431–442
Mowll JL, Gadd GM (1984) Cadmium uptake by *Aureobasidium pullulans*. Microbiology 130 (2):279–284
Oller AR (2005) *Aureobasidium pullulans* morphology: two adapted polysaccharide stains. Can J Microbiol 51(12):1057–1060
Park JK, Kim WS, Chang HN (2001) Specific Cd^{2+} uptake of encapsulated *Aureobasidium pullulans* biosorbents. Biotech lett 23(17):1391–1396
Patil YB, Paknikar KM (1999) Removal and recovery of metal cyanides using a combination of biosorption and biodegradation processes. Biotech Lett 21(10):913–919
Paul D (2017) Research on heavy metal pollution of river Ganga: a review. Ann Agrar Sci 1–9
Radulović MĐ, Cvetković OG, Nikolić SD, Đorđević DS, Jakovljević DM, Vrvić MM (2008) Simultaneous production of pullulans and biosorption of metals by *Aureobasidium pullulans* strain CH-1 on peat hydrolysate. Biores Technol 99(14):6673–6677
Reeslev M, Jørgensen BB, Jørgensen OB (1993) Influence of Zn2 + on yeast-mycelium dimorphism and exopolysaccharide production by the fungus *Aureobasidium pullulans* grown in a defined medium in continuous culture. Microbiology 139(12):3065–3070
Regulez P, Ponton J, Dominguez JB, Goñi FM, Uruburu F (1980) Lipid composition and the transition from yeast-like to chlamydospore cells of *Pullularia pullulans*. Can J Microbiol 26 (12):1428–1437
Shabtai Y, Mukmenev I (1995) Enhanced production of pigment-free pullulan by a morphogenetically arrested *Aureobasidium pullulans* (ATCC 42023) in a two-stage fermentation with shift from soy bean oil to sucrose. Appl Microbiol Biotechnol 43(4):595–603
Suh JH, Kim DS, Oh SJ, Park YS, Song SK (1997) The biosorption rate of lead by *Aureobasidium pullulans*. Environ Eng Res 2(4):287–290
Suh JH, Yun JW, Kim DS (1999) Effect of extracellular polymeric substances (EPS) on Pb^{2+} accumulation by *Aureobasidium pullulans*. Bioprocess Biosyst Eng 21(1):1–4
Tobin JM, Cooper DG, Neufeld RJ (1984) Uptake of metal ions by *Rhizopus arrhizus* biomass. Appl Environ Microbiol 47(4):821–824
Tunali S, Cabuk A, Akar T (2006) Removal of lead and copper ions from aqueous solutions by bacterial strain isolated from soil. Chem Eng J 115(3):203–211
Veglio F, Beolchini F (1997) Removal of metals by biosorption: a review. Hydrometallurgy 44 (3):301–316
Viraraghavan T, Srinivasan A (2011) Fungal biosorption and biosorbents. In: Microbial biosorption of metals, Springer, Netherlands pp 143–158
Wang J, Chen C (2006) Biosorption of heavy metals by *Saccharomyces cerevisiae*: a review. Biotechnol Adv 24(5):427–451
Zalar P, Gostinčar C, De Hoog GS, Uršič V, Sudhadham M, Gunde-Cimerman N (2008) Redefinition of *Aureobasidium pullulans* and its varieties. Stud Mycol 61:21–38
Zodape GV (2014) Metal contamination in commercially important prawn and shrimp species collected from malad market of Mumbai suburb of India. Nat Environ Pollut Technol 13 (1):125

Author Biographies

Prof. Ekambaram Nakkeeran is working as Professor and Head of Biotechnology Department at Sri Venkateswara College of Engineering, Chennai, India. He holds M.Tech. (2005) and Ph.D. degree in Biotechnology (2011). He authored 12 publications and presented 34 research papers.

He received travel grants from SERC, CSIR and SERB, New Delhi to present research papers at International Conferences held in Malaysia (2008), Egypt (2009) and USA (2014). His research expertise are Bioprocess and Downstream Processing of biomolecules and synthesis and application nanoparticles and value-added products from cactus pear. He is guiding 1 PDF, 2 Ph. D. and 2 M.S. (By Research) Scholars.

Ms. Ravichandran Rathna is pursing M.S (by Research) in the department of biotechnology under the guidance of Prof. E. Nakkeeran, Head of Biotechnology Department at Sri Venkateswara college of Engineering, Chennai. She holds B.Tech (2014) degree in Biotechnology. She authored 1 publication and presented research papers in international/national conferences. Her area of specialization are Bioprocess and Downstream Processing, synthesis of phospholipid based nanoparticles and Molecular Modeling.

Ms. Ravichandran Viveka holds B.Tech (2015) and M.Tech (2017) degree in Biotechnology. She authored 1 publication and presented research papers in international/national conferences. Her area of specialization are Bioprocess and Downstream Processing.

Chapter 12
Landfill biodegradation process and leachate

Rajkumar Joshi and Deepak Pant

Abstract Increasing landfill sites leads to serious environmental problems due to the generation of leachates. Leachates are the potentially hazardous waste from landfill sites and consist of many organic and inorganic, dissolved suspended, biodegradable or non-biodegradable matters. In this chapter, we are discussing various process involved during landfill biodegradation, generation of leachate and factor affecting biodegradation process. Microbes played an important role in the biodegradation and bioremediation of landfill waste and leachates. The different kind of microbes used organic material present in landfill and leachate for their metabolic activities and resulted in the production of gases like CH_4 and CO_2. The biodegradation processes of waste material depend on the various factors including characteristics of waste materials, moisture content, temperature, pH and the presence of inhibitors.

Keywords MSW · Landfill · Leachate · Biodegradation · Waste management

1 Introduction

Landfills are usually found operative for a span of 25–35 years in developing countries and often accept heterogeneous throughout. The waste present can be in various stages of decomposition with different rate, i.e. most recently dumped waste had different phase of decomposition than the previously dumped waste in given area. Common types of landfills include (i) a ship's system, usually arranged in cells wastes, (ii) storage space, (iii) a leachate collector, (iv) gas collector and (v) cap (Andrés et al. 2007; Chandra et al. 2005; Erses et al. 2008). Leachates are an

R. Joshi
Department of Science and Technology, New Delhi, India
e-mail: rkjoshi.serb@gmail.com

D. Pant (✉)
Department of Environmental Sciences, Central University of Himachal Pradesh, Dharamshala 176215, Himachal Pradesh, India
e-mail: dpant2003@yahoo.com; dpant2003@gmail.com

S. J. Varjani et al. (eds.), *Waste Bioremediation*, Energy, Environment, and Sustainability, https://doi.org/10.1007/978-981-10-7413-4_12

aqueous effluent generated as a result of precipitation that passes through landfill site and decomposition of waste within landfill. Leachates are rich in organic and inorganic matter including ammonia-nitrogen, heavy metals, chlorinated organic and inorganic salts (Wiszniowski et al. 2006). The discharge of landfill leachate causes serious environmental problems leading to proper treatment of leachate. Some of the conventional leachate treatment methods are air stripping, coagulation, flocculation, reverse osmosis, active carbon adsorption and settling. Landfill leachates are rich in ammonium nitrogen, and air stripping is the most common method for removing a high concentration of NH_4^+–N. However, if ammonia is not properly absorbed, then NH_3 is released into the atmosphere and causes severe air pollution. Coagulation and flocculation are used as pre- and post-treatment of leachates to remove non-biodegradable organic matter. Active carbon adsorption reduces COD levels, and the main drawback is the need for frequent regeneration of columns. All these treatment methods are costly and create environmental pollution (Renou et al. 2008).

The bacteria present in these landfills decompose the waste in four major steps, and leachate as well as gas produced varies in all the stages/steps. These phases are (i) hydrolysis; (ii) acedogenesis; (iii) acetogenesis and methanogenesis; and (iv) settlement phase. Figure 1 shows decomposition phases and dominant chemicals transformation at these stages.

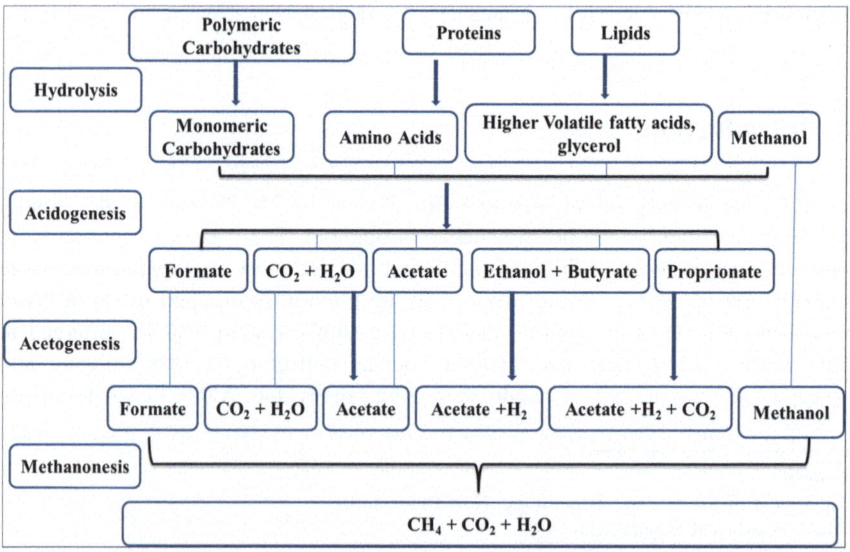

Fig. 1 Phases of waste decomposition in landfills

1.1 Hydrolytic Phase (Phase I)

The first stage of waste decomposition in landfill depends on the amount of oxygen present. Phase I remains to continue until the available oxygen is depleted, and decomposition of waste varies according to the compactness of the substrate. The first phase is therefore called the hydrolytic phase which occurs under aerobic conditions and in the presence of aerobic bacteria that convert the complex carbohydrates, protein, lipid and other long-chain molecular constituents of organic waste into simple constituents via consumption of oxygen. Carbon dioxide is the primary by-product of this process. Nitrogen content is high at this stage, but when the available oxygen is completely consumed, landfill moves to another phase.

1.2 Acidogenic Phase (Phase II)

This phase is primarily acid fermentation stage which is usually established after about 2 weeks to 2 months after the depletion of oxygen. In this phase, cellulose, proteins and fats were biodegraded into semi-decomposed substances in the absence of oxygen that results in production of short-chain fatty acids, CO_2 and H_2. Maximum production of CO_2 and H_2 is observed during this period which is an anaerobic phase. Under this condition, the bacteria consume organic constituents and convert them into simpler compounds, which are acidic in nature such as lactic, acetic and formic acids and also alcohols like methanol and ethanol. Under this condition, the landfill is highly acidic in nature and acts as acid bath. However, if disturbance occurs in landfill and oxygen is introduced, the phase reverts to the hydrolytic phase.

1.3 Acetogenic and Methanogenic Phase (Phase III)

This stage takes approximately 3–4 months for its establishment. Methane and acid-producing bacteria are in symbiosis during this phase. Methanogenic bacteria consumed compounds from acid-producing bacteria. Following are the major reactions:-

$$4H_2 + 2HCO_3^- + H^+ \rightarrow CH_3COO^- + 4H_2O \text{ (Acetogenesis)}$$

$$4H_2 + 2HCO_3^- + H^+ \rightarrow CH_4 + 3H_2O \text{ (Hydrogen based methanogenesis)}$$

$$CH_3COO^- + H_2O \rightarrow CH_4COO^- + HCO_3^- \text{ (Acetate based methanogenesis)}$$

The main degradation products during this phase are methane (CH_4) and carbon dioxide (CO_2). CH_4 and CO_2 are the major landfill gases (LFG) with relative amounts of 40–45 and 55–60% by volume, respectively, for CH_4 and CO_2. The factors that influence methane generation in landfill are the composition of the waste and availability of readily biodegradable organic matter, the age of the waste, moisture content, pH and temperature. Three processes that lead to the formation of landfill gas are bacterial decomposition, volatilization and chemical reactions. Gas production rates get stabilized within 2–3 years and the fourth phase begins. It may be noted that natural gas has about 90–95% of methane and landfill gas about 45–60% of methane. Landfill gas has ignition point at 700 °C with dry density 1.2 kg/m^3 and heat value 5.0–7.5 kWh/m^3 (Balat and Balat 2009).

1.4 Settlement Phase (Phase IV)

In the settlement phase, the decomposition rate and production of landfill gas remain relatively stable according to prevailing condition. In this phase, the landfill gas volume is usually about 45–60% methane, 35–45% carbon dioxide and 2–10% other gases such as sulphides, water vapour. The rate of production of gases remains constant in phase IV usually for about 20 years in conventional landfill. If waste contains sufficient quantity of organics, then the gases can continue to be emitted for more than 50 years in the landfill.

2 Biological Treatment of Leachates

Depending on requirement of O_2, biological treatment processes are classified as aerobic or anaerobic. In aerobic processing, atmospheric O_2 is utilized for conversion of organic pollutants into CO_2 and sludge. In anaerobic treatment, organic matter is converted into biogas. Conventional activated sludge processes and trickling filters have been widely studied and adopted. Trickling filters reduces biological nitrogen from municipal landfill leachate. Above 90% nitrification of leachate in laboratory and also on site pilot was achieved with loading rate of 100 to 130 mg L^{-1} d^{-1} of NH3–N at 25 °C by Abbas et al. (2009). Activated sludge processes are widely applied for the elimination of organic carbon, nutrients and ammonia content and produced effluent with 150–500 mg L^{-1} of COD, BOD < 7 mg L^{-1}, NH_3–N < 13 mg L^{-1} (Jokela et al. 2002).

Biodegradation which occurs through wide variety of microorganisms in landfills involves organic compounds. The different kind of microbes uses biodegradable material for their metabolic activities and thereby produces CH_4 and CO_2 and other trace gases. To adjust, control and optimally design a recirculation landfill, it is important to know the type of microbial consortia present and its behaviour in landfill, especially those microbes which have explicit roles. The bacterial metabolisms play a crucial role in the stabilization of solid wastes through their complicated interactions with various physical and chemical reactions.

The most credible work done by microbes for the carbon cycle via biodegradation of solid waste is disintegration of organic material into carbon dioxide, methane, hydrogen, ammonia, sulphide and water. Initially, degradation of cellulose, lignin, proteins and polysaccharides are accomplished by extracellular enzymes, followed by fermentation of the monomeric constituent. Number of reactions to decompose organic fraction and metals are carried by microorganisms in the landfills. The studies have been conducted to get the results of group of consortia to achieve specialized function. Microbial metabolisms play a vital role to stabilize substrate through their complex interfaces with the organic and inorganic matter through different physicochemical responses. With the passage of time and degree of waste stabilization, the role and types of microbes taking part in different activities required to be analysed to achieve optimal performance. The types of microbes taking part in degradation of decomposable solid waste are discussed in the proceeding section.

2.1 Archea

In landfills, three types of processes for methanogenesis exists to produce methane, which are conversion of acetate present in waste to methane (acetoclastic), reduction of carbon dioxide and hydrogen to methane (under hydrogenotrophic process with the help of methanomicrobiales and methanobacteriales) and reduction of methyl hydroxide, methyl amines and dimethyl sulphur oxide to methane (under methylotrophic process with the help of methanosarcinales). In anaerobic regime, the dominance of *Crenarchaeota*, *Methanosarcinales*, *Methanomicrobiales, Methanobrevibacter*, *Methanospirillum*, *Methanococcus sp. Methanoculleus sp.*, *Methanomethylovoransbn*, *Methanosphaerula Methanoculleusbourgensis*, *Methanofollisformosanus*, *Methanosphaerastadtmanae* are found. *Methanoculleuspalmolei, Symbiont trimyema compress and Methanosarcina dominance* are found in hybrid landfill while methanosphaera, *M. palmolei* and methanosarcina are found in aerobic landfill (Calli et al. 2006; Thauer et al. 2008; Sang et al. 2009; Cardinali-Rezende et al. 2009; Li et al. 2009; Nayak et al. 2009; Bareither et al. 2013). Figure 2 depicts common archea present in landfill.

2.2 Bacteria

Under anaerobic regime, different kinds of bacteria play an important role to degrade waste. The prevailing bacteria's are *Actinobacteria*, *Bacteroidetes*, *Bacillus*, *Chloroflexi*, *Clostridium*, *Escherichia*, *Enterobacter*, *Empedobacter*, *Flavobacterium Firmicute Lactobacillus*, *Leuconostoc, Listonella*, *Klebsiella*,

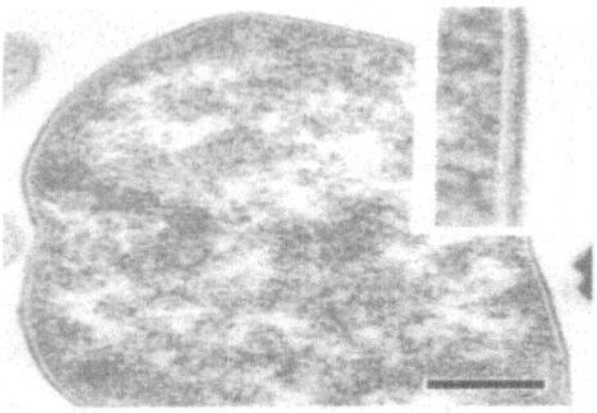

Methanoculleuspalmolei (Zellner *et al.*, 1998)

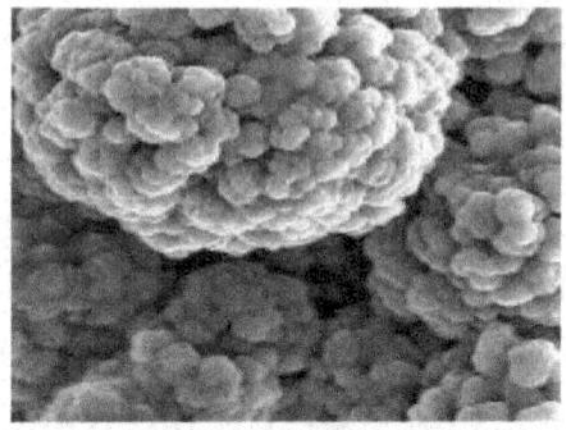

Methanosarcina dominance (Liu et al., 1985)

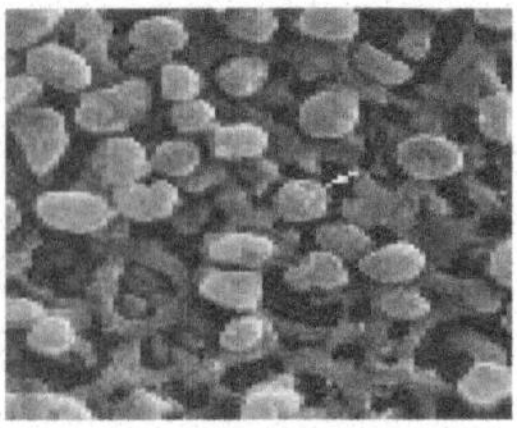

Methanofollis formosanus(Wu et al., 2005)

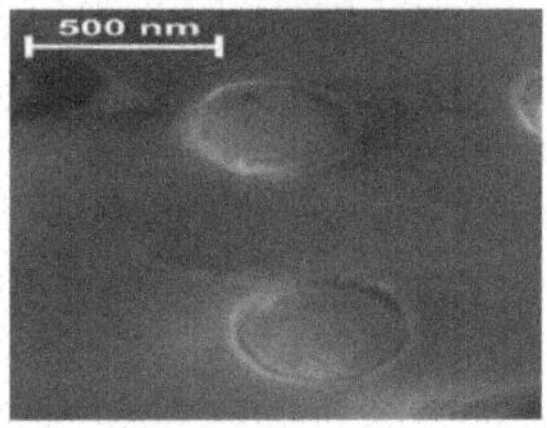

Methanoculleus bourgensis (Strapoc et al., 2008)

Fig. 2 Various species of archea

Proteobacteria, *Providentia*, *Pseudomonas Rhodospirillum Sporanaerobacter*, *Schineria*, *Thiomicrospira*, *Thermacetogenium*. For protein-rich biomass, the active bacterial consortia include *Flavobacterium* and *Rhodospirillum*. Acids and sugar content were used by *Clostridium* and *Bacillus* for their metabolic activities. Bacteria which reduce sulphur content are *Desulfobacter*, *Desulfococcus* and *Desulfonema*. Bacteria of the genus *Azospirillum*, which degrade various nitrogen-containing organic matters, were also found in anaerobic regime. Uncultured species *Bacteroidetes bacterium* and *Thermoactinomyces dichotomicus* can degrade fatty acids, sugars and amino acids. *Methanotrophs* have attracted a great deal of attention because they consume 10–90% of methane, a greenhouse gas, before it is emitted to the atmosphere. *Methanotrophs* help to reduce greenhouse gases like methane and nitrous oxide. Due to the unavailability of proper database, many bacteria found in the recirculatory regime and landfill still remain unidentified (Sei et al. 2006; Liamleam and Annachhatre 2007; De Visscher et al. 2007; Lee et al. 2009; Sang et al. 2009; Scheutz et al. 2009; Sawamura et al. 2010). Figure 3 depicts the microscopic view of common bacteria present in landfill. Common reactions by landfill bacteria are indicated below, where M = Zn, Cu, Co, Ni and Pb:

$$2FeSO_4 + \frac{1}{2}O_2 + H_2SO_4 \xrightarrow{\text{bacteria}} Fe_2SO_4$$

$$MS + Fe_2(SO_4)_3 \xrightarrow{\text{bacteria}} MSO_4 + 2FeSO_4 + Si$$

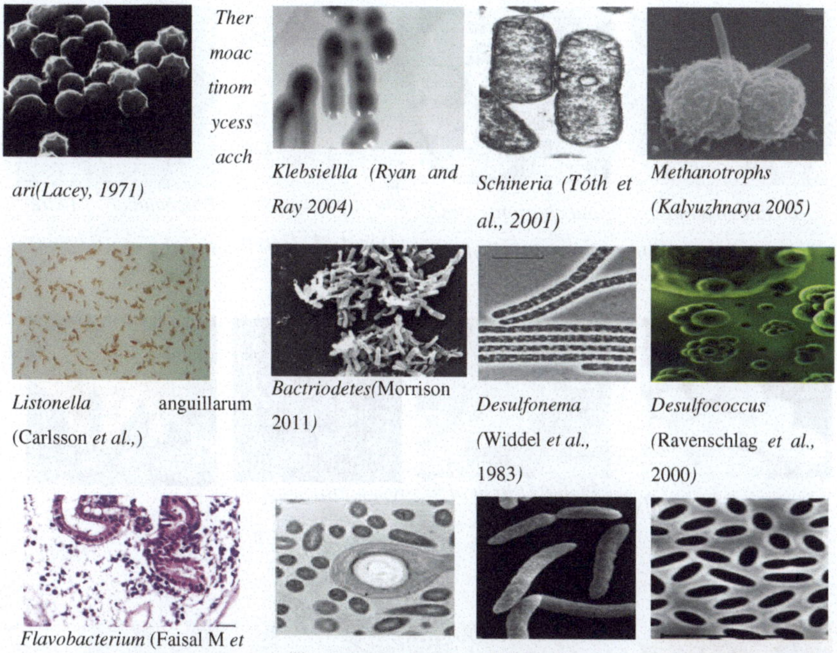

Thermoactinomycess acchari(*Lacey, 1971*)

Klebsiellla (*Ryan and Ray 2004*)

Schineria (*Tóth et al., 2001*)

Methanotrophs (*Kalyuzhnaya 2005*)

Listonella anguillarum (Carlsson *et al.,*)

Bactriodetes(Morrison 2011)

Desulfonema (Widdel *et al.*, 1983)

Desulfococcus (Ravenschlag *et al.*, 2000)

Flavobacterium (Faisal M *et al.*, 2011)

Thermacetogenium (*Hattori et al., 2000*)

Rhdodsprillium (*Munk et al., 2011*)

Desulfobacter (*Kuever et al., 2015*)

Fig. 3 Various species of bacteria

$$CaCo_3 + H_2SO_4 \xrightarrow{\text{bacteria}} CaSO_4 + H_2O + CO_2$$

$$MO + H_2SO_4 \xrightarrow{\text{bacteria}} MSO_4 + H_2O$$

2.3 Fungi

Fungi play an important role in degrading cellulosic and other biodegradable species in recirculatory and conventional regime. Fungi can survive in inadequate nutrient levels, low moisture and low pH conditions. Under aerobic regime, fungi's metabolic rate is enhanced due to degradation in organic matter. Development of the fungal population like *Basidiomycota Saccharomycetes*, *Basidiomycota Eurotiomycetes Rhizomucor*, *Geotrichum Pseudeurotium Pichia*, *Saccharomyces, Aspergillus, Mucor, Neocallimastigales* and *Candida* can putrefy miscellaneous carbon-based biomolecules (Lockhart et al. 2006) and enhance the stabilization of landfills. The anaerobic regime fungi play an important role in degradation of active cellulose. *Subphyla agaricomycotina, Mucoromycotina, Pezizomycotina, Pucciniomycotina* and *Saccharomycotina* and *Neocallimastigomycetes* plays

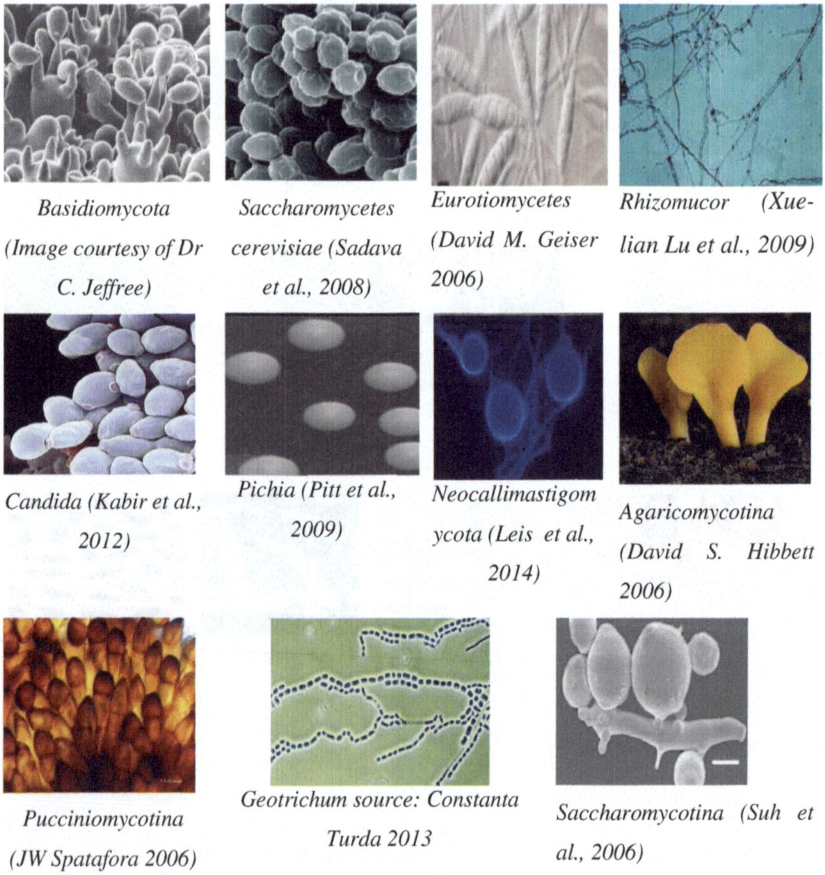

Fig. 4 Various species of fungi

important role to degrade cellulose present in the substrate (Lockhart et al. 2006). Figure 4 shows common fungi present in the landfill.

Microorganisms took part in the stabilization of heavy metals like Fe, Cd, Cu, Zn, Ni and organic matters in landfills directly or indirectly through different reactions and pathways. At low pH, leachate enhances metal solubility and reactions observed. Metals and metalloids also exert toxicity to some microorganisms and reduce response to biochemical activities; it changes cell morphology and affects growth. In the presence of sulphate-reducing bacteria, heavy metals sulphides are formed during acidic phase of the biodegradation under anaerobic conditions. Under aerobic conditions, too redox potential positive affects metal speciation and mobility.

Different responses of various biological microorganisms are due to heavy metals in landfills. Aerated landfill with/without recirculation of leachate has also driven various actions to stabilize organic waste and enhance metal solubility, adsorptions and absorption.

3 Composting

Composting is a natural biological phenomenon of breaking down the organic matter present in MSW under aerobic and humid environment and the product formed has high nutrient value known as humus (compost). According to Bhide and Shekdar (1998), composting is either labour-intensive or mechanical. Labour-intensive composting is carried out in smaller towns; while in big Indian cities, power-driven composting units are installed. An MSW composting centre installed at Indore city, Madhya Pradesh, is one of the best-maintained facilities. In Bengaluru, Vadodara, Mumbai, Delhi and Kanpur, mechanical composting units of 150–300 tonnes/day capacity were also installed (Sharholy et al. 2007).

3.1 Vermi-Composting

The compost produced by the help of worms is known as vermi-compost and the process is vermi-composting. Earthworms were well known for vermi-composting and can consume five times of organic matter per day in comparison to their body weight. Initially, biodegradable organic matter is decomposed through microbial enzymatic activity. In India, largest vermi-composting plant is located in Bengaluru having capacity of 100 million tonnes/day, and some small plants were located in Hyderabad, Bangalore, Mumbai and Faridabad. Details of composting and vermi-composting plants installed in different states are shown in Table 1.

Table 1 Number of composting/vermi-composting plants in some states

State	Number of plants (composting/ vermi-composting)	State	Number of plants (composting/ vermi-composting)
Andhra Pradesh	32	Madhya Pradesh	4
Chhattisgarh	15	Maharashtra	125
Delhi	3	Meghalaya	2
Goa	5	Orissa	3
Haryana	2	Punjab	2
Gujarat	86	Rajasthan	2
Himachal Pradesh	13	Tripura	13
Karnataka	5	Uttarakhand	3
Kerala	29	West Bengal	9

Source CPCB 2013

4 Aerobic and Anaerobic Digestions

Degradation of MSW in landfills occurs through various anaerobic processes. There are four processes in MSW degradation, which are assisted by microbes; they are hydrolysis, acid fermentation, acetogenesis and methanogenesis. Different kinds of microbes play different role in each of these processes. Fermentative microbes degrade the fresh MSW, which contains complex organic matter, into hydrolyzed products. The latter is then converted into volatile fatty acids and alcohols by acidogenic microbes. Acetogenic microbes cause fermentation of the partially stabilized MSW into acetic acid, which is eventually converted into gaseous products (methane and carbon dioxide).

Methanogenesis is the ultimate process reached by a stabilizing MSW. Methanogenesis of MSW wet mass occurs quickly when moisture content reaches 60% or above. Methanogenesis is favoured in the range of 6.0–8.0 pH. Alkalinity of around 2000 mg/L also facilitates methanogenesis. The temperatures between 35 and 45 °C enhance population growth of methanogenic microbes. Biomass maintaining nutrients are easily available in MSW (Yuen et al. 1995).

Production of methane gas is the indicators of stabilization phase of MSW. Different enhancing technical processes have been developed from time to time to quicken the stabilization of MSW. These processes include pretreatment, covering, compaction, buffering, seeding and liquid addition as well as water or leachate. The surface area available for biodegradation can be increased by shredding. Early composting with aeration before the anaerobic degradation can help in easy oxidation of organic matter. This pretreatment can result in the development of an intensive acid phase. Shredding can lead to the increase in the degradation rate, but its reverse effect has also been reported (Reinhart and Townsend 1997). Anaerobicity can be maintained in the system with the help of cover layer, and the volume of the waste in a particular landfill space can be increased by compaction. The advantages of setting up a cover layer over MSW fills are reported by Hurst et al. (2005). Excess production of volatile acid is prevented by adding buffer to the waste since its high levels stop the process of methanogenesis. It has been reported that low organic strength leachate is produced by controlling the pH value (Tittlebaum 1982). In case when methanogenic microbial biomass is present in low quantities in landfill environment, anaerobic sewage sludge and cow dung are being added to the waste to introduce it (Xi et al. 2005). Such cases require proper checking of the seed to prevent any inhibitor. Moisture is maintained by adding a liquid in the form of water or leachate because it is important for maintaining biological activity, and this has been reported in countries like Germany, USA, UK, Australia and Denmark where experiments on recirculatory MSW are being carried out (Reinhart and Townsend 1997).

4.1 Factors Affecting Degradation Process

The biodegradation processes of waste material in an anaerobic condition depends on the various factors including characteristics of waste materials, moisture content, temperature, pH, the availability of ingredients and the presence of inhibitors such as oxygen, nitrogen, sulphates and metals (EI-Fadel and Khoury 2000; Wall and Zeiss 1995; Edil et al. 1990; Barlaz et al. 1990). These factors have been indicated in Fig. 5.

The density of refuse in municipal solid wastes materials relies on the number of components which include the different nature of the solid waste materials, condition of available data, daily cover and the degree of soil compaction of landfill during placement (Zekkos et al. 2005). When waste becomes older, the density becomes more dependent on the climatic conditions, nature and degree of decomposition. All these factors rise to a spatial and natural variation of density and give weight to the landfill (Dixon and Jones 2005). From the literature survey, the density is found to be in the range of 321–694 kg/m^3 during the stage of filling of waste and between 1186 and 1653 kg/m^3 after landfill stabilization (Chakma and Mathur 2012).

Similarly, the amount of moisture present in waste depends on the initial composition of waste, local climatic conditions, rate of decomposition, operating

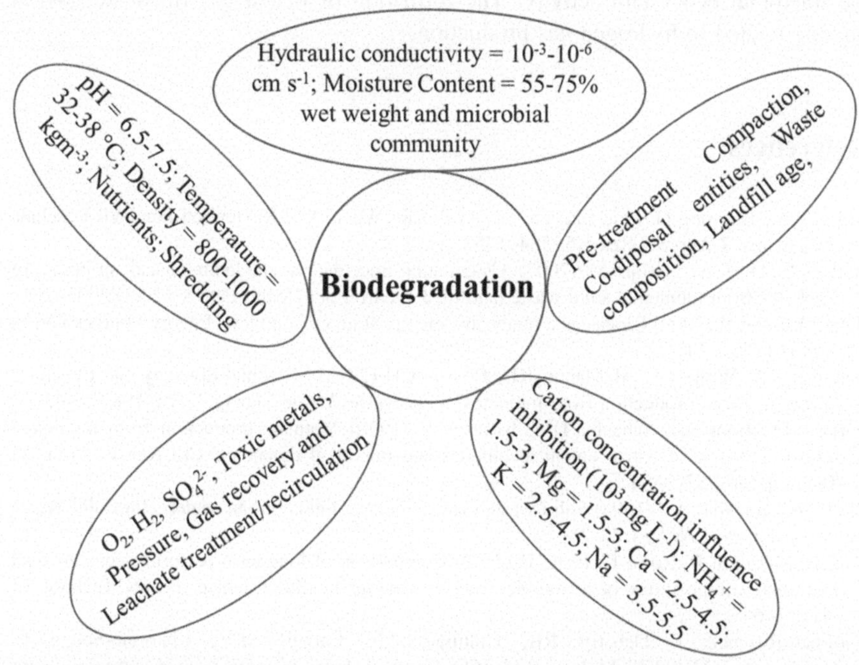

Fig. 5 Factors influencing biodegradation in landfills

condition and the available organic content in the refuse. Moisture content is an important factor that has the greatest influence on the biodegradation phenomena because it provides aqueous environment for easy transportation of minerals and microbes in the landfill. Also, the rate of production of methane enhances with an increase in water content of the waste. However, Pohland (1980) found that the presence of high water content does not confirm to either increase or decrease the rate of production of methane gas. On the other hand, it is also mentioned that high addition of moisture has a bad effect of over-cooling the process, which results in lowering the degradation rate because it depends on the nature of climate (Rovers and Farquhar 1973). Moreover, knowledge of the amount of moisture-holding capacity of the refuse is required to measure the amount of moisture that is required to be added in a landfill before any leachate is produced or removed through the bottom of the landfill. Based on the available source of literature, survey data of the field capacity of MSW landfills are in range of 14–44% (Zeiss and Major 1993).

Furthermore, the availability of oxygen also inhibits the process of anaerobic bacteria in a landfill. The atmospheric oxygen enters into the landfill with disposal of fresh waste, diffusion, rainwater infiltration and due to landfill slide failure also inhibits anaerobic process and aerobic bacteria present in the topmost layer again started waste oxidization with the help of aerobic bacteria. However, diffusion of atmospheric oxygen is limited up to the uppermost 1 m layer of the waste (Pohland 1980). Hydrogen works as both substrate as well as product of biodegradation phenomena in the various anaerobic processes; it plays vital role in the balance of the microbial ecosystem activity. The formation of marsh gas (methane gas) is directly related to hydrogen gas production.

References

Abbas AA., Jingsong G, Ping LZ, Ya PY, Al-Rekabi WS (2009) Review on LandWll Leachate Treatments. J App Sci Res 5(5):534–545

Andrés P, Diaz A, Cortijo M (2007) Coagulation-flocculation and ammoniacal stripping of leachates from municipal solid waste landfill. J Environ Sci Health Part A 42(13):2033–2038

Balat M, Balat H (2009) Biogas as a renewable energy source—a review. Energy Sources Part A 31(14):1280–1293

Bareither CA, Wolfe GL, McMahon KD, Benson CH (2013) Microbial diversity and dynamics during methane production from municipal solid waste. Waste Manag 33(10):1982–1992

Barlaz MA, Ham RK, Schaefer DM, Isaacson R (1990) Methane production from municipal refuse: a review of enhancement techniques and microbial dynamics. Crit Rev Environ Sci Technol 19(6):557–584

Bhide AD, Shekdar AV (1998) Solid waste management in Indian urban centers. Int Solid Waste Assoc Times (ISWA) 1:26–28

Calli B, Mertoglu B, Roest K, Inanc B (2006) Comparison of long-term performances and final microbial compositions of anaerobic reactors treating landfill leachate. Biores Technol 97 (4):641–647

Cardinali-Rezende J, Debarry RB, Colturato LF, Carneiro EV, Chartone-Souza E, Nascimento AM (2009) Molecular identification and dynamics of microbial communities in reactor treating organic household waste. Appl Microbiol Biotechnol 84(4):777–789

Carlsson O, Fernström LL, Johansson KE. Listonellaanguillarum, http://www.vetbact.org/popup/image.php?imgtable=vetbact_images&imgid=149

Census of India (2011) Registrar general of India. http://censusindia.gov.in/

Chakma S, Mathur S (2012) Postclosure long-term settlement for MSW landfills. J Hazard Toxic Radioact Waste 17(2):81–88

Chandra S, Chauhan LKS, Murthy RC, Saxena PN, Pande PN, Gupta SK (2005) Comparative biomonitoring of leachates from hazardous solid waste of two industries using Allium test. Sci Total Environ 347(1):46–52

De Visscher A, Boeckx P, Van Cleemput O (2007) Artificial methane sinks, 184200, In: Reay DS, Hewitt CN, Grace J (eds) Greenhouse gas sinks. Oxfordshire, Wallingford. https://doi.org/10.1079/9781845931896.0184

Dixon N, Jones DRV (2005) Engineering properties of municipal solid waste. Geotext Geomembr 23(3):205–233

Edil TB, Ranguette VJ, Wuellner WW (1990) Settlement of municipal refuse. In: Geotechnics of waste fills—theory and practice. ASTM STP, 1070, pp 25–239

El-Fadel M, Khoury R (2000) Modeling settlement in MSW landfills: a critical review. Crit Rev Environ Sci Technol 30(3):327–361

Erses AS, Onay TT, Yenigun O (2008) Comparison of aerobic and anaerobic degradation of municipal solid waste in bioreactor landfills. Biores Technol 99(13):5418–5426

Faisal M, Loch TP, Fujimoto M, Woodiga SA, Eissa AE, Honeyfield DC, Wolgamood M, Walkerand ED, Marsh TL (2011) Characterization of novel flavobacterium spp. Involved in the mortality of Coho salmon (Oncorhynchus kisutch) in their early life stages. J Aquac Res Dev S 2:2

Hattori S, Kamagata Y, Hanada S, Shoun H (2000) Thermacetogenium phaeum gen. nov., sp. nov., a strictly anaerobic, thermophilic, syntrophic acetate-oxidizing bacterium. Int J Syst Evol Microbiol 50(4):1601–1609

Hibbett DS (2006) A phylogenetic overview of the Agaricomycotina. Mycologia 98(6):917–925

Hurst C, Longhurst P, Pollard S, Smith R, Jefferson B, Gronow J (2005) Assessment of municipal waste compost as a daily cover material for odour control at landfill sites. Environ Pollut 135 (1):171–177

Jokela JP, Kettunen RH, Sormunen KM, Rintala JA (2002) Biological nitrogen removal from municipal landfill leachate: Low-cost nitrification in biofilters and laboratory scale insitu denitrification. Water Res 36:4079–4087. https://doi.org/10.1016/S0043-1354(02)00129-X

Kabir MA, Hussain MA, Ahmad Z (2012) Candida albicans: a model organism for studying fungal pathogens. ISRN microbiol 2012:1–15

Kalyuzhnaya MG, Stolyar SM, Auman AJ, Lara JC, Lidstrom ME, Chistoserdova L (2005) Methylosarcina lacus sp. nov., a methanotroph from Lake Washington, Seattle, USA, and emended description of the genus Methylosarcina. Int J Syst Evol Microbiol 55(6):2345–2350

Kuever J, Rainey FA, Widdel F (2015) Desulfobacter proteobacteria delta proteobacteria desulfo bacterales. In: Bergey's Manual of Systematics of Archaea and Bacteria. doi:https://doi.org/10.1002/9781118960608.gbm01011

Lacey J (1971) Thermoactinomyces sacchari sp. nov., a thermophilic actinomycete causing bagassosis. Microbiology 66(3):327–338

Lee SW, Im J, DiSpirito AA, Bodrossy L, Barcelona MJ, Semrau JD (2009) Effect of nutrient and selective inhibitor amendments on methane oxidation, nitrous oxide production, and key gene presence and expression in landfill cover soils: characterization of the role of methanotrophs, nitrifiers, and denitrifiers. Appl Microbiol Biotechnol 85(2):389–403

Li T, Mazéas L, Sghir A, Leblon G, Bouchez T (2009) Insights into networks of functional microbes catalysing methanization of cellulose under mesophilic conditions. Environ Microbiol 11(4):889–904

Liamleam W, Annachhatre AP (2007) Electron donors for biological sulfate reduction. Biotechnol Adv 25(5):452–463

Liu Y, Boone DR, Sleat R, Mah RA (1985) Methanosarcina mazei LYC, a new methanogenic isolate which produces a disaggregating enzyme. Appl Environ Microbiol 49(3):608–613

Lockhart RJ, Van Dyke MI, Beadle IR, Humphreys P, McCarthy AJ (2006) Molecular biological detection of anaerobic gut fungi (Neocallimastigales) from landfill sites. Appl Environ Microbiol 72(8):5659–5661

Lu XL, Liu ZH, Shen YN, She XD, Lu GX, Zhan P, Fu MH, Zhang XL, Ge YP, Liu WD (2009) Primary cutaneous zygomycosis caused by Rhizomucor variabilis: a new endemic zygomycosis? A case report and review of 6 cases reported from China. Clin Infect Dis 49(3):e39–e43

Morrison B (2011) Inside the microbiome: balancing bacteria for fat loss. http://liftbigeatbig.com/inside-the-microbiome-balancing-bacteria-for-fat-loss/

Munk AC, Copeland A, Lucas S, Lapidus A, Del Rio TG, Barry K, Detter JC, Hammon N, Israni S, Pitluck S, Brettin T (2011) Complete genome sequence of Rhodospirillum rubrum type strain (S1 T). Stan Genomic Sci 4(3):293–302

Nayak BS, Levine AD, Cardoso A, Harwood VJ (2009) Microbial population dynamics in laboratory-scale solid waste bioreactors in the presence or absence of biosolids. J Appl Microbiol 107(4):1330–1339

Pitt JI, Hocking AD (2009) Fungi and food spoilage, vol 519. Springer, New York

Pohland FG (1980) Leachate recycle as landfill management option. J Sanitary Eng Div 106 (6):1057–1069

Ravenschlag K, Sahm K, Knoblauch C, Jørgensen BB, Amann R (2000) Community structure, cellular rRNA content, and activity of sulfate-reducing bacteria in marine arctic sediments. Appl Environ Microbiol 66(8):3592–3602

Reinhart DR, Townsend TG (1997) Landfill bioreactor design & operation. CRC press

Renou S, Givaudan JG, Poulain S, Dirassouyan F, Moulin P (2008) Landfill leachate treatment: review and opportunity. J Hazard Mater 150(3):468–493

Rovers FA, Farquhar GJ (1973) Infiltration and landfill behavior. J Environ Eng Div 99(5):671–690

Ryan KJ, Ray CG (2004) Medical microbiology. McGraw Hill 4:370

Sadava DE (2008) Fungi: recyclers, pathogens, parasites, and plant partners. In: Life, the Science of Biology. Sunderland, MA: Sinauer Associates, p 665

Sang NN, Soda S, Inoue D, Sei K, Ike M (2009) Effects of intermittent and continuous aeration on accelerative stabilization and microbial population dynamics in landfill bioreactors. J Biosci Bioeng 108(4):336–343

Sawamura H, Yamada M, Endo K, Soda S, Ishigaki T, Ike M (2010) Characterization of microorganisms at different landfill depths using carbon-utilization patterns and 16S rRNA gene based T-RFLP. J Biosci Bioeng 109(2):130–137

Scheutz C, Kjeldsen P, Bogner JE, De Visscher A, Gebert J, Hilger HA, Huber-Humer M, Spokas K (2009) Microbial methane oxidation processes and technologies for mitigation of landfill gas emissions. Waste Manage Res 27(5):409–455

Sei K, Uchikawa Y, Sang NN, Ike M, Fujita M, Ishigaki T, Inanc B, Inoue Y, Mitsui K, Maeda S, Suzuki M, Kadokami K, Koezuka T (2006) Demonstration test of accelerated stabilization of coastal reclamation waste disposal model reactor applying leachate circulation. Environ Eng Res 43:319–325

Sharholy M, Ahmad K, Vaishya RC, Gupta RD (2007) Municipal solid waste characteristics and management in Allahabad, India. J Waste Manag 27(4):490–496

Spatafora JW (2006) A five-gene phylogeny of Pezizomycotina. Mycologia 98(6):1018–1028

Strąpoć D, Picardal FW, Turich C, Schaperdoth I, Macalady JL, Lipp JS, Lin YS, Ertefai TF, Schubotz F, Hinrichs KU, Mastalerz M (2008) Methane-producing microbial community in a coal bed of the Illinois Basin. Appl Environ Microbiol 74(8):2424–2432

Suh SO, Blackwell M, Kurtzman CP, Lachance MA (2006) Phylogenetics of Saccharomycetales, the ascomycete yeasts. Mycologia 98(6):1006–1017

Thauer RK, Kaster AK, Seedorf H, Buckel W, Hedderich R (2008) Methanogenic archaea: ecologically relevant differences in energy conservation. Nat Rev Microbiol 6(8):579–591

Tittlebaum ME (1982) Organic carbon content stabilization through landfill leachate recirculation. J Water Pollut Control Fed 54(5):428–433

Tóth E, Kovács G, Schumann P, Kovács AL, Steiner U, Halbritter A, Márialigeti K (2001) Schineria larvae gen. nov., sp. nov., isolated from the 1st and 2nd larval stages of Wohlfahrtia magnifica (Diptera: Sarcophagidae). Int J Syst Evol Microbiol 51(2):401–407

Wall DK, Zeiss C (1995) Municipal landfill biodegradation and settlement. J Environ Eng 121 (3):214–224

Widdel F, Kohring GW, Mayer F (1983) Studies on dissimilatory sulfate-reducing bacteria that decompose fatty acids. Arch Microbiol 134(4):286–294

Wiszniowski J, Robert D, Surmacz-Gorska J, Miksch K, Weber JV (2006) Landfill leachate treatment methods: a review. Environ Chem Lett 4(1):51–61

Wu SY, Chen SC, Lai MC (2005) Methanofollis formosanus sp. nov., isolated from a fish pond. Int J Syst Evol Microbiol 55(2):837–842

Xi B, Zhang G, Liu H (2005) Process kinetics of inoculation composting of municipal solid waste. J Hazard Mater 124(1):165–172

Yuen STS, Styles JR, McMahon TA (1995) An active landfill management by leachate recirculation–a review and an outline of a full-scale project. In: Proceedings sardinia, vol 95, pp 1–17)

Zeiss C, Major W (1993) Moisture flow through municipal solid waste: patterns and characteristics. J Environ Syst 22(3):211–231

Zekkos DP, Bray JD, Kavazanjian E, Matasovic N, Rathje E, Riemer M, Stokoe KH (2005) Framework for the estimation of MSW unit weight profile. In: Proceedings, Sardinia '05, 10th international waste management and landfill symposium, October, Santa Margherita di Pula, Cagliari, Italy, pp 3–7

Zellner G, Messner P, Winter J, Stackebrandt E (1998) Methanoculleuspalmolei sp. nov., an irregularly coccoid methanogen from an anaerobic digester treating wastewater of a palm oil plant in North-Sumatra, Indonesia. Int J Syst Bacteriol 48(4):1111–1117

Author Biographies

Dr. Rajkumar Joshi is working in Department of Science and Technology (DST), New Delhi; a Government Department which has a mandate to promote and manage basic scientific research in India. His broad areas of interest are management of research, innovations and its development. The author also worked on solid waste management. He has about 25 years working experience.

Dr. Deepak Pant is currently the Dean and Head of Environmental Science, Central University of Himachal Pradesh, Dharamshala. He received the Silver Jubilee Research Fellowship Award in 2003 from Kumaun University, Nainital, India; the UCOST Young Scientist Award 2009, INSA Fellow (2010), DST-SERC, Visiting Fellow (2010) and DST-SERC Young Scientist Award (2011). He has 5 patents in the area of waste management using green techniques and published 10 books, 45 research papers in various national and international journals. He recently received Visitor's Award 2017 from the President of India for his innovations in the area of waste management.

Chapter 13
Microbial Transformation of Heavy Metals

E. Raja Sathendra, R. Praveen Kumar and G. Baskar

Abstract In natural environments, the average abundance of heavy metals is generally low and much of that sequestered in sediments, soil and mineral deposits may be biologically unavailable. Microorganisms have ability to adapt and live in all ecological condition. In natural habitat, the cause for microbes on heavy metal depends on the physico-chemical properties of the environmental condition. Microbes can metabolize the metal ion and yield energy through oxidation and reduction process by dissolving them. Many trace metals are necessary for growth and metabolism at low concentrations, (e.g. Co, Cu, Ni, Mo, Fe, Zn), and microorganism acquires mechanisms of varying specificity for the intracellular increase from the external environment. The molecular mechanism of microorganism and plants in the removal of toxic heavy metals into nontoxic form using plants and microorganisms is well studied, and this has many biotechnology implications in the bioremediation of heavy metal contaminated sites.

Keywords Heavy metal · Biotransformation · Biodegradation
Bioremediation

1 Introduction

There are nearly sixty-five elements, which may be termed as heavy metal as they exhibit metallic properties. In the ecosystem, organic toxic contaminants cause harmful effect because of releasing hazardous waste such as lead, cadmium and chromium by the increasing technological advancement. A wide variety of activi-

E. Raja Sathendra (✉) · R. Praveen Kumar
Department of Biotechnology, Arunai Engineering College,
Tiruvannamalai 606603, India
e-mail: e.rajasathendra@gmail.com

G. Baskar
Department of Biotechnology, St. Josephs College of Engineering,
Chennai 600119, India

S. J. Varjani et al. (eds.), *Waste Bioremediation*, Energy, Environment, and Sustainability, https://doi.org/10.1007/978-981-10-7413-4_13

ties by industries have generated the mobilization of large heave metal in the elevated point to the natural geochemical system (Gaur 2014; Dixit 2015; Tak 2013).

The non-degradable nature of toxic heavy metal pollution is the important environmental trouble caused by metal ions that continue to present in the environment that is controlled biological system. Based on the dose of adsorption and course of exposure, toxic heavy metal causes a serious lethal effect at very high concentration and the toxic effect depends on the amount present to organisms (Sand et al. 1992). Bioremediation techniques are used to reduce or completely remove the toxic heavy metals from the environment (Dixit 2015; Akcil et al. 2015).

Microbes and plants are used to remediate the polluted environment through sustained technology and bring back natural environment condition (Doelman 1994). The change in the microbial make up is due to the degradation of function group in the heavy metal or alteration in the conformation of the molecules (Wood 1983; Li 1994).

The reaction of microbes to heavy metals is because of the concentration, and accessibility of toxic metal is a completed process which is managed by many factors like metal chemical property, medium type and the microbes (Goblenz et al. 1994). A perceptive of microbial responses towards heavy metals is also significant for the preservation and protection of natural and synthetic materials, since inorganic and organometallic compounds are still widely used as microbicidal agent in domestic and agricultural pest and pathogen control operations.

2 Toxic Heavy Metals in the Environment

Industrial actions have increased the mobilization of many heavy metals above rates of natural geochemical cycling, and there is improved deposition in aquatic and terrestrial ecosystem, as well as release into the atmosphere. The heavy metal contamination of the soil and water is mainly due to manmade activity and by natural. Table 1 shows the sources of certain heavy metals in the environment and its impact.

3 Bioremediation

Bioremediation is process of using microorganisms to remove polluted environment or to prevent pollution. The basis of bioremediation is the inherent natural capacity of microorganism to degrade organic compound present in the environment.

Table 1 Shows certain heavy metals in the environment with their significance (Donald 2003)

Elements	Source	Effects and significance
Arsenic	Pesticides, chemical	Toxic and carcinogenic
Beryllium	Nuclear power and space Industries and coal	Acute, chronic toxicity and Carcinogenic
Boron	Industrial wastes and coal	Plant toxic
Cadmium	Waste from mining, metal plating, water pipes and industrial discharge	Replaces zinc biochemically, lead to high blood pressure and damage to kidney
Chromium	Cooling-tower water additive (chromate), normally found as Cr (VI) in polluted water and metal plating	Important trace element, (glucose tolerance factor), and carcinogenic as Cr(VI)
Copper	Metal plating, industrial and domestic wastes, mining, mineral leaching	Necessary trace element, not very toxic to animals, cause plant toxicity and algae
Fluorine (Fluoride)	Industrial wastes, water additive and natural geological sources	Avoid decaying of tooth at above 1 mg/L, lead to mottled teeth and bone harm at around 5 mg/L in water
Iodine (Iodide)	Seawater intrusion, industrial wastes and natural brines	Prevents goitre
Iron	Industrial wastes, acid mine drainage, low pE water in contact with iron minerals and corroded metal	Essential nutrient (component of haemoglobin), not very toxic, damages material (bathroom fixtures and clothing)
Lead	Plumbing, coal, gasoline, industry, mining	Toxicity (anaemia, kidney disease, nervous system), wildlife destruction
Manganese	Acid mine drainage, microbial action on manganese minerals at low pE, mining, industrial waste	For animal nontoxic, but harmful to plants at elevated point and cause material to strain
Mercury	Mining, pesticides, coal, industrial waste	Acute and chronic toxicity
Molybdenum	Natural sources, cooling-tower water additive, industrial waste	Possible toxic to animals, essential for plants
Selenium	Coal, natural geological sources, sulphur	Low levels are necessary, at higher levels toxic, causes "alkali disease" and "blind staggers" to cattle
Silver	Disinfection of water, from natural sources and also mining	Causes blue-grey discolouration of skin, mucous membranes, eyes
Zinc	Metal plating, plumbing, industrial wastes	Important metalloenzymes, aids wound healing, harmful to plants at elevated levels

4 Bioremediation Methods

The toxic metal may remain in the form of solid or liquid phase; bioremediation technology will vary accordingly the waste material remains in its natural form or is taken to the fermentor. Bioremediation of toxic waste material treatment has two methods: in situ bioremediation and ex situ bioremediation.

5 Microbes in Metal-Containing Habitats

A wide variety of organism from all the major groups may be found in the metal-polluted habitats such as bacteria, fungi, yeast and algae (White et al. 1997; Vieira and Volesky 2000). Microorganisms can be isolated from almost any environmental conditions. In contaminated environments, there are some dominant microbes, which can tolerate the metal stress and grow more or less normally or even better, due to easy availability of resources.

Microorganisms can be subdivided into the following groups:

a. Aerobic

Pseudomonas, Alcaligenes, Sphingomonas, Rhodococcus and *Mycobacterium.* These microbes have ability to degrade pesticides and hydrocarbons, both alkanes and polyaromatic compounds. Microorganisms use the contaminant as the prime source of carbon and energy.

b. Anaerobic

Anaerobic bacteria are used for bioremediation of dechlorination of the solvent trichloroethylene (TCE) and chloroform, polychlorinated biphenyls (PCBs) in river sediments.

c. Lignilolytic fungi

Phanerochaete chrysosporium has the capability to remove an extremely diverse range of persistent or toxic pollutants from environment.

d. Methylotrophs

Aerobic bacteria take methane for carbon and energy. The primary enzyme in the metabolic pathway for aerobic degradation is the methane monooxygenase, has a wide range of substrate and is active against a large amount of compounds, together with the chlorinated aliphatic trichloroethylene and 1,2-dichloroethane. Table 2 shows some microbes utilizing heavy metals.

Table 2 Microbes utilizing toxic heavy metals

Microorganism	Elements	References
Bacillus spp. *Pseudomonas aeruginosa*	Cu, Zn	Philip et al. (2000), Gunasekaran et al. (2003)
Zooglea spp. *Citrobacter* spp.	U, Cu, Ni Co, Ni, Cd	Sar and D'Souza (2001)
Citrobacter spp. *Chlorella vulgaris*	Cd, U, Pb Au, Cu, Ni, U, Pb, Hg, Zn	Gunasekaran et al. (2003)
Aspergillus niger	Cd, Zn, Ag, Th, U	Gunasekaran et al. (2003)
Pleurotus ostreatus	Cd, Cu, Zn	Gunasekaran et al. (2003)
Rhizopus arrhizus	Ag, Hg, P, Cd, Pb, Ca	Favero et al. (1991), Gunasekaran et al. (2003)
Stereum hirsutum	Cd, Co, Cu, Ni	Gabriel et al. (1994, 1996)
Phormidium valderium	Cd, Pb	Gabriel et al. (1994, 1996)
Ganoderma applanatum	Cu, Hg, Pb	Gabriel et al. (1994, 1996)

6 Metal and Microbe Interaction

The capability of microbes to survive and grow in a contaminated metal habitat may depend on genetical and physiological adaptation. The mechanisms of metal ions binding to the cell surface include electrostatic interactions, van der Waals forces, covalent bonding, redox interactions and extracellular precipitation or combination of these processes (Blanco 2000). The negatively charged groups (carboxyl, hydroxyl and phosphoryl) of bacterial cell wall adsorb metal cations, which are then retained by mineral nucleation. Biosorption of metal such as U, Zn, Pb, Cd, Ni, Cu, Hg, Th, Cs, Au, Ag, Sn and Mn shows the extent of sorption which varies clearly with the metal and microorganisms.

7 Microbes with Metal Tolerance

Some microorganisms are supposed to have developed metal resistance because of their contact with toxic metals in an already metal-polluted world (Gupta et al. 1993). In response to metal resistance in the environment, microorganisms have evolved ingenious mechanisms of metal resistance and detoxification. Some resistance mechanisms are plasmid encoded and tend to be specific for a particular metal. Others are general resistance to a variety of toxic heavy metals. There are two mechanisms of metal resistance:

- General mechanism of metal resistance,
- Metal-dependent mechanism of metal resistance.

7.1 General Mechanism of Metal Resistance

Metal binding on to extracellular materials immobilizes the metals and prevents the entry into the cell. Metal binding to anionic cell surfaces occurs with a large number of cationic metals, including cadmium, lead, zinc and iron. For example, algal cell surfaces contain carboxylic, amino, thio, hydroxo and hydroxyl carboxylic groups that strongly bind metals. Phosphoryl groups and phospholipids in bacterial lipopolysaccharides in the outer membrane also strongly interact with cationic metals.

7.1.1 Exopolymer Binding

The major molecules which constitute the exopolymeric substances include nucleic acids, carbohydrates, fatty acids and polysaccharides. These extracellular polymeric substances (EPSs) are very general in natural environments, and they provide protection against desiccation, phagocytosis and parasitism in addition to their role in heavy metal binding. They bind metal such as lead, cadmium and uranium.

7.1.2 Siderophore Complexation

It is an iron-chelating organic molecule. Their biological function is to concentrate iron in the environment where their concentration is very low and to transport iron into the cell. The interactions of siderophore with other metals are chemically similar to iron, viz., aluminium, gallium and chromium; for example, siderophores in cyanobacteria reduce copper toxicity.

7.1.3 Biosurfactants Complexation

They have ability to complex metals such as lead, cadmium and zinc. This apparently enhances the solubility of metals.

7.1.4 Precipitation by Metal Reduction

In this, soluble metals are reduced to less soluble metal salts, as well as sulfidic and phosphatidic metal salts. For example, *Citrobacter* spp. under aerobic condition can enzymatically produce phosphates which results in the precipitation of lead and copper. In anaerobic condition, *Desulfovibrio* spp., can have ability to precipitate sulphate metal.

7.2 Metal-Dependent Mechanism of Metal Resistance

Metallothioneins are proteins that play pivotal role in heavy metal metabolism and in the management of various forms of stress. Metallothioneins are similar to phytochelatins of plants and have high number of cysteine residues in the protein and also in the fact that both are responsible for the detoxification of heavy metals. Metallothioneins have three characteristics, low molecular weight proteins, and have high cysteine content and high metal with organization of metal ions in clusters.

The metallothioneins have two categories. The class-I MTs are found in all the animal metallothioneins, and the class-II is present in microorganisms and plants.

Table 3 Microbial metal resistance mechanisms

Metals	Mechanism of resistance	Microorganisms	References
As	Reduction, efflux, intracellular sequestration	*Lactobacillus* sp. As-1	Nair and Pradeep (2002)
Cd	Intracellular and extracellular sequestration, efflux, reduction	*Gloeothece magna*, *Staphylococcus aureus*, *Fusarium oxysporum*, *Pseudomonas azotoformans*, *Pseudomonas* sp. H1	Mohamed (2001), Nies and Silver (1989), Ahmad et al. (2002), Nair et al. (2007), Roane et al. (2001)
Cr	Extracellular sequestration and reduction	*Streptomyces* spp.	Amoroso et al. (2001), Lioyd (2003)
Cu	Intracellular and extracellular sequestration	*Enterobacter* sp., *Saccharomyces cerevisiae*	Culotta et al. (1994)
Hg	Volatilization	*Clostridium glycolicum*AS1-1	Meyer et al. (2007)
Pb	Biosurfactant, extracellular sequestration	*Pseudomonas aeruginosa* BS2, *Enterobacter* sp. J1	Nair et al. (2007)
Zn	Efflux	*Ralstonia metallidurans CH34*	Nies (2003)

In mammalian archetype, the class-I MTs have been very well characterized and represent. The invariant arrangement of Cys within the protein provides the thiol for mercaptide bonds in arrangements typical of metal-thiolate clusters. These proteins are isolated from the cyanobacterium *Synechococcus* spp., *Escherichia coli* and *Pseudomonas putida* (Gupta et al. 1993). Table 3 shows microbial metal resistance mechanisms.

8 Microbial Immobilization and Transformation of Metals

Bioremediation strategy involves mechanisms like precipitation and solubilization which shows the microbial transformations of metal as a result of metal resistance mechanisms.

8.1 Extracellular Complexation

Extracellular polysaccharides have anionic properties which show the purpose of efficient biosorbents for metal cations. Metal ions interaction is generally considered a direct effect of the presence of negatively charged functional groups on these exopolymers; they are pyruvate, phosphate, hydroxyl succinyl and uronic acids. A pH reliant binding of positively charged cations can quickly occur, with stability constants in excess of those generally measured for humic substances and other naturally occurring ligands.

Many organic metabolites can be significant on detoxification of metals because of the chelating properties or complexation properties. Organic or inorganic acids produced by microorganism, including *Thiobacillus, Serratia, Pseudomonas, Bacillus, Penicillium and Aspergillus,* are competent to extract metals from solid substrates. The oxalic acid can precipitate metals as insoluble oxalates around cell wall efficiently done by metal chelator citric acid.

8.2 Extracellular Precipitation and Crystallization

Microbes like sulphate-reducing bacteria, e.g. *Desulfovibrio,* are concerned in the formation of sulphide deposits which contain large amounts of metals. Metal-resistant strains of *Klebsiella aerogenes* precipitate Pb, HG or Cd as insoluble sulphide granules on outer surfaces of cells, and particles of AgS are deposited on *thiobacilli* when growth in silver-containing sulphide leaching systems. Strains of the green algae *Cyanidium caldarium* can grow in acidic waters at 45 °C containing high concentrations of metal ions. Iron, copper, nickel, aluminium and chromium can be removed from solution by precipitation, at cell surfaces, as metal sulphides. Cells can contain up to 20% metal on a dry weight basis. Yeasts can also precipitate metals as sulphides in and around cells walls, and colonies may appear dark brown in the presence of copper (Gadd 1990).

Algae and fungi can promote Mn^{2+} oxidation in a variety of habitats and can become coated with manganic oxides. Other bacteria can become coated with oxidized iron compounds by metabolism-dependent and -independent processes. In *Thiobacillus ferrooxidans* and certain algae, Au^{3+} can be adsorbed to cell walls and the plasma membrane and reduced to particles of elemental Au. Reducing compounds produced by silver-resistant bacteria can reduce Agy^{2+} to metallic AgO, which results in silver deposition on glass surfaces of culture vessels or in growing colonies.

8.3 Transport-Related Metal Resistance

Metabolism-dependent intracellular transport may be a slower process than binding and is inhibited or halted by metabolic inhibitors, low temperatures, the absence of an energy source (e.g. glucose in heterotrophs, light in phototrophs) and uncouplers. In several algae, bacteria and yeasts, amounts accumulated by transport may greatly exceed amounts taken up by general biosorption. Rates of transport are also affected by the metabolic state of the cell and nature and composition of the growth medium. Integral to heavy metal transport systems are ionic gradients across the cell membrane, e.g. H^+ and/or K^+ together, generating the electrochemical gradient across the cell membrane, and the membrane potential.

A relationship between heavy metal transport into microbial cells and toxicity is often observed, with sensitive strains taking up more metal ions than resistant strains. When transport rates do not vary, the sensitive organisms have more intracellular toxic metal species than the resistant ones. It is therefore not surprising that decreased transport, impermeability or the occurrence of metal efflux systems constitute resistance mechanisms in many organisms. In several bacteria, including *Staphylococcus aureus* and *Bacillus subtilis*, active transport of Cd^{2+} depends on the cross-membrane electrical potential and this uptake system is highly specific for Mn^{2+}, as well as Cd^{2+}. Resistance may arise from reduced Cd^{2+} transport in gram-negative bacteria. The efflux of cadmium from the cells of actively growing *S. aureus* has been found to be an energy-consuming process, involving cell membrane ATPase. This $Cd^{2+}/2H^+$ antiport is presented in resistant strains which results in a net reduction of Cd^{2+} uptake. Cadmium-resistant *P. putida* also exhibits reduced uptake, and this may also be a result of defective transport and/or the presence of a Cd^{2+} efflux system. In an *Alcaligenes* sp., adaptation to Cd^{2+} resistance was associated with the synthesis of a new membrane protein, which may be involved in the prevention of Cd^{2+} influx or the cause of Cd^{2+} efflux. Membrane proteins that reduce bacterial uptake of Hg^{2+} have also been described. Some Cu^{2+}

resistance strains of *E. coli* exhibit reduced uptake and/or exclude Cu^{2+}, due to the disappearance of outer membrane proteins which may be involved in Cu^{2+} transport.

8.4 Intracellular Compartmentation and Detoxification

Certain bacteria, yeasts, fungi and algae can actively accumulate intracellular metal ions against a gradient. Once inside cells, metal ions may be compartmentalized and/or converted to more innocuous forms. Such processed can be effective detoxification mechanism, and microbes expressing them may be able to accumulate metals to high intracellular concentrations. It has been suggested that such mechanisms may be temporary and precede other means of expulsion of accumulated metals from the cells, possibly by means of vacuoles in eukaryotic microbes. A variety of electron-dense deposits, many of unknown compositions, have been recorded in cyanobacteria, bacteria, algae, protozoa and fungi after exposure to heavy metals. Polyphosphate has been implicated as a metal sequestering agent in *P. putida*, *Anabaena cylindrica* and *Plectonema boryanum*, a variety of eukaryotic algae and certain fungi and yeasts. In eukaryotic microbes, e.g. algae, yeasts and fungi, there may be compartmentation of accumulated Co^{2+}, Mn^{2+}, Zn^{2+} and K^{+} that is located in the vacuoles, where it may be in ionic form or bound to low molecular weight polyphosphates.

A common metal-induced response in many microorganisms is the synthesis of intracellular metal-binding proteins which function in detoxification and also the storage and regulation of intracellular metal ion concentrations. Metallothioneins are the most widely studied, and these are low molecular weight, cysteine-rich proteins that can bind metal like Cd, Zn and/or Cu. A Cd-binding prokaryotic metallothionein (molecular weight approx. 8100) was first described in the cyanobacterium *Synechococcus* sp. and others have since been reported in several bacteria including P. *putida* inducible Cd-binding proteins also occur in E. coli, but these are larger than metallothioneins and may be more related to recovery from cadmium toxicity.

8.5 Enzymatic Transformation of Metals

Microbes transform heavy metals, such as oxidation, reduction, methylation and demethylation, and these are significant not only to biogeochemical cycling, but may also constitute mechanisms of resistance. Metal transformation has concentrated on bacterial involvement in the mercury cycle, which can be simply represented as:

$$CH_3Hg^+ \underset{\text{methylation}}{\overset{\text{Demethylation}}{\rightleftarrows}} Hg^{2+} \underset{\text{oxidation}}{\overset{\text{reduction}}{\rightleftarrows}} Hg^0 \quad (1)$$

Mercury resistance to gram-positive and gram-negative bacteria is a common property, the determinants usually being plasmid encoded, particularly in gram-negative bacteria. The most ubiquitous mechanism of mercury resistance in bacteria is the enzymatic reduction of Hg^{2+}, by cytoplasmic mercuric reductase, to metallic HgO, which is less toxic than Hg^{2+}, volatile and rapidly lost from the environment. This has also been recorded in certain fungi and yeasts. Organomercurial compounds are enzymatically detoxified by organomercurial lyase which cleaves the Hg–C bond of, e.g. methyl-, ethyl- and phenyl-mercury, to form Hg^{2+} and methane, ethane and benzene, respectively. The Hg^{2+} can then be volatile by mercuric reductase. More than one kind of organomercurial lyase may be present in a given bacterium. In organomercurial resistance *E. coli*, two enzymes are active against methyl- and ethylmercury chloride. In a Pseudomonas strain, also there are two enzymes of similar specifies, as well as being able to cleave ρ-hydroxy-mercury benzoate.

In FAD-containing mercuric reductase is a flavoprotein has been studies for metal detoxification enzyme this, and organomercurial lyase, require an excess of thiols for activity. NADPH is the preferred reducing cofactor for mercuric reductase as well as organomercurial lyase, but in some bacterial strains, NADH is effective. The thiols prevent the formation of NADPH-Hg^{2+} complex and ensure that the Hg^{2+} is present as a dimercaptide. The reaction catalysed by mercuric reductase in vivo can be represented as:

$$RS - Hg \quad SR + NADPH + H^+ \longrightarrow HgO + NADP + 2RSH \quad (2)$$

Structurally and mechanically mercuric reductase is related to glutathione reductase and lipoamide dehydrogenase. The mechanism of mercuric reductase probably involves electrons being transferred from NADPH via FAD to reduce the active site cystine, converting it to two cysteine residues with titratable SH groups. One Cys residue forms a charge transfer complex with FAD. The active site cysteines then reduce Hg^{2+}, bound to the C-terminal cysteines, forming HgO. Mercuric reductase has been reported to exist as a monomer, dimer and trimer with a subunit molecular weight between 54,000 and 69,000, depending on the source, though it seems probable that the dimeric structure is active in vivo.

9 Genetic Aspects of Heavy Metal Resistance

The metal microbe genetic relations have been inadequate for mutant strains, primarily because understanding of the physiology and biochemistry of resistance is fragmentary in many cases. The areas, for which significant genetic advances have been made to date, are in heavy metal resistance in bacteria, cyanobacteria and the metallothionein of *Saccharomyces cerevisiae*.

9.1 Genetic Aspects of Heavy Metal Resistance in Bacteria

Many bacteria isolated from metal-contaminated environment shown mercury resistance. The mercury resistance determinants in bacteria (gram-positive and gram-negative) are usually plasmid encoded, predominantly in gram-negative bacteria. A common feature of the Hg^{2+} resistance determinants of bacteria from the environment is the absence of association to drug resistance markers in contrast to clinical bacterial isolates, where Hg^{2+} resistance is usually linked to genes for antibiotic resistance. Several mercury resistance transposons could be responsible for the widespread occurrence of Hg^{2+} resistance in gram-negative bacteria. To date, the most recent advances have stemmed from DNA sequencing studies of the mer genes located on transposons Tn501 and Tn21 from gram-negative bacteria, and such studies have been carried out with broad spectrum determinants, in both gram-negative and gram-positive bacteria (Gadd 1990).

9.2 Genetic Aspects of Heavy Metal Resistance in Fungi and Other Eukaryotes

In fungi, the copper-induced metallothionein of *S. cerevisiae* (molecular weight approx, 6573) has received most attention although a copper inducible metallothionein has also been described from *Neurospora crassa*. The yeast protein is inducible only by copper not cadmium or zinc and is therefore often referred to as Cu-MT or yeast MT.

Inducible Cd-binding proteins that are structurally different to Cu-MT have been isolated from *Schizosaccharomyces pombe*. A common name for these sulphur-rich metal-binding polypeptides is *phytochelatin*. The metal-binding peptides are composed of only three amino acids, L-cysteine, L-Glutamic and glycine and are analogous to similar peptides found in plant cell exposed to have metals like Cd, Cu, Hg, Pb and Zn. The general structure of two peptides found in Cd-exposed *S. pombe* is (g-Glu-Cys) Gly (n = 2 and 3). The unit peptides are called cadystin, and the phytochelatins $n = 2$ and $n = 3$ are called cadystin A and B, respectively. In addition to these, five homologous peptides with chain lengths from $n = 4$ and $n = 8$ are also observed in *S. pombe,* indicating that these are synthesized by elongation of the peptide with one (g-Glu-Cys) unit at the expense of and possibly, starting from glutathione. The cysteine-rich peptides of *S. pombe* can confiscate several transition metals and therefore share some basic features with metallothioneins. Multiple enzymes may be involved in phytochelatin synthesis, and therefore, mechanisms must exist for metal-dependent induction or activation of such enzymes.

10 Conclusion

The main strategy is to reduce the bioavailability, mobility and toxicity of the metal. Toxic heavy metals have adverse effects on aquatic and terrestrial ecosystems. All though various microorganisms are commonly found in polluted habitats and possess a range of morphological and physiological attributes that enable survival. Certain kinds of resistances may be genetically determined and transferrable, whereas, in other cases, tolerance may depend on intrinsic properties of the organism and/or detoxification due to environmental components. Many organisms can express a variety of survival strategies, and besides providing excellent model systems for scientific endeavour in fields ranging from ecology to molecular biology and genetics, they are also of biotechnological relevance to the detoxification of metals and for containing environmental metal recovery and reclamation for preservation of natural, synthetic materials and treatment of infection.

In relation to industrial application of microbial metal removal/recovery, most attention has so far been focused on high value elements of commercial importance. With non-precious metals and radionuclides, the economics are frequently unattractive and there is the problem of contaminated biomass. However, the containment of low-volume waste or elute is preferable to its entry into the environment. At the moment, environmental protection receives only limited political and industrial attention. However, with increasing pollution and public awareness of this, and the dangers of heavy metals and radionuclides to all living components of the biosphere, action may be taken and the fuller involvement of microbe-based technologies may be part of this.

References

Ahmad A, Mukherjee P, Mandal D, Senapati S, Khan MI, Kumar R, Sastry M (2002) Enzyme mediated extracellular synthesis of CdS nanoparticles by the fungus, *Fusarium oxysporum*. J Am Chem Soc 124:12108–12109

Akcil A, Erust C, Ozdemiroglu S, Fonti V, Beolchini F (2015) A review of approaches and techniques used in aquatic contaminated sediments: metal removal and stabilization by chemical and biotechnological processes. J Clean Prod 86:24–36

Amoroso MJ, Castro GR, Durain A, Peroud O, Oliver G, Hill RT (2001) Chromium accumulation by two *Streptomyces* spp. isolated from riverine sediments. J Ind Microbiol Biotechnol 26: p210–p215

Blanco A (2000) Immobilization of nonviable cyanobacteria and their use for heavy metal adsorption from water. In Oluguin EJ, Sanchez, Hernandez E (eds) Environmental biotechnology and cleaner bioprocesses. Taylor and Amp Francis, Philadelphia, p 135

Culotta VC, Howard WR, Liu XF (1994) CRS5 encodes a metallothionein-like protein in *Saccharomyces cerevisiae*. J Biol Chem 269:25295–25302

Dixit R, Malaviya D, Pandiyan K, Singh UB, Sahu A, Shukla R, Singh BP, Rai JP, Sharma PK, Lade H (2015) Bioremediation of heavy metals from soil and aquatic environment: an overview of principles and criteria of fundamental processes. Sustainability 7:2189–2212

Doelman P, Jansen E, Michels M, van Til M (1994) Effects of heavy metals in soil on microbial diversity and activity as shown by the sensitivity-resistance index, an ecologically relevant parameter. Biol Fertil Soil 17:177–1784

Favero N, Costa P, Massimino ML (1991) In vitro uptake of cadmium by basidiomycete Pleurotusostreatus. Biotechnol Lett 10:701–704

Gabriel J, Mokrejs M, Bily J, Rychlovsky P (1994) Accumulation of heavy metal by some Woodrooting fungi. Folia Microbiol 39:115–118

Gabriel J, Kofronova O, Rychlovsky P, Krenzelok M (1996) Accumulation and effect of cadmium in the wood rotting basidiomycete, Daedaleaquercina. Bull Environ Contam Toxicol 57:383–390

Gadd GM (1990) Heavy metal accumulation by bacteria and other microorganisms, vol 46, no 8. Springer, Berlin, pp 834–840

Gaur N, Flora G, Yadav M, Tiwari A (2014) A review with recent advancements on bioremediation-based abolition of heavy metals. Environ Sci Process Impacts 16:180–193

Goblenz A, Wolf K, Bauda P (1994) The role of glutathione biosynthesis in heavy metal resistance in the fission yeast *Schizosaccharomyces pombe*. FEMS Microbiol Rev 14:303–308

Gunasekaran P, Muthukrishnan J, Rajendran P (2003) Microbes in heavy metal remediation. Indian J Exp Biol 41(1):935–944

Gupta A, Morby AP, Turner JS, Whitton BA, Robinson NJ (1993) Deletion within the metallothionein locus of cadmium-tolerant *Synechococcus* PCC-6301 involving a highly iterated palindrome (hip1). Mol Microbiol 7:189–195

Li F, Tan TC (1994) Monitoring BOD in the presence of heavy metal ions using a poly (4-vinylpyridine) coated microbial sensor. Biosens Bioelectron 9:445–455

Lioyd JR (2003) Microbial reduction of metals and radionuclides. FEMS Microbiol Rev 27:411–425

Meyer J, Schmidt A, Michalke K, Hensel R (2007) Volatilization of metals and metalloids by the microbial population of an alluvial soil. Syst Appl Microbiol 31:81–87

Mohamed ZA (2001) Removal of cadmium and manganese by a non-toxic strain of the fresh water Cyanobacterium *Gloeothece magna*. Water Res 35(18):p4405–p4409

Nair B, Pradeep T (2002) Coalescence of nanoclusters and formation of submicron crystallites assisted by Lacto strains. Cryst Growth Des 2:293–298

Nair A, Juwarkar AA, Singh SK (2007) Production and characterization of siderophores and its application in arsenic removal from contaminated soil. Water Air Soil Pollut 180:p199–p212

Nies DH (2003) Efflux mediated heavy metal resistance in prokaryotes. FEMS Microbiol Rev 27: p313–p339

Nies DH, Silver S (1989) Plasmid determined inducible efflux is responsible for resistance to cadmium, zinc and cobalt in *Alcaligenes eutrophus*. J Bacteriol 171:896–900

Philip L, Iyengar L, Venkobacher L (2000) Site of interaction of copper on. Water Air Soil Pollut 119:11–21

Roane TM, Josephson KL, Pepper IL (2001) Dual-bioaugmentation strategy to enhance remediation of cocontaminated soil. Appl Environ Microbiol 67:p3208–p3215

Sand W, Rohde K, Sabotke B, Zenneck C (1992) Evaluation of *Leptospirillum ferrooxidans* for leaching. Appl Environ Microbiol 58:85–92

Sar P, D'Souza SF (2001) Biosorptive uranium uptake by Pseudomonas strain: characterization and equilibrium studies. J Chem Technol Biotechnol 76:1286–1294

Smedley PL, Kinniburgh DG (2002) A review of the source behaviour and distribution of arsenic in natural waters. Appl Geochem 17:517–568

Spark DL, Chemistry Environmental soil (2003) Academic Press. San Diego, California

Tak HI, Ahmad F, Babalola OO (2013) Advances in the application of plant growth-promoting rhizobacteria in phytoremediation of heavy metals. In: Reviews of environmental contamination and toxicology. Springer, New York, pp 33–52
Vieira R, Volesky B (2000) Biosorption: a solution to pollution? Int Microbiol 3:17–24
White C, Sayer JA, Gadd GM (1997) Microbial solubilization and immobilization of toxic metals: key biochemical processes for treatment of contamination. FEMS Microbiol Rev 20:503–516
Wood JM, Wang HK (1983) Microbiol resistance to heavy metals. Environ Sci Technol 17:582–590

Author Biographies

E. Raja Sathendra currently working as Assistant Professor in Biotechnology, Arunai Engineering College, Tiruvannamalai. He has published 5 papers in national journals. His research area is biofuel from lignocellulosic materials and Antibacterial activity using biologically synthesized nanoparticles.

Dr. R. Praveen Kumar is graduate in Chemical Engineering and postgraduation in Industrial Biotechnology from Annamalai University. He is working as Head, Department of Biotechnology, Arunai Engineering College, Tiruvannamalai, Tamil Nadu, India. His area of research includes Renewable Energy from Biomass and Municipal Waste. He has done provisional registration of four patents and published more than 52 papers in peer-reviewed journals. He was awarded with Outstanding Young Investigator Award for excellence in research and teaching in a rural setting, awarded by Raise.Rural.

Prof. G. Baskar is currently working as Professor of Biotechnology at St. Joseph's College of Engineering, Chennai, India. He has published more than 100 papers in national and international journals and 5 book chapters. His research expertise is in biofuels, therapeutic proteins, microbial enzymes, nanomedicine and nanocatalysis. He has received the International Bioprocessing Association's Young Scientist Award (2015) and the Indian Society for Technical Education (ISTE)-Syed Sajed Ali National Award (2016).

Chapter 14
Bioremediation of Metals from Lithium-Ion Battery (LIB) Waste

Tenzin Dolker and Deepak Pant

Abstract Technological advancement has greatly increased the demand for newer lithium-ion batteries (LIBs) due to the more use of advanced energy storage devices like electric vehicles, consumer electronics, renewable energy storage, backup power, medical devices. The existing methods for metal recovery from LIB recycling involved: (i) aqueous stream-based limited recycling using the liquid stream mixed with soda ash with the hammer mill and shaker table (ii) supercritical CO_2-based recycling of cathode and anode (iii) pyro- and hydrometallurgical processes. Microbial participation for the recovery of metals from waste (LIBs) was found to be an attractive method due to its environmental-friendly approaches. *Lysinibacillus, Micrococcus, Sporosarcina, Empedobacter, Barrientosiimonas, Lysinibacillus, Paenibacillus, Bacillus, Acidithiobacillus* are among the species involved in the recycling of metal form LIB.

Keywords Lithium-ion batteries · Bioleaching · Bioprecipitation Bioaccumulation · Biosorption

1 Introduction

Energy is the indispensable need of the today's technological advancing era. Batteries are electrochemical devices which stored energy in electrical form to be used when needed. But with every asset comes certain disadvantages and is same for batteries (Luo et al. 2015). Sony Co. in 1991 introduced rechargeable lithium-ion batteries (LIBs) into the market (Zhu et al. 2012; Sonoc and Jeswiet 2014) and is widely used in portable electronic gadgets (Guo et al. 2016).

LIBs had enviable characteristics, such as high energy density, high power density, good cycle life; less memory effect and low self-discharge due to which

T. Dolker (✉) · D. Pant
Department of Environmental Sciences, Central University of Himachal Pradesh, Dharamshala 176215, Himachal Pradesh, India
e-mail: dpant2003@gmail.com

S. J. Varjani et al. (eds.), *Waste Bioremediation*, Energy, Environment, and Sustainability, https://doi.org/10.1007/978-981-10-7413-4_14

they were widely used in portable electronic devices (Xin et al. 2016; Ku et al. 2016; Nan et al. 2005; Nayaka et al. 2015). The main constituents of lithium-ion battery are anode, cathode, electrolyte, and separator (Wang et al. 2014; Barik et al. 2016). Al foil and Cu foils were used as a current collector in cathode and anode, respectively (Dewulf et al. 2010). During charge and discharge in Fig. 1, lithium-ions migrate from cathode which is oxides of transition metals to the carbonaceous material of the anode, respectively (Alper 2002).

Percentage composition of cobalt, nickel, lithium, and plastics in LIBs consist of 5–20, 5–10, 5–7, 7–15%, respectively (Zeng et al. 2014; Xu et al. 2008). London metal exchange for August 2017 shows that cobalt is a relatively more expensive material than other battery constituents (Co > Ni > Cu > Al), so, its recovery is economically beneficial. Lithium also plays a crucial role in many industrial applications (Mantuano et al. 2006; Mishra et al. 2008).

Spent LIBs are not only the waste that needs to be disposed of but it can also be the resource that can be reused for the production of other products if recycled properly. Spent LIBs were produced in huge quantity in each year. About 36,000 tons of waste LIBs were produced only in the year 2014 (Guo et al. 2016). Mantuano et al. 2006 reported that the waste LIBs were composed (w/w) of 36 ± 9% Co, 5 ± 6% Li, 13% Cu, 8 ± 3% Al, 0.02% Ni, and by the recovery of metals from spent LIBs can save 51% of natural resources (Horeh et al. 2016). These wastes LIBs can be a secondary source for metals and can provide economic, environmental and social benefits if properly recovered (Fig. 2).

Worldwide production of LIB is about 36,000 tons in 2014 (Guo et al. 2016), and it is still growing. The increasing production of LIBs demands new methods for its management (Wang et al. 2014; Guo et al. 2016; Chen et al. 2015a, b, c). Toxicity due to LIBs waste is due to its flammable nature (Castilo et al. 2002; Vanitha and Balasubramanian 2013), also it can cause soil and groundwater pollution due to the leakage of the organic component (Meshram et al. 2015a, b; Ku et al. 2016). Recycling and recovery of the spent LIBs are an attractive means to avert environmental pollution and natural resource depletion. Pyrometallurgical,

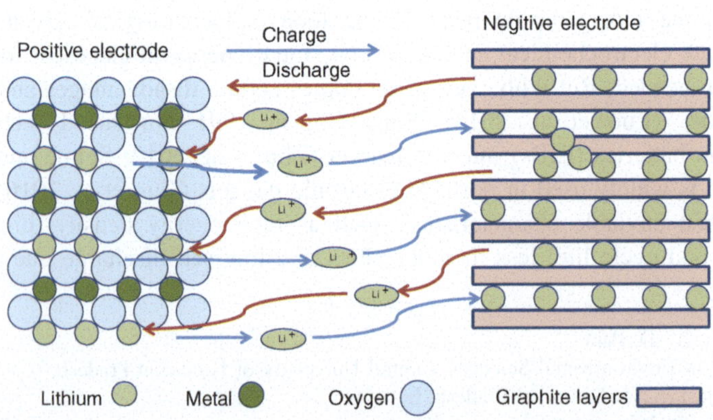

Fig. 1 Charge and discharge process in LIB (Alper 2002)

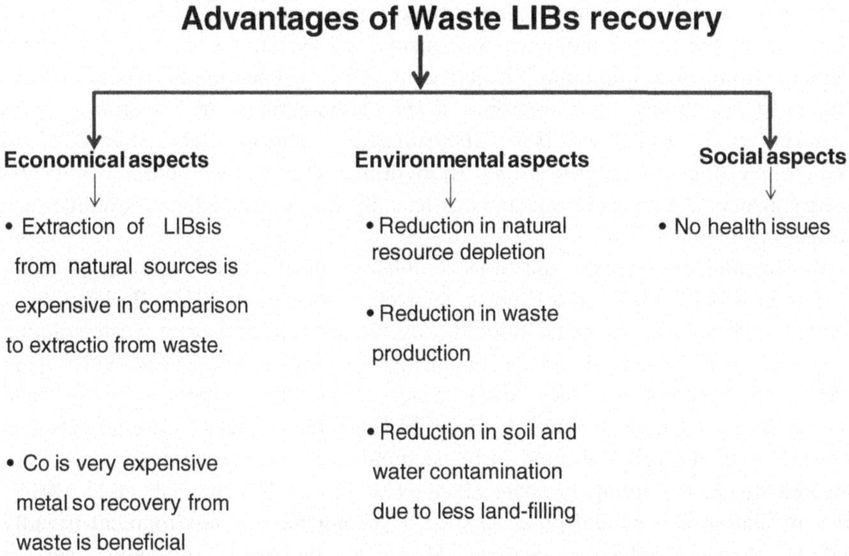

Fig. 2 LIBs recycling

hydrometallurgical and biometallurgical recycling processes were currently being used to recover metals from batteries (Dorella and Mansur 2007; Barik et al. 2016; Li et al. 2009; Freitas et al. 2010).

SUMIMOTO and INMETCO process is based on the pyrometallurgical process and were industrially used for the recycling of waste LIBs (Bernardes et al. 2004; Ferreira et al. 2009). Pyrometallurgy process mainly involves thermal treatment which involves burning off the organic compound and recovery of Co from spent LIBs (Zeng et al. 2014; Meshram et al. 2015a, b). However, the recovery of lithium is not possible and is a major drawback of pyrometallurgical recycling processes (Horeh et al. 2016; Georgi-Maschler et al. 2012). Other drawback of pyrometallurgical process involves loss of materials, high energy utilization and emission of hazardous gases like dioxins, mercury and chloride compounds (Nayaka et al. 2016a, b; Freitas et al. 2007, 2010) disadvantage.

The hydrometallurgical process of recycling is a widely used process and is very efficient in comparison to pyrometallurgy (Xin et al. 2016; Chagnes and Pospiech 2013). Hydrometallurgy is stepwise process which consists of pre-treatment, secondary treatment and purification processes. In pre-treatment, batteries were discharged, dismantled, sorted and shredded followed by secondary treatment like leaching, extraction, crystallization and precipitation (Hanisch et al. 2015). Leaching process is carried out in acid or alkaline medium followed by precipitation to recover metal from leaching liquor. The shortcomings of pyro- and hydrometallurgical processes advocate environment-friendly biotechnological strategies for metal recovery.

Biohydrometallurgy is the branch of biotechnology applied for efficient recovery of metal by the use of biological organisms and system to produce extractable elements from solid compounds (Anjum et al. 2012; Willner and Fornalczyk 2013). The most commonly used microbes in bioremediation were: *Acidithiobacillus ferrooxidans, Acidithiobacillus thiooxidans, Leptospirillum ferrooxidans, Penicillium* sp. and *Aspergillus niger*. Biohydrometallurgical methods consume less energy; hence, it is an economic and eco-friendly way for remediation (Willner and Fornalczyk 2013).

Bioleaching, biosorption and bioaccumulation are the three methods of bioremediation widely employed for metals recovery (Jaafar et al. 2015). Bioleaching is defined as dissolution of metal sulphides by microbes in ores or in waste solution (Xin et al. 2009; Sand et al. 2001). Bioleaching mechanism is well studied for spent LIBs, and microbes mostly used were acidophilic sulphur-oxidizing and iron-oxidizing bacteria or fungi *A. niger*. In biosorption, metals were adsorbed to the surface of the cell wall, and in bioaccumulation process, pollutants get accumulated inside the living biomass (Pant et al. 2012; Nancharaiah et al. 2016). Bioremediation is a natural process, so it is sustainable and environment-friendly and is also cost-effective process. However, biological processes were a time-consuming process and were not able to tolerate change in environment. So, advanced and sustainable recycling technologies need to be developed.

2 Extraction of Metals from LIB

Increasing spent battery in waste streams and decreasing natural resources divert public attention towards recycling of waste batteries to meet the increasing demand of metals. LIBs consists of metals like cobalt, nickel and lithium and its composition varies with different brands (Zeng et al. 2014; Xu et al. 2008). Also spent electric vehicle LIBs quantity is estimated to reach 500 thousand metric tons by year in 2020 (Zeng et al. 2014; Xin et al. 2016; Changnes and Pospiech 2013). Therefore, the recovery of metals from spent LIBs is the demand of developing technology (Richa et al. 2014).

Recycling processes were broadly divided into two categories: primary treatment and secondary treatment. The primary treatment or physical processes are the initial process of recycling. It involves skinning, removing of crust, shredding, crushing, sieving, thermal processes, dissolution process and mechanical separation techniques. Pre-treatment of materials is done to separate the cathode materials from spent LIBs with which hydrometallurgical or a pyrometallurgical recycling process was conducted. The different techniques that can be utilized for the extraction process are given in flowchart (Fig. 3).

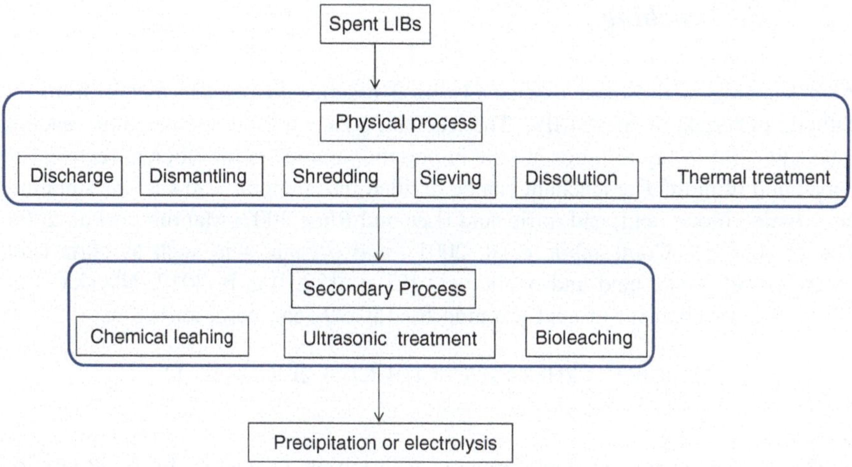

Fig. 3 Different techniques of LIB recycling

2.1 Chemical Processes

In chemical process, either acid or base was employed to extract metals into the solution from LIBs scraps and then was recovered via solvent extraction, precipitation, crystallization or by the electrolysis process. In precipitation, pH of the leaching solution was altered and was carried out by adding some reaction agent. In the crystallization tests, leached liquor were placed in an oven at 60 °C until 80–95% of liquor is evaporated (Ferreira et al. 2009). Figure 4 represents the flowchart of these technologies.

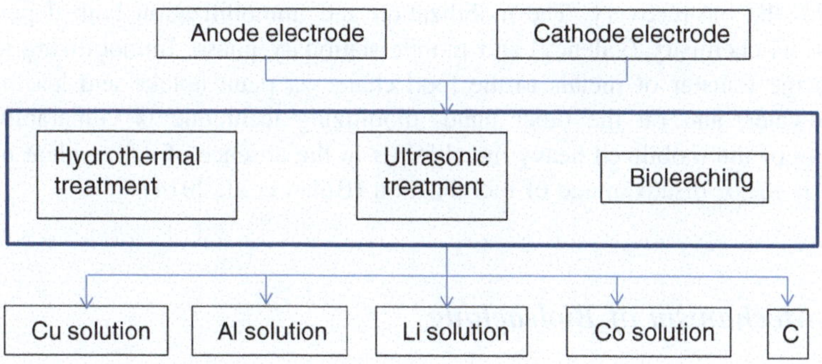

Fig. 4 Chemical process for LIB recycling (Zeng et al. 2014)

2.2 Acid Leaching

Acid leaching is the most extensively used practice to recover metals from the cathode materials of spent LIBs. The leachant concentration, temperature, reaction time and solid-to-liquid ratios are the important parameters for efficient recovery of cobalt and lithium. The leachant can be divided into mineral acid such as sulphuric acid, hydrochloric acid, and nitric acid (Lee and Rhee 2002; Mantuano et al. 2006; Zou et al. 2013; Contestabile et al. 2001), and organic acid such as citric acid, succinic acid, malic acid and oxalic acid (Li et al. 2010a, b, 2012; Nayaka et al. 2015). The mechanism of acid leaching for $LiCoO_2$ can be described as:

$$2LiCoO_2 + 8HCl \Leftrightarrow 2CoCl_2 + Cl_2 + 2LiCl + 4H_2O$$

(Zou et al. 2013).

The overall techniques and process of leaching of metals from lithium-ion battery are presented in Table 1.

3 Bioleaching of Spent LIB

A biohydrometallurgical process is a natural way for the recovery of metals (Brandl and Faramarzi 2006) and viable alternative for chemical and physical waste treatment technique (Pant et al. 2012; Ilyas et al. 2010; Ehrlich 2015; Rohwerder et al. 2003; Cerruti et al. 1998; Mishra et al. 2008). Table 2 gives general idea about metal–microbe interaction that occurs either by mobilization of solid metals or by immobilization of solubilized metals (biosorption, bioaccumulation) (Brandl and Faramarzi 2006; Nancharaiah et al. 2016). Also Fig. 5 illustrates the microbe–metal interaction by different mechanisms of metal solubilization and immobilization used for the bio-recovery. The mobilization and immobilization both depend on metals, its chemistry (valency) and mobile-stationary phase. Immobilizing agents reduce the transfer of metals to the food chain via plant uptake and leaching to groundwater, and on the other hand, mobilizing technique is vulnerability to leaching of the mobilized heavy metal(loid)s in the absence of active plant uptake which is major disadvantage of mobilization (Bolan et al. 2014).

3.1 Mechanism of Bioleaching

The microbes exploited for the metals recovery from LIBs waste include *Aspergillus niger* (fungi) (Horeh et al. 2016) and mixed culture of acidophilic sulphur-oxidizing and iron-oxidizing bacteria (Xin et al. 2009, 2016). Xin et al. (2009) explained the non-contact bioleaching mechanism for recovery of Co and Li

Table 1 Metal leaching

S. no.	Leaching chemicals	Techniques	Reagent	% recovery of metals		References
				Li	Co	
1.	Citric acid, H_2O_2	–	–	100	90	Li et al. (2010a)
2.	Ascorbic acid	Ultrasonic washing and calcination	–	98.5	94.8	Li et al. (2012)
3.	Tartaric and ascorbic acids	Reductive-complexing mechanism	–	95	95	Nayaka et al. (2016b)
4.	Iminodiacetic acid Maleic acid	–	–	99 100	91 97	Nayaka et al. (2016a)
5.	Malic acid	–	–	100	90	Li et al. (2010b)
6.	H_2SO_4 + H_2O2 + NaOH	Precipitation		80	100	Zou et al. (2013)
7.	HNO_3 + H_2O_2	Precipitation	Citric acid	95	95	Lee and Rhee (2002)
8.	H_2SO_4, H_2O_2	Liquid–liquid extraction	Cyanex 272	74	68	Mantuano et al. (2006)
9.	HCL	–	–			Contestabile et al. (2001)
10.	NaOH, H_2SO_4	Crystallization	–	100	97	Ferreira et al. (2009)
11.	H_2SO_4, H_2O_2	Solvent extraction	D2EHPA	81	98.2	Chen et al. (2015a)
12.	Citric acid and ascorbic acid	Precipitation	Oxalic acid and NH_4F	100	80	Nayaka et al. (2015)
13.	NH_4OH, H_2SO_4 + H_2O_2	–	–	97	97	Nayl et al. (2014)
14.	H_2SO_4 + H_2O_2	Extraction and sedimentation	P507, oxalic acid and carbonic acid	86.04	90.02	Jian et al. (2012)
15.	EDTA	Bipolar membrane electrodialysis	–	99	99	Iizuka et al. (2013)
16.	Oxalate, H_2O_2, HCl and HNO_3	Vacuum pyrolysis	Oxalate			Sun and Qiu (2012)
17.	Citric acid and D-glucose	Selective precipitation	MnO_2, $KMnO_4$	99		Chen et al. (2016)

(continued)

Table 1 (continued)

S. no.	Leaching chemicals	Techniques	Reagent	% recovery of metals		References
				Li	Co	
18.	Oxalic acid, H_2O_2	–	–	98	97	Zeng et al. (2015)
19.	Citric acid, H_2O_2	Selective precipitation	Dimethylglyoxime reagent	89	97	Chen et al. (2015b)
20.	H_2SO_4			93.4	66.2	Meshram et al. (2015a)
21.	H_2SO_4 $NaHSO_3$.	–	–	96.7	91.6	Meshram et al. (2015b)
22.	H_2SO_4, H_2O_2	Solvent extraction	Cyanex 272	94	93	Swain et al. (2007
23.	H_2SO_4, HNO_3 and HCl	Oxidative precipitation	Co_2O_3, $3H_2O$		100	Joulié et al. (2014)
24.	Succinic acid, H_2O_2	Grinding, calcination	–	96	100	Li et al. (2015)
25.	H_3Cit + tea waste $H_3Cit + H_2O_2$ H3Cit + *Phytolacca americana*	–	–	98 99 96	96 98 83	Chen et al. (2015c)

Table 2 Microbial reactions in LIB bioleaching (Brandl and Faramarzi 2006)

Process	Mechanism	Description of the reaction
Mobilization	Redoxolysis	Metals were either oxidized or reduced by microbe, resulting in increase of metal mobility
	Acidolysis	Proton was induced by microbial secretion resulting in changes of the metal mobility
	Complexolysis	Through complexation, microbes lead to an increase metal solubilization
	Alkylation	Alkyl groups are enzymatically transferred to the metal and covalently bound. After bioalkylation, metals show high volatility incomparision to their elemental form
Immobilization	Biosorption	Biosorption is defined as a passive process of metal sequestering and concentration by chemical sites (functional groups such as carboxyl, sulfonate, phosphate, hydroxyl, amino or imino residues) naturally present on the surface of living or dead microbial biomass. Metal sorption can be more or less selective depending on the organisms used and environmental conditions (e.g. pH, salinity)
	Bioaccumulation	Soluble metals are actively transported through the cell membrane and accumulated within the cells as solid particles or in vacuoles
	Redox reaction	Metals were either oxidized or reduced by microbe, resulting in increase of metal mobility
	Complex formation	Soluble metal species were precipitated by microbially formed complexing agents, e.g. sulphide. Microbial activities reduces the acidity resulting to the precipitation of metals as hydroxides

from spent LIBs by using sulphur-oxidizing and iron-oxidizing bacteria. According to them, Co and Li were extracted from spent batteries by the formation of inorganic acid, H_2SO_4 by bio-oxidation of elemental sulphur. Li shows highest bioleaching efficiency in S system while Co shows in FeS_2 or $S + FeS_2$ system. Figure 6 shows some general mechanism involved in bioleaching process.

3.2 *Mechanism of Biosorption and Bioaccumulation*

Immobilization of metal is the main process involved in the biosorption and bioaccumulation. In biosorption, metal ions were adsorbed to the cell wall of the microbe by the process of chelation, complexation, ion exchange and physical adsorption (Jaafar et al. 2016; Das et al. 2008). The cell wall provides number of active metal binding sites such as polysaccharides and proteins (Das et al. 2008). Figure 7 shows the adsorption of cadmium on the cell wall of bacteria (Manasi

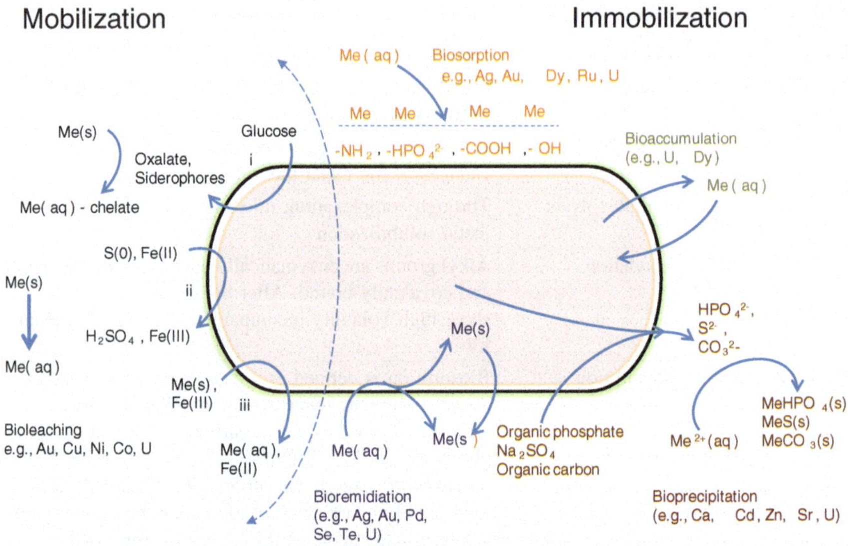

Fig. 5 Metal solubilization and immobilization used for bio-recovery (Nancharaiah et al. 2016)

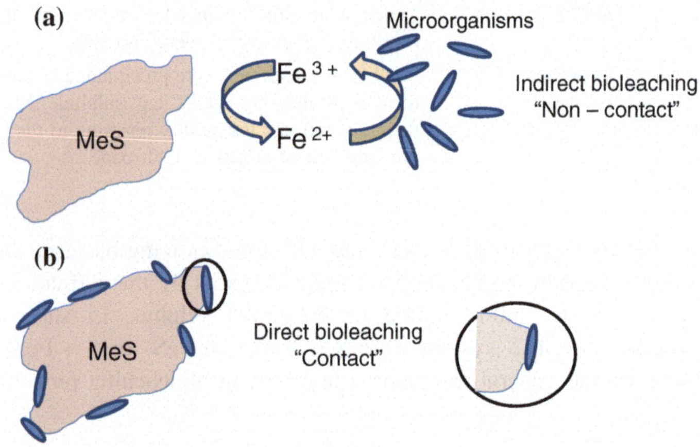

Fig. 6 Bioleaching leaching mechanism (Anjum et al. 2012)

et al. 2014). The cell wall has many functional groups (amines, hydroxyl, carboxyl and phosphate groups), and these functional groups are responsible for Cd binding. The cell surfaces of microbe change after reacting with Cd signifying an enlargement in cell surface to improve the interaction with toxic substances (Manasi et al. 2014).

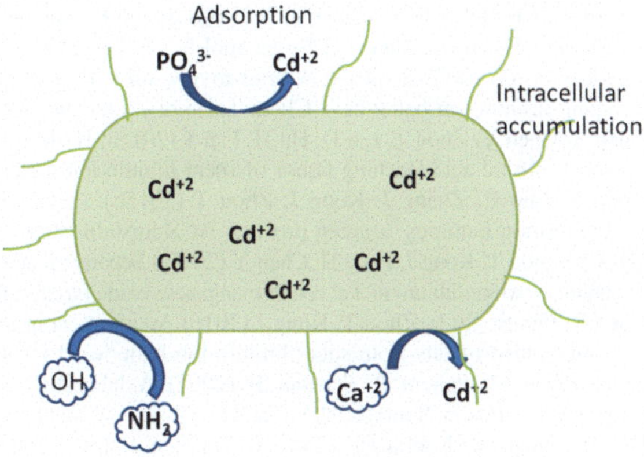

Fig. 7 Adsorption of cadmium on the cell wall of bacteria (Manasi et al. 2014)

4 Conclusion

Use of LIBs is expanding day by day leading to the various environmental problems. For the protection of these resources from depletion, efficient and sustainable technologies for the recycling and recovery of metals are in high demand (Anjum et al. 2012). Conventional methods such as hydrometallurgy are efficient but not eco-friendly, and on the other hand, bioremediation is eco-friendly but time-consuming. However, in order to meet strict quality standards for direct discharge of leachate into the surface water, a development of integrated methods of treatment, i.e. a combination of chemical, physical and biological steps are required. A compatible combination of chemical and biological method can improve the efficiency of the process both in terms of time and environment. Microbial in cooperation make the recycling more environmentally sound and viable. The improved processes ameliorate the drawbacks of individual processes and contributing to a higher efficacy of the overall treatment.

References

Alper J (2002) The battery: not yet a terminal case. Science 296(5571):1224–1226

Anjum F, Shahid M, Akcil A (2012) Biohydrometallurgy techniques of low grade ores: a review on black shale. Hydrometallurgy 117:1–12

Barik SP, Prabaharan G, Kumar B (2016) An innovative approach to recover the metal values from spent lithium-ion batteries. Waste Manage 51:222–226

Bernardes AM, Espinosa DCR, Tenório JS (2004) Recycling of batteries: a review of current processes and technologies. J Power Sources 130(1):291–298

Bolan N, Kunhikrishnan A, Thangarajan R, Kumpiene J, Park J, Makino T, Kirkham MB, Scheckel K (2014) Remediation of heavy metal (loid) s contaminated soils—to mobilize or to immobilize? J Hazard Mater 266:141–166

Brandl H, Faramarzi MA (2006) Microbe-metal-interactions for the biotechnological treatment of metal-containing solid waste. China Particuology 4(2):93–97

Castillo S, Ansart F, Laberty-Robert C, Portal J (2002) Advances in the recovering of spent lithium battery compounds. J Power Sources 112(1):247–254

Cerruti C, Curutchet G, Donati E (1998) Bio-dissolution of spent nickel–cadmium batteries using *Thiobacillus ferrooxidans*. J Biotechnol 62(3):209–219
Chagnes A, Pospiech B (2013) A brief review on hydrometallurgical technologies for recycling spent lithium-ion batteries. J Chem Technol Biotechnol 88(7):1191–1199
Chen X, Chen Y, Zhou T, Liu D, Hu H, Fan S (2015a) Hydrometallurgical recovery of metal values from sulfuric acid leaching liquor of spent lithium-ion batteries. Waste Manage 38:349–356
Chen X, Luo C, Zhang J, Kong J, Zhou T (2015b) Sustainable recovery of metals from spent lithium-ion batteries: a green process. ACS Sustain Chem Eng 3(12):3104–3113
Chen X, Zhou T, Kong J, Fang H, Chen Y (2015c) Separation and recovery of metal values from leach liquor of waste lithium nickel cobalt manganese oxide based cathodes. Sep Purif Technol 141:76–83
Chen X, Fan B, Xu L, Zhou T, Kong J (2016) An atom-economic process for the recovery of high value-added metals from spent lithium-ion batteries. J Clean Prod 112:3562–3570
Contestabile M, Panero S, Scrosati B (2001) A laboratory-scale lithium-ion battery recycling process. J Power Sources 92(1):65–69
Das N, Vimala R, Karthika P (2008) Biosorption of heavy metals—an overview. Ind J Biotechnol 7:159–169
Dewulf J, Van der Vorst G, Denturck K, Van Langenhove H, Ghyoot W, Tytgat J, Vandeputte K (2010) Recycling rechargeable lithium ion batteries: critical analysis of natural resource savings. Resour Conserv Recycl 54(4):229–234
Dorella G, Mansur MB (2007) A study of the separation of cobalt from spent Li-ion battery residues. J Power Sources 170(1):210–215
Ehrlich HL, Newman DK, Kappler A (eds) (2015) Ehrlich's geomicrobiology. CRC Press, Boca Raton
Ferreira DA, Prados LMZ, Majuste D, Mansur MB (2009) Hydrometallurgical separation of aluminium, cobalt, copper and lithium from spent Li-ion batteries. J Power Sources 187(1):238–246
Freitas MBJG, Garcia EM (2007) Electrochemical recycling of cobalt from cathodes of spent lithium-ion batteries. J Power Sources 171(2):953–959
Freitas MBJG, Celante VG, Pietre MK (2010) Electrochemical recovery of cobalt and copper from spent Li-ion batteries as multilayer deposits. J Power Sources 195(10):3309–3315
Georgi-Maschler T, Friedrich B, Weyhe R, Heegn H, Rutz M (2012) Development of a recycling process for Li-ion batteries. J Power Sources 207:173–182
Guo Y, Li F, Zhu H, Li G, Huang J, He W (2016) Leaching lithium from the anode electrode materials of spent lithium-ion batteries by hydrochloric acid (HCl). Waste Manage 51:227–233
Hanisch C, Schünemann JH, Diekmann J, Westphal B, Loellhoeffel T, Prziwara PF, Haselrieder W, Kwade A (2015) In-production recycling of active materials from lithium-ion battery scraps. ECS Trans 64(22):131–145
Horeh NB, Mousavi SM, Shojaosadati SA (2016) Bioleaching of valuable metals from spent lithium-ion mobile phone batteries using *Aspergillus niger*. J Power Sources 320:257–266
Iizuka A, Yamashita Y, Nagasawa H, Yamasaki A, Yanagisawa Y (2013) Separation of lithium and cobalt from waste lithium-ion batteries via bipolar membrane electrodialysis coupled with chelation. Sep Purif Technol 113:33–41
Ilyas S, Ruan C, Bhatti HN, Ghauri MA, Anwar MA (2010) Column bioleaching of metals from electronic scrap. Hydrometallurgy 101(3):135–140
Jaafar R, Al-Sulami A, Al-Taee A, Aldoghachi F, Napes S (2015) Biosorption and bioaccumulation of some heavy metals by *Deinococcus radiodurans* isolated from soil in Basra Governorate-Iraq. J Biotechnol Biomater 5:190
Jaafar R, Al-Sulami A, Al-Taee A, Aldoghachi F, Suhaimi N, Mohammed S (2016) Biosorption of some heavy metals by *Deinococcus radiodurans* isolated from soil in Basra Governorate-Iraq. J Bioremediat Biodegrad 7(332):2
Jian G, Guo J, Wang X, Sun C, Zhou Z, Yu L, Kong F, Qiu JR (2012) Study on separation of cobalt and lithium salts from waste mobile-phone batteries. Proc Environ Sci 16:495–499
Joulié M, Laucournet R, Billy E (2014) Hydrometallurgical process for the recovery of high value metals from spent lithium nickel cobalt aluminum oxide based lithium-ion batteries. J Power Sources 247:551–555

Ku H, Jung Y, Jo M, Park S, Kim S, Yang D, Rhee K, An EM, Sohn J, Kwon K (2016) Recycling of spent lithium-ion battery cathode materials by ammoniacal leaching. J Hazard Mater 313:138–146

Lee CK, Rhee KI (2002) Preparation of $LiCoO_2$ from spent lithium-ion batteries. J Power Sources 109(1):17–21

Li J, Shi P, Wang Z, Chen Y, Chang CC (2009) A combined recovery process of metals in spent lithium-ion batteries. Chemosphere 77(8):1132–1136

Li L, Ge J, Chen R, Wu F, Chen S, Zhang X (2010a) Environmental friendly leaching reagent for cobalt and lithium recovery from spent lithium-ion batteries. Waste Manage 30(12):2615–2621

Li L, Ge J, Wu F, Chen R, Chen S, Wu B (2010b) Recovery of cobalt and lithium from spent lithium ion batteries using organic citric acid as leachant. J Hazard Mater 176(1):288–293

Li L, Lu J, Ren Y, Zhang XX, Chen RJ, Wu F, Amine K (2012) Ascorbic-acid-assisted recovery of cobalt and lithium from spent Li-ion batteries. J Power Sources 218:21–27

Li L, Qu W, Zhang X, Lu J, Chen R, Wu F, Amine K (2015) Succinic acid-based leaching system: a sustainable process for recovery of valuable metals from spent Li-ion batteries. J Power Sources 282:544–551

Luo X, Wang J, Dooner M, Clarke J (2015) Overview of current development in electrical energy storage technologies and the application potential in power system operation. Appl Energy 137:511–536

Manasi RV, Kumar ASK, Rajesh N (2014) Biosorption of cadmium using a novel bacterium isolated from an electronic industry effluent. Chem Eng J 235:176–185

Mantuano DP, Dorella G, Elias RCA, Mansur MB (2006) Analysis of a hydrometallurgical route to recover base metals from spent rechargeable batteries by liquid–liquid extraction with Cyanex 272. J Power Sources 159(2):1510–1518

Meshram P, Pandey BD, Mankhand TR (2015a) Recovery of valuable metals from cathodic active material of spent lithium ion batteries: leaching and kinetic aspects. Waste Manage 45:306–313

Meshram P, Pandey BD, Mankhand TR (2015b) Hydrometallurgical processing of spent lithium ion batteries (LIBs) in the presence of a reducing agent with emphasis on kinetics of leaching. Chem Eng J 281:418–427

Mishra D, Kim DJ, Ralph DE, Ahn JG, Rhee YH (2008) Bioleaching of metals from spent lithium ion secondary batteries using *Acidithiobacillus ferrooxidans*. Waste Manage 28(2):333–338

Nan J, Han D, Zuo X (2005) Recovery of metal values from spent lithium-ion batteries with chemical deposition and solvent extraction. J Power Sources 152:278–284

Nancharaiah YV, Mohan SV, Lens PNL (2016) Biological and bioelectrochemical recovery of critical and scarce metals. Trends Biotechnol 34(2):137–155

Nayaka GP, Manjanna J, Pai KV, Vadavi R, Keny SJ, Tripathi VS (2015) Recovery of valuable metal ions from the spent lithium-ion battery using aqueous mixture of mild organic acids as alternative to mineral acids. Hydrometallurgy 151:73–77

Nayaka GP, Pai KV, Manjanna J, Keny SJ (2016a) Use of mild organic acid reagents to recover the Co and Li from spent Li-ion batteries. Waste Manage 51:234–238

Nayaka GP, Pai KV, Santhosh G, Manjanna J (2016b) Dissolution of cathode active material of spent Li-ion batteries using tartaric acid and ascorbic acid mixture to recover Co. Hydrometallurgy 161:54–57

Nayl AA, Elkhashab RA, Badawy SM, El-Khateeb MA (2014) Acid leaching of mixed spent Li-ion batteries. Arab J Chem

Pant D, Joshi D, Upreti MK, Kotnala RK (2012) Chemical and biological extraction of metals present in E waste: a hybrid technology. Waste Manage 32(5):979–990

Richa K, Babbitt CW, Gaustad G, Wang X (2014) A future perspective on lithium-ion battery waste flows from electric vehicles. Resour Conserv Recycl 83:63–76

Rohwerder T, Gehrke T, Kinzler K, Sand W (2003) Bioleaching review part A. Appl Microbiol Biotechnol 63(3):239–248

Sand W, Gehrke T, Jozsa PG, Schippers A (2001) (Bio) chemistry of bacterial leaching—direct vs. indirect bioleaching. Hydrometallurgy 59(2):159–175

Sonoc A, Jeswiet J (2014) A review of lithium supply and demand and a preliminary investigation of a room temperature method to recycle lithium ion batteries to recover lithium and other materials. Procedia Cirp 15:289–293

Sun L, Qiu K (2012) Organic oxalate as leachant and precipitant for the recovery of valuable metals from spent lithium-ion batteries. Waste Manage 32(8):1575–1582

Swain B, Jeong J, Lee JC, Lee GH, Sohn JS (2007) Hydrometallurgical process for recovery of cobalt from waste cathodic active material generated during manufacturing of lithium ion batteries. J Power Sources 167(2):536–544

Vanitha M, Balasubramanian N (2013) Waste minimization and recovery of valuable metals from spent lithium-ion batteries–a review. Environ Technol Rev 2(1):101–115

Wang X, Gaustad G, Babbitt CW, Bailey C, Ganter MJ, Landi BJ (2014) Economic and environmental characterization of an evolving Li-ion battery waste stream. J Environ Manage 135:126–134

Willner J, Fornalczyk A (2013) Extraction of metals from electronic waste by bacterial leaching. Environ Prot Eng 39(1):197–208

Xin B, Zhang D, Zhang X, Xia Y, Wu F, Chen S, Li L (2009) Bioleaching mechanism of Co and Li from spent lithium-ion battery by the mixed culture of acidophilic sulfur-oxidizing and iron-oxidizing bacteria. Biores Technol 100(24):6163–6169

Xin Y, Guo X, Chen S, Wang J, Wu F, Xin B (2016) Bioleaching of valuable metals Li Co, Ni and Mn from spent electric vehicle Li-ion batteries for the purpose of recovery. J Clean Prod 116:249–258

Xu J, Thomas HR, Francis RW, Lum KR, Wang J, Liang B (2008) A review of processes and technologies for the recycling of lithium-ion secondary batteries. J Power Sources 177(2):512–527

Zeng X, Li J, Singh N (2014) Recycling of spent lithium-ion battery: a critical review. Crit Rev Environ Sci Technol 44(10):1129–1165

Zeng X, Li J, Shen B (2015) Novel approach to recover cobalt and lithium from spent lithium-ion battery using oxalic acid. J Hazard Mater 295:112–118

Zhu SG, He WZ, Li GM, Xu Z, Zhang XJ, Huang JW (2012) Recovery of Co and Li from spent lithium-ion batteries by combination method of acid leaching and chemical precipitation. Trans Nonferrous Metals Soc China 22(9):2274–2281

Zou H, Gratz E, Apelian D, Wang Y (2013) A novel method to recycle mixed cathode materials for lithium ion batteries. Green Chem 15(5):1183–1191

Author Biographies

Miss Tenzin Dolker Miss Tenzin Dolker currently perusing her Ph.D in Environmental Sciences under the supervision of Prof. Deepak Pant from, Central University of Himachal Pradesh, Dharamshala India. She did her post-graduation in Environmental Sciences from Central University of Himachal Pradesh, Dharamshala, in year of 2013. She was the recipient of best young research award in 2015 (RTCM-15) and expert in metal recycling techniques.

Dr. Deepak Pant Dr. Deepak Pant is currently the Dean and Head of Environmental Science, Central University of Himachal Pradesh, Dharamshala. He received the Silver Jubilee Research Fellowship Award in 2003 from Kumaun University, Nainital, India; the UCOST Young Scientist Award 2009, INSA Fellow (2010), DST-SERC, Visiting Fellow (2010) and DST-SERC Young Scientist Award (2011). He has 5 patents in the area of waste management using green techniques and published 10 books, 45 research papers in various national and international journals. He recently received Visitor's Award 2017 from the President of India for his innovations in the area of waste management.

Chapter 15
Characterization of Leachate and Groundwater in and Around Saduperi Municipal Solid Waste Open Dump Site, Vellore District, Tamil Nadu, India

N. Manoj Kumar and M. Chaithanya Sudha

Abstract In India, a large portion of the landfill is open or unlined. The administration of municipal solid waste (MSW) requires proper infrastructure, upkeep in all actions. This turns out to be extremely costly and complex due to the unconstrained improvement of urban dominion. The landfill has been concerned with air contamination, soil contamination, surface and groundwater contamination. The origin of landfill gases is subjective to various factors such as the composition of solid waste product, decomposition of waste, oxygen availability, moisture and rain percolation, pH, organic amount and microorganism population. Dioxins are exceedingly dangerous and can cause reproductive and developmental problems. The waste put in the landfills influences the groundwater stream, and rainwater may permeate through the waste. The water gets mixed with organic and inorganic compounds and accumulated at the bottom of the landfill. The present study represents a real-time case study of a solid waste dump yard located at Saduperi, Vellore District, Tamil Nadu, India. Groundwater samples are collected in and around landfill site to analyse the possible impact of leachate on the quality of groundwater. Various physicochemical parameters and heavy metal concentration of groundwater and leachate sample are analysed and reported. Leachate analysis showed a neutral pH (7.4) and BOD concentration of 9100 mg/L. Total hardness and alkalinity were found to be 5500 and 10,000 mg/L, respectively. The chloride concentration was found to be higher (5317.5 mg/L). The concentration of heavy metals such as nickel, cadmium and chromium was found in concentrations of 0.05, 0.09 and 2.84 mg/L, respectively. Groundwater samples showed slightly acidic to

N. Manoj Kumar
Department of Civil Engineering, Builders Engineering College,
Kangayam, Tamil Nadu, India

M. Chaithanya Sudha (✉)
CO_2 Research and Green Technologies Centre, VIT University,
Vellore, Tamil Nadu, India
e-mail: mchaithanyasudha@gmail.com

S. J. Varjani et al. (eds.), *Waste Bioremediation*, Energy, Environment, and Sustainability, https://doi.org/10.1007/978-981-10-7413-4_15

neutral pH values with higher concentration in TDS (9690 mg/L) and chloride (2153 mg/L) parameters. Further leachate pollution index was calculated to know the potential of impact from the dump site leachate.

Keywords Municipal solid waste · Leachate · Groundwater contamination
Water quality

1 Introduction

In the past decades, the collection, conveyance and disposal of municipal solid waste (MSW) have become a crucial issue especially in major cities all over the world. Due to rapid urbanization, almost half of the world's population was residing in cities and future projection shows that population may reach 5 billion by 2030 (Omar et al. 2012). The increase in population eventually increases the amount of waste generated in the major cities. In 2012, MSW generation in world cities is estimated to be 1.3 billion tonnes per annum, and it is expected to reach 2.2 billion tonnes per annum by 2025 (Hoornweg and Bhada-Tata 2012). The per capita MSW generation in developed countries and developing countries ranges from 522–759 and 109–526 kg per person per year (Karak et al. 2012). The MSW is usually collected from each home, academic institutions, offices and commercial complexes and is composed of organic materials (food waste, market waste, yard leaves, wood, etc.), plastic, glass, metals and other refuses (Albores et al. 2016). The quantity, composition and proportion of MSW vary from one part of the world to another. This variation in the composition depends on culture, location, climatic conditions, lifestyle, type of energy source and economic status. However, a study by Hoornweg and Bhada (Hoornweg and Bhada-Tata 2012) shows that in global scale, 46% consists of organic waste in MSW followed by paper (17%), plastic (10%), glass (5%) and metal (4%). In the same study, it is explained that OECD member countries have a low organic fraction of 27% when compared to East Asia and Pacific region countries which have a high organic fraction of 62%.

The cities are unable to manage increased MSW generation due to their lack of regulatory, financial, knowledge, institutional, and public participation. As a result of improper MSW management, degradation of the environment and human health effects occur. The major impacts could be contamination of soil, surface water, groundwater by heavy metals in the leachate (Xiaoli et al. 2007; Prechthai et al. 2008), toxic emissions while burning (McKay 2002), human health problems (Giusti 2009; Rovira et al. 2015) and methane emissions (Boeckx et al. 1996; Das et al. 2016). Hence by knowing these effects, there should be a proper MSW management for safe disposal of wastes. The management of waste involves the collection of waste, resource recovery and recycling, transportation and processing or disposal. Of these, the most important one is processing/disposal of waste (Reddy 2011). The common processing/disposal practices adopted in various countries are open burning, landfilling, composting, incineration and recycling or recovery from waste. Open burning and unsystematic landfilling are carried out widely in low-income and developing countries because they are very cheap–cum-rapid process. On the other hand, it will degrade the environment and affect human health, whereas, in developed countries systematic landfilling is carried out extensively. But for systematic landfilling, a vast land resource is required which is a major constraint in cities. In such places, incinerators are preferred which require less place and has the benefit of heat recovery and waste reduction by volume. If the MSW has high organic content, low calorific value and high moisture content (especially in low-income and developing countries) then incinerators will not be a suitable option. Those wastes which have high organic content are preferred for composting in which wastes are transformed into stabilized product. Only after the segregation of organic and inorganic wastes, the composting will become the alternative for the incinerator and landfill (Karak et al. 2012). Recycling and recovery of useful material from MSW at its source leads to waste reduction and recovery of valuable materials (Jha et al. 2011). A study by Lavee (2007) showed that reduction in direct cost up to 11% can be achieved when adopting recycling of MSW.

1.1 MSW Management in India

India is the second highest populated country (1.21 billion) in the world with annual urban population growth rate of 3.35%. In 2011 census, there are 7935 towns and it has increased by 2774 since 2001. The number of Class I towns were also increased from 394 to 468 which includes 53 million-plus cities. The Indian urban population has increased to 31.16% (377 million) in 2011 from 17.29% in 1951, and correspondingly rural population has come down to 68.84% in 2011 from 82.7% in 1951 (Census 2001 and 2011). The projection shows that by next 10 years nearly 50% of Indian population will reside in urban areas (Vij 2012). Activities associated with such population create solid waste in large quantity (Zhu et al. 2008). The solid waste includes solid or semi-solid domestic waste, agriculture, dairy waste, horticulture waste, commercial waste, sanitary waste, institutional waste, street sweepings, silt from the surface drains, catering and market waste and other non-residential wastes, treated biomedical waste (Solid Waste Management Rules 2016). In developing countries like India, the management of solid waste has become crucial since it is a major source of contamination of air, soil and water. The common activities involved in MSW management are waste generation, storage of waste in individual or community bins, waste collection, transport, processing and waste disposal. Generally, the waste generated in urban areas are of large quantities when compared to rural areas.

The wastes generated are collected separately as biodegradable and non-biodegradable waste. This segregation will help in the recovery of useful products for processing and eventually reduce the amount of waste to be handled. However, this segregation practice is not done properly by local authorities and individuals (Vij 2012). The collected wastes are stored in either movable or fixed bin. For transportation, movable bins are flexible while the fixed bins are time-consuming (Shekdar 2009). The stored wastes are transported to transfer station or taken directly to processing/disposal sites. The segregation step can also be carried out in the transfer station. In India, the waste management system rarely comprises the waste processing unit because of the cost involved in setting up of such units. Further, the unpredictable urban growth makes it more complex process. Without processing unit, the volume of waste to be dumped in landfills will be nearly equal to generated volume. In India, more than 90% of solid waste is disposed of unscientifically in landfills and open dump sites. Apart from landfilling, some of MSW disposal mechanisms such as composting, incineration, refuse-derived fuel and biomethanation are adopted (Sharholy et al. 2008; Kalyani and Pandey 2014).

To make the MSW management more effective in India, a revised and much-defined version of The Municipal Solid Wastes (Management & Handling) Rules, 2000, named Solid Waste Management Rules (SWMR) came into effect from April 2016. Some of the important highlights are discussed here. In SWMR, 2016, the jurisdiction is unfolded beyond municipal area to encompass outgrowths in agglomerated urban areas, notified industrial townships, census towns, areas under the control of Indian airports, railways, port and harbour, defence establishments, central and state government organizations, places of pilgrims, places of historical importance and special economic zones. The SWMR, 2016, gives priority in promoting waste to energy plant by (1) encouraging industries to use refuse-derived fuel, (2) non-recyclable combustible waste having calorific value of 1500 kcal/kg or more shall be utilized for generating energy through refuse-derived fuel, and (3) high calorific wastes shall be used for co-processing in cement or thermal power plants. The SWM Rules, 2016, mandate all local bodies for (1) setting up MSW processing facilities when the population is 1 lakh or more, (2) setting up common or stand-alone sanitary landfills and (3) carrying out bioremediation or capping of old and abandoned dump sites. It also provides specific criteria for site selection of sanitary landfills, setting up development facilities in landfill sites like MSW processing and its treatment facility, SWM in hilly areas, specifications for operation, closure of landfill and rehabilitation of old dump sites (Solid Waste Management Rules 2016).

1.2 MSW Generation in India

The Central Pollution Control Board estimated that 1, 41,064 tonnes of municipal solid waste was generated per day during 2014–2015. Of the total generated waste, 90% of wastes are collected by local bodies and only 34,752 tonnes of collected waste are treated. Maharashtra state generated highest MSW of 22,570 tonnes per day (TPD) followed by Uttar Pradesh (19,180 TPD) and Tamil Nadu (14,500 TPD) (Annual Status Report on Municipal Solid Waste Management 2014–2015). It is estimated that urban India will generate 2,76,342 TPD by 2021, 4,50,132 TPD by 2031 and 11,95,000 TPD of MSW by 2050 (Planning Commission's Report of the Task Force on Waste to Energy 2014). Another estimate shows that MSW generation will reach 300MT by 2047 and about 169.6 km^2 of land is required for disposal (Pappu et al. 2007; Management of Municipal Solid Waste 2010). The average per capita waste generation was found to be 0.11 kilograms per day (Mani and Singh 2016). The MSW composition at the source of generation and waste collection points was calculated on a wet weight basis, and it comprises 40–60% of organic fraction, 30–40% of ash and fine earth, 3–6% of paper and each less than 1% of plastic, glass and metals. The C/N ratio of MSW ranges from 20 to 30, and the calorific value ranges from 800 to 1000 kcal/kg (Sharholy et al. 2008; Gupta et al. 2015).

1.3 Environmental Issues of MSW

Almost all anthropogenic activities will have an impact on the environment and so MSW disposal. Even though proper waste management does reduce the magnitude of impact, it will not eliminate the impact totally. The assessment of the environmental impacts is important to protect environmental settings (Chandrappa and Das 2012). Groundwater is a substantial and invaluable resource for human beings. Groundwater contamination is accelerated after the establishment of industrial development and urbanization (Maiti et al. 2016). MSW unlined and lined landfills are considered to be primary sources of groundwater pollution due to the leachate migration from waste (Reyes-López et al. 2008; Sizirici and Tansel 2015). In India, most of the landfills do not have a barrier or leachate collection system to restrict the migration of leachate into groundwater (Naveen et al. 2016). Leachate is a complex mixture of pollutants having high biochemical oxygen demand, chemical oxygen demand, suspended particles, ammonium nitrogen and toxic characteristics (Kurakalva et al. 2016; Han et al. 2016; Fatta et al. 1999; Regadio et al. 2012). The leachate composition depends upon the nature of MSW, chemical and biochemical processes responsible for the decomposition of waste materials and total water content in waste (Naveen et al. 2016; Fatta et al. 1999). Therefore, groundwater contamination resulting from the landfill leachate shall be considered as a major environmental concern (Singh et al. 2008). Various studies have indicated that total dissolved solids, total hardness, organic matter, sodium, chloride and heavy metals are the important groundwater contaminants emanating from landfill leachate (Akinbile 2012; Smahi et al. 2013; Marzougui and Ben Mammou 2006). The major potential environmental effects associated with leachate are contamination of groundwater and surface water (Kjeldsen et al. 2002). Physicochemical and heavy metal parameters reported in the Indian literatures are presented in Table 1. Apart from leachate, MSW impacts include air pollution and global warming, fires and explosions, unpleasant odours, vegetation damage and landfill settlement (Shenbagarani 2013; Raman and Narayanan 2008).

Table 1 Physicochemical and heavy metal parameters reported in the Indian literature

Location of dump site/landfill	Reported water parameters	References
Perungudi, Chennai, Tamil Nadu, India	pH, EC, TDS, TH, Ca^{2+}, Mg^{2+}, Cl^-, Zn, Cd, Ni, Fe, Cu, Cr and Pb	Shenbagarani (2013)
Pallavaram, Chennai, Tamil Nadu, India	Colour, odour, taste, pH, EC, TDS, TSS, TA, TH, Ca^{2+}, Mg^{2+}, Cl^-, NO_3^-, SO_4^{2-}, K^{+}, Na^+, Cu and Mn	Raman and Narayanan (2008)
Raipur, Chhattisgarh, India	Temperature, pH, EC, TDS, Turbidity, TA, TH, Cl^-, DO and MPN Test	Agrawal et al. (2011)
Erode, Tamil Nadu, India	pH, EC, TDS, TA, TH, Ca^{2+}, Mg^{2+}, Cl^-, NO_3^-, SO_4^{2-}, K^{+}, Na^+, and F^-	Nagarajan et al. (2012)
Hyderabad, Telangana, India	pH, EC, TDS, Ca^{2+}, Mg^{2+}, Cl^-, NO_3^-, SO_4^{2-}, K^{+}, Na^+, F^-, HCO_3^-, As, Cd, Cr, Cu, Fe, Mn, Pb and Zn	Kurakalva et al. (2016)
Dhapa, Kolkata, West Bengal, India	pH, EC, TDS, TH, Cl^-, PO_4^{3-}, SO_4^{2-}, BOD_5, COD, As, Cu, Cr, Cd, Hg, Pb and Zn	Maiti et al. (2016)
Ahmedabad, Gujarat, India	pH, EC, TDS, Ca^{2+}, Mg^{2+}, Cl^-, NO_3^-, SO_4^{2-}, K^{+}, Na^+, Fe, Cr, Zn, Mn, Cd, Pb, Ni and Cu	Singh et al. (2008)
Vijayawada city, Andhra Pradesh, India	Odour, turbidity, pH, EC, dissolved solids, TA, TH, Ca^{2+}, Mg^{2+}, Cl^-, NO_3^-, NO_2^-, SO_4^{2-}, Fe and F^-	Raju (2012)
Navi Mumbai, Maharashtra, India	pH, EC, TDS, SS, TA, BOD_5, COD, Cl^- and SO_4^{2-}	Rathod et al. (2013)
Nanded, Maharashtra, India	pH, EC, TDS, TA, Ca^{2+}, Mg^{2+}, Cl^-, K^{+}, Na^+, PO_4^{3-}, SO_4^{2-}, Fe, Cr, MPN test, salinity	Shaikh et al. (2012)
New Delhi, India	pH, TDS, Cl^-, Fe, As, Cn^-, Pb, Total Cr, Hg, Ni, Cu and Zn	Gupta and Arora (2016)
Thiruvananthapuram, Kerala, India	TDS, TA, Cl^-, TH, Ca^{2+}, Mg^{2+}, NO_3^-, SO_4^{2-} and F^-	Anilkumar et al. (2015)

1.4 *MSW in Vellore City*

Vellore is a Sprawling city situated on the banks of River Palar in the north-eastern part of Tamil Nadu. Vellore City has an area of 87.91 km^2 with a population of 5.02 lakhs. In the past 20 years, the normal rainfall per year ranges from 917 to 1030.3 mm. Vellore City has a semi-arid climate with very high temperature. Dry and hot weather exists throughout the year. The open dump site is located at the Saduperi village which is 5 km from Vellore City, and it has coordinates of 12° 90′ N and 79° 09′E.

The area of open dump site is about 11 acres. For the past 50 years, MSW was dumped here. In dump site, segregation of organic and inorganic waste was not

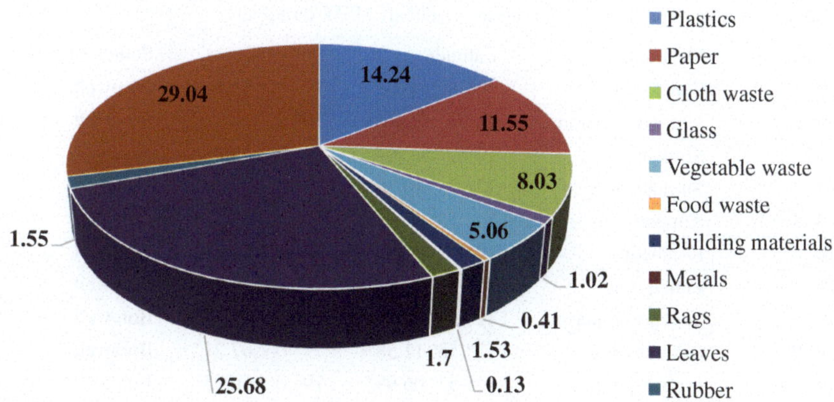

Fig. 1 Weight percentage of components in Vellore MSW (Saravanan and Bhagavanulu 2004)

practised. Earlier manure preparation was done and now because of large proportion of plastics waste, separation of plastic process becomes tedious and manure production was stopped. MSW generation per head in Vellore City was 400 grams per day, and total solid waste dumped per day is of 200 tonnes. 563 sanitation workers are employed in Vellore Corporation. A total of 276 vehicles are engaged in sanitary works. With the help of these vehicles, only 150 tonnes of waste are collected daily. The ultimate analysis of Vellore MSW was done by Saravanan et al. (Saravanan and Bhagavanulu 2004), and it was represented in Fig. 1.

2 Materials and Methodology

2.1 Sampling of Groundwater and Leachate

The sampling wells were identified in and around the solid waste dump site using a random sampling method with 3 km as the periphery. Twenty-two samples are collected for characterization. Of 22 samples, one is leachate sample and remaining 21 are groundwater samples. Within the stretch, we identified and collected 20 borewell samples and 1 open well sample. The open well was situated very closely to landfill. In 20 borewells, one well is situated inside landfill site and others are scattered in 3 km stretch (Table 2). The samples were collected once in 15 days for about six months and transferred to Environmental Laboratory at VIT University, Vellore. The leachate sample is collected from the leachate collecting pit at the dump site.

Table 2 Sampling locations in and around Saduperi MSW dump site

Sample no.	Place name	Latitude	Longitude	Source of water
GW-1	Fort Road	12° 55′ 06.08″	79° 07′ 30.96″	Borewell
GW-2	Virupakshipuram	12°54′ 12.86″	79° 07′ 14.91″	Borewell
GW-3	Sirukanchi 1	12°53′ 52.07″	79° 05′ 48.25″	Borewell
GW-4	Sirukanchi 2	12° 53′ 40.70″	79° 05′ 27.07″	Borewell
GW-5	Kuppam	12° 53′ 09.14″	79° 04′ 48.05″	Borewell
GW-6	Kembedu	12° 53′ 31.98″	79° 05′ 17.25″	Borewell
GW-7	Jamalpuram	12° 54′ 08.12″	79° 06′ 32.29″	Borewell
GW-8	Saduperi dumpsite	12° 54′ 06.71″	79° 06′ 00.97″	Borewell
GW-9	Palavansathu	12° 53′ 11.58″	79° 06′ 07.21″	Borewell
GW-10	Kamaraj Nagar	12° 53′ 06.95″	79° 06′ 56.95″	Borewell
GW-11	Gandhi Nagar	12° 55′ 22.29″	79° 07′ 09.21″	Borewell
GW-12	Bajanaikov	12° 55′ 14.89″	79° 06′ 20.31″	Borewell
GW-13	Aavarampalayam	12° 55′ 12.16″	79° 05′ 40.87″	Borewell
GW-14	Abdullah Puram 1	12° 54′ 49.69″	79° 04′ 46.59″	Borewell
GW-15	Abdullah Puram 2	12° 54′ 48.53″	79° 05′ 08.76″	Borewell
GW-16	Sirukanchi 3	12° 53′ 50.80″	79° 05′ 24.08″	Borewell
GW-17	Saduperi 1	12° 53′ 56.84″	79° 06′ 12.89″	Borewell
GW-18	Jamalpuram Road	12° 54′ 09.95″	79° 06′ 44.64″	Borewell
GW-19	Ariyur	12° 52′ 31.90″	79° 06′ 10.78″	Borewell
GW-20	Palavansathu	12° 53′ 39.89″	79° 07′ 31.89″	Borewell
GW-21	Saduperi 2	12° 53′ 56.33″	79° 05′ 58.40″	Open well

2.2 Analytical Methods

The collected samples are taken to laboratory and are immediately stored at 4 °C. All groundwater samples were analysed for pH, total dissolved solids (TDS), electrical conductivity (EC), dissolved oxygen (DO), chemical oxygen demand (COD), turbidity, alkalinity, total hardness, calcium, magnesium, chloride, sulphate, nitrate, nitrite, potassium (K), sodium (Na), and heavy metals like nickel (Ni), cadmium (Cd) and chromium (Cr). The experimental analyses are carried out as per Bureau of Indian Standard and American Public Health Association Standard methods (Table 3).

Table 3 Details of measured parameters, its adopted methods and instruments used

Parameter	Adopted method	Instrument/apparatus used
pH	IS:3025 Part 11, electrometric method	pH meter
Turbidity	IS:3025 Part 10, nephelometric method	Turbidimeter
TDS	IS:3025 Part 16, gravimetric method	Desiccator and analytical balance
EC	IS:3025 Part 14, laboratory method	Conductivity meter
Alkalinity	IS:3025 Part 23, indicator method	–
Total hardness	IS:3025 Part 21, EDTA method	–
Magnesium	IS:3025 Part 46, EDTA method	–
Chloride	IS:3025 Part 32, argentometric method	–
Calcium	IS:3025 Part 40, EDTA titrimetric method	–
DO	IS:3025 Part 38, Winkler method	Incubator
COD	IS:3025 Part 58, reflux Method	Reflux apparatus and digestion vessels
Sodium and potassium	IS:3025 Part 45, flame photometry method	Flame photometer
Nitrates	APHA 4500-NO_3^-, ultraviolet spectrophotometric screening method	Spectrophotometer
Nitrites	APHA 4500-NO_2^-, colorimetric method	Spectrophotometer
Sulphates	IS:3025 Part 24, turbidity method	Turbidimeter
Ni	IS:3025 Part 54, atomic absorption method	Atomic absorption spectrophotometer
Cd	IS:3025 Part 41, atomic absorption method	Atomic absorption spectrophotometer
Cr	IS:3025 Part 52, atomic absorption method	Atomic absorption spectrophotometer

3 Results and Discussion

3.1 Leachate Characteristics

MSW composition, temperature, time, moisture and oxygen are the major factors influencing the quality of leachate (Naveen et al. 2017). Various physiochemical parameters of dump site leachate are presented in Table 4. In this study, leachate has pH value of 7.2, and it indicates the mature stage of dumping site (Jorstad et al. 2004), whereas pH values ranging from 6.9 to 9.8 are observed in similar studies (Raju 2012; Rathod et al. 2013; Shaikh et al. 2012). In MSW, organic matter is degraded to CO_2 and NH_3 and it further leads to the production of carbonic acid and ammonium ions. This carbonic acid is dissociated to form hydrogen cations and bicarbonate anions (Mahapatra et al. 2011).

BOD_5 of leachate was 9100 mg/L and value of COD was 13,200 mg/L. This high level of BOD indicates the presence of organic matter in the leachate (Rathod

Table 4 Dump site leachate characteristics

Parameters	Leachate concentration
pH	7.2
TDS	12,820
COD	13,200
BOD	9100
Chlorides	5317.5
NO_3^-	4.12
Cd	0.09
Cr	2.84
Ni	0.053
Sodium	230
Potassium	20.56

Note Except pH, all values are in mg/L

et al. 2013). From the BOD and COD values, the BOD_5/COD ratio was found to be 0.68. BOD and COD values are well correlated with (Rathod et al. 2013) and (Archana and Dutta 2014). The BOD5/COD ratio is the good indicator of leachate age. The young leachate (3–12 months) has BOD_5/COD ratio of 0.6–1, followed by a medium leachate (1–5 years) of 0.30–0.60 and old leachate (greater than 5 years) of 0–0.30 (Alvarez-Vazquez et al. 2004). Hence, this dump sites leachate is found to be in young age. The young leachate is mainly composed of organic compounds that will not easily decompose and biodegrade. It also produces refractory compounds that are resistant to biochemical degradation (Agrawal et al. 2011; Abd El-Salam and Abu-Zuid 2015).

The concentration of TDS was recorded as 12,820 mg/L, and this extremely high value of TDS indicates the existence of inorganic materials. In the literature, TDS values ranges from 2027 to 81,000 mg/L (Agrawal et al. 2011; Rathod et al. 2013; Naveen et al. 2017; Jorstad et al. 2004; Bhalla et al. 2012; Aderemi et al. 2011). The chloride content of 5317.5 mg/L in leachate may be due to the mixing of domestic waste. The chloride content is well correlated with (Regadio et al. 2012; Jorstad et al. 2004) similar studies were done in the past. In leachate, the nitrogen cycle is dominated by microbial decomposition of organic carbon. As the time progresses, the nitrogen concentration decreases because of microbial utilization of nitrate compounds and denitrified as ammonia gas. Nitrate concentration was recorded as 4.12 mg/L, which is low in comparison to the reference values (Bhalla et al. 2012). In our study, chromium is abundant with a concentration of 2.84 mg/L.

The metallic elements such as nickel and cadmium are found to be 0.05 and 0.09 mg/L, respectively. Heavy metals concentration in landfills will be high in initial stages because the higher metal solubility is higher as a result of low pH (Kulikowska and Klimiuk 2008). The decrease in solubility of metal occurs in later stages, and a sharp decrease in heavy metal's concentration is observed (Harmsen 1983). Na^+ and K^+ in the leachate are recorded with a concentration of 230 and

20.56 mg/L. These ions are not affected due to microbiological activities inside the dump site. Both Na^+ and K^+ are derived from domestic waste and vegetable residues (Christensen et al. 2001).

3.2 Groundwater Characteristics

Various physicochemical characteristics of leachate-contaminated groundwater samples are given in Tables 5 and 6. The collected groundwater samples were tested for pH and have an average value of 7.07 with 6.58 (GW-8) being the lowest and highest being 7.34 (GW-14). The least and the highest pH values are below the BIS standard values. In earlier studies, the pH of the leachate-contaminated groundwater varied between 4 and 8.16 (Rathod et al. 2013; Anilkumar et al. 2015; Alvarez-Vazquez et al. 2004; Lee et al. 2010; Moody and Townsend 2017; Banar et al. 2006). Dissolved gases and materials influence the pH of the water and shift it to alkaline or acidic side. Acidity in water is because of the presence of carbonic, fulvic, humic and organic acids (Mahapatra et al. 2011). The alkaline pH values can sustain with a high amount of dissolved substances and were good in supporting plant life.

Total dissolved solids represent both dissolved and suspended matter in a water sample. All groundwater samples exceeded the permissible limit of 500 mg/L. The highest TDS values are recorded in GW-8 and GW-21 with a concentration of 9690 and 7620 mg/L, respectively. The very high TDS observed in the groundwater samples suggests percolation of leachate into groundwater and indicates the presence of inorganic materials (Mor et al. 2006). In the literature, TDS values of the groundwater ranges from 1440 to 25,514 mg/L (Bhalla et al. 2012; Alvarez-Vazquez et al. 2004; Kulikowska and Klimiuk 2008). High concentration of TDS makes water unpalatable and also causes irritation in gastrointestine in humans. Electrical conductivity (EC) indicates the presence of metals and amount of materials dissolved in water. The EC values in this study range from 757 to 6492 μS/cm. EC values up to 11,560 μS/cm are observed in the literature (Smahi et al. 2013). Except for GW-1 and GW-16, all groundwater samples exceeded the BIS limits which make them hard water and unfit for drinking. The highest values of calcium and magnesium hardness are 2050 and 5950 mg/L, respectively. Even the lowest hardness recorded in GW-16 (582.5 mg/L) is very close to BIS limit of 600 mg/L. By Piper plot (Fig. 2), calcium and magnesium are found to be major ions in groundwater sample. Hardness values from 1070 to 2890 mg/L are reported in the literature (Kulikowska and Klimiuk 2008; Harmsen 1983). Alkalinity is caused by the presence of carbonate, bicarbonate and hydroxide compounds.

Dissolution of CO_2 and carbonate minerals supplies bicarbonate into nearby groundwater. MSW in unlined dump site and oxidation of organic materials are the potential sources of alkalinity in groundwater (Mor et al. 2006). In present study the alkalinity concentration of the groundwater samples are recorded in the range of 660–2400 mg/L. The maximum value 2400 mg/L is 12 times higher than the desirable limit, i.e. 200 mg/L.

Table 5 Physicochemical characteristics of leachate-contaminated groundwater samples

Parameter	GW1	GW2	GW3	GW4	GW5	GW6	GW7	GW8	GW9	GW10	GW11	GW12	GW13	GW14	GW15	GW16	GW17	GW18	GW19	GW20	GW21
pH	7.29	7.2	7.29	7.16	7.17	7.05	6.8	6.58	7.15	6.9	7.18	7.03	7.08	7.34	7.18	7.15	6.91	7.16	6.71	7.03	6.74
TDS	1440	3810	1130	2380	2790	2690	3290	9690	3050	2670	2410	2070	1410	1820	3230	3090	4080	4330	1720	2210	7620
EC (µS/cm)	965	2553	757	1595	1869	1802	2204	6492	2044	1789	1615	1387	945	1219	2164	2070	2734	2901	1152	1481	5105
COD	128	96	32	32	32	160	320	960	64	192	416	40	34	256	64	32	128	288	160	160	1600
Turbidity	21.05	12.6	24.3	22.7	11.8	15.55	3.75	6.25	4.75	5.2	5.45	27.2	5.05	3.9	5.75	15	27.05	4.6	42.85	8.65	10.3
Alkalinity	875	2400	660	1287	1512	1137	1372	1850	750	1050	937.5	960	827.5	1082	1870	670	777.5	1015	1875	1277	1125
TH	590	920	692.5	895	1330	750	1705	8000	1262	1060	782.5	792	687	857	1025	582.5	2950	2075	1227	907	2677.5
Chloride	758	877	386	638	665	638	796	2154	678	573	558	638	465	346	784	578	1157	1037	572	532	2034
Sulphate	7.47	14.5	11.69	12.05	10.71	12.02	11.24	9.26	7.19	11.41	13.36	10.05	12.02	11.38	13.03	7.12	10.88	12.36	10.45	11.48	10.69
Nitrate	0.15	0.07	0.02	1.32	0.14	0.02	0.22	0.03	0.03	0.16	0.03	0.32	0.05	0.01	0.37	0.06	0.05	0.04	0.01	0.57	1.32
Nitrite	3.87	4.29	4.47	3.87	4.47	4.78	4.29	4.12	4.47	4.12	4.12	3.99	3.77	4.29	4.12	4.12	3.87	4.12	3.99	3.79	3.87
K	1.28	26.06	2.8	1.9	3.45	2.43	4.28	9.53	1.43	4.18	2.45	4.51	1.42	2.49	18.42	2.64	1.49	2.09	1.39	42.74	64.76
Na	180.9	765.5	110.6	299.8	430.8	358.2	424.6	1966	502.4	425.2	387.2	467.7	222.1	34.9	481.7	522.7	190.3	497.4	193.7	263.9	1083.7
Ni	0.485	0.109	0.027	0.011	0.04	0.105	0.007	0.067	0.048	0.07	0.059	0.06	0.067	0.125	0.039	0.006	0.146	0.02	0.048	0.123	0.067
Cd	0.013	0.024	0.023	0.018	0.022	0.017	0.009	0.041	0.026	0.032	0.017	0.028	0.027	0.012	0.47	0.057	0.051	0.012	0.045	0.039	0.01
Cr	0.249	0.678	0.204	0.339	0.415	0.56	0.764	1.025	0.647	0.845	0.658	0.951	0.874	0.894	1.082	0.957	1.041	0.745	0.977	1.103	1.064

Table 6 Minimum, maximum and average values of water quality parameters

Parameters (mg/L)	Minimum	Maximum	Average	BIS standards
pH	6.58	7.40	7.07	6.5–8.5
TDS	1410.00	9690.00	3187.14	2000
EC (μS/cm)	945	6492	2135	–
COD	32	1600	247.33	NA
Turbidity	3.75	42.85	13.51	10
Alkalinity	670	2400	1205.36	600
Total Hardness	582.5	8000	1512.86	600
Chloride	346	2153.39	802.91	1000
Sulphate	7.12	14.50	10.97	400
Nitrate	0.01	1.32	0.24	45
Nitrite	3.77	4.78	4.13	NA
K	1.39	64.76	9.61	NA
Na	34.9	1966.00	413.582	NA
Ni	0.01	0.49	0.08	NA
Cd	0.01	0.47	0.05	0.01
Cr	0.42	1.10	0.77	0.05

Note In Tables 5 and 6, all values are in mg/L except pH and EC

An excess of Cl^- in water is usually taken as an index of pollution and considered to be a tracer for groundwater contamination (Loizidou and Kapetanios 1993). The range of chloride concentration is between 345.64 and 2153.39 mg/L. High concentration of chloride in groundwater is in correlation with leachate chloride concentration. The only source of anthropogenic contamination near sampling area is dump site indicating migration of leachate into groundwater. The chloride content, exceeding the concentration level of 250 mg/L, causes odour and taste problems.

Chemical oxygen demand (COD) is a measure of oxygen equivalent to the organic matter content of the water susceptible to oxidation by a strong chemical oxidant and thus is an index of organic pollution. Previous studies used COD as organic indicators to assess the groundwater pollution caused by the landfill (Mor et al. 2006). In this study, the average COD values of all samples are estimated to be 247 mg/L. Hence, it shows the strong correlation of organic matter in the leachate and shows the state of pollution level. Nitrate is a common contaminant of groundwater that originates from septic systems, manure storage and fertilizers. Nitrates, most highly oxidized form of nitrogen compounds, are the end product of the aerobic decomposition of organic nitrogenous matter (Moody and Townsend 2017; Al Sabahi et al. 2009). Nitrate concentration in samples was in the range between 0.01 and 1.32 mg/L which is within the standard limits.

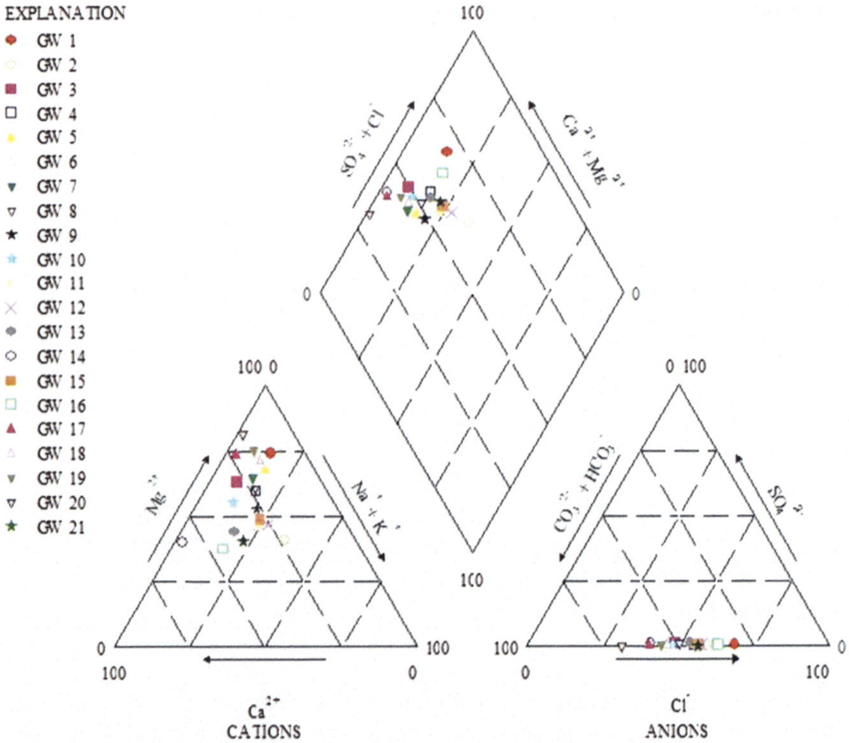

Fig. 2 Piper trilinear plot of major ions in the groundwater samples

As per World Health Organization (WHO), the concentration of potassium should be within 200 mg/L. Potassium is weakly hazardous in water, but it does spread pretty rapidly, because of its relatively high mobility and low transformation potential (Bali and Devi 2013). Potassium has been reported being an indication of the leachate effect, and their concentration in the sample is between 1.28 and 64.76 mg/L. All samples are within desirable limits. Maximum sulphate concentration was found in GW-2. The sulphate ion concentration in groundwater is due to the presence of domestic waste. In similar studies, maximum potassium and sulphate concentrations are 76 and 300 mg/L (Nagarajan et al. 2012).

Chromium in drinking water should be less than 0.05 mg/L. In most of the samples, the concentration is exceeding the standard value. Maximum concentration is found in sample GW-20 (1.1 mg/L), and lowest concentration is 0.2 mg/L in sample GW-3. Cadmium has a standard limit of 0.003 mg/L. In sample GW-21, the concentration is 0.05 mg/L which is highest. In the case of nickel, the maximum concentration observed was 0.48 which exceeds the standard limit.

3.3 Piper Plot

Piper plot represents the chemistry of water samples graphically. In general, we can classify the sample points in the piper diagram into six fields (Kumar 2013). They are as follows:

(i) Ca–HCO_3 type,
(ii) Na–Cl type,
(iii) Ca–Mg–Cl type,
(iv) Ca–Na–HCO_3 type,
(v) Ca–Cl type,
(vi) Na–HCO_3 type.

Piper plot was created for collected samples using results obtained from analytical tests. Piper diagram consists of anion triangle, cation triangle and a diamond apex which shows combined plot of anions and cations.

Anions are plotted as a percentage of $SO_4{}^{2-}$, Cl^-, $HCO_3{}^-$ and $CO_3{}^{2-}$. Cations are plotted as a percentage of Na^+, K^+, Ca^{2+} and Mg^{2+} (Shenbagarani 2013). From the plot, it is concluded that most of the samples are in Ca–Mg–Cl and Ca–HCO_3 type. Similar result was observed in the previous study (Nagarajan et al. 2012). This result indicates Ca^{2+}, Mg^{2+} of cations and Cl^-, $HCO_3{}^-$ of anions are dominant in water samples. It is well correlated with higher concentration of hardness and chlorides in groundwater samples.

3.4 Leachate Pollution Index (LPI)

Leachate pollution potential varies according to geographical area, and for comparing potentials of the different landfill site, 80 panellists were surveyed. The survey was conducted using multiple questionnaires to formulate LPI based on Rand Corporation's Delphi Technique (Kumar and Alappat 2005). LPI is a number ranges from 5 to 100 that expresses the overall leachate contamination potential of a landfill based on several leachate pollution parameters at a given time. It is an increasing scale index, wherein a higher value indicates a poor environmental condition. The LPI can be used to report leachate pollution changes in a particular landfill over time (Mor et al. 2006; Umar et al. 2010). LPI can be used as an environmental monitoring tool and indicator for remedial measures. LPI depends on various parameters like m, P_i, W_i which represents the number of leachate pollutant parameters, the sub-index score of the ith leachate pollutant variable and the weight for the ith pollutant variable, respectively. The stepwise detail of formulation of LPI is presented in (Kumar and Alappat 2005).

3.5 Parameters Considered for LPI Calculation

In this study, the parameters available for calculating LPI are pH, TDS, COD, BOD, chlorides, chromium and nickel. LPI for these parameters are calculated in Tables 7 and 8. For calculation, the following formula was used.

$$\mathrm{LPI} = \frac{\sum_{i=1}^{m} w_i p_i}{\sum_{i=1}^{m} w_i}$$

where

LPI Leachate pollution index
m Number of leachate pollutant parameters available
P_i The sub-index score of the ith leachate pollutant variable
W_i The weight for the ith pollutant variable

The sub-index scores of the ith leachate pollutant and the weight for the ith pollutant variable were taken from the literature (Bali and Devi 2013; Umar et al. 2010). LPI of leachate from Saduperi dump site was found to be 34.14 which is higher than untreated leachate in Jamalpur landfill site of Ludhiana City (Bhalla et al. 2014). LPI ranging from 19.5 to 45.01 was observed in previous studies (Bali and Devi 2013; Umar et al. 2010). The LPI calculation for standards given in municipal solid waste management and handling rules 2013 was also done. LPI value was found to be 7.88 for inland surface water disposal and 6.64 for land disposal. When compared with suggested LPI standards (7.88 and 6.64), the leachate from Saduperi dump site has high pollution potential with LPI of 34.14 and hence immediate remedial measures are recommended.

Table 7 Leachate pollution index of Saduperi dump site sample

Parameters	Concentration at Saduperi landfill site	Significance	Pollutant weight (W_i)	P_i	$P_i \times W_i$
pH	7.2	3.509	0.055	5	0.275
TDS	12,820	3.196	0.050	30	1.5
COD	13,200	3.963	0.062	75	4.65
BOD	9100	3.902	0.061	60	3.66
Chlorides	5317.5	3.078	0.048	50	2.4
Cr	2.84	4.057	0.064	10	0.64
Ni	0.053	3.321	0.052	5	0.26
Total			0.392		13.385
LPI value					34.14

Table 8 LPI calculation for standards (solid waste management rules)

Parameters	W_i	LPI standard for inland surface water disposal			LPI standard for land disposal		
		Standard	P_i	$P_i \times W_i$	Standard	P_i	$P_i \times W_i$
pH	0.055	5.5–9	5	0.275	5.5–9	5	0.275
TDS	0.050	2300	7	0.35	2300	7	0.35
COD	0.062	250	10	0.62	–	–	–
BOD	0.061	30	6	0.366	100	9	0.549
Chlorides	0.048	1000	8	0.384	600	7	0.336
Cr	0.064	2	9	0.576	2	9	0.576
Ni	0.052	3	10	0.52	3	10	0.52
Total	0.392			3.091			2.606
LPI Value				7.88			6.64

4 Conclusions

This work represents a real-time case study of MSW open dump site located at Saduperi, Vellore District, Tamil Nadu, India. Leachate and groundwater samples are collected in and around open dump site to analyse the possible impact of leachate on the quality of groundwater. Various physicochemical parameters and heavy metal concentration are carried out and reported. Results showed that leachate collected from dump site is of young age and has a high BOD, COD and TDS values. To quantify leachate pollution potential, LPI was calculated for standard concentration given in MSW rules and dump sites leachate. The LPI value was found to be four times greater than the corresponding standard LPI values. By this result, the leachate emanated from dump site should have contaminated nearby resources. The water quality analysis showed that wells situated near the dump site (e.g. GW-8) are highly contaminated than the wells that are far away. This is due to the fact that leachate being a viscous fluid is hindered due to the mass of solid soil matter. One of the significant findings is that the concentration of TDS, alkalinity and total hardness in most of the groundwater sample exceeds the BIS limits. The heavy metal concentrations present in the samples are found to have a potential threat to public health. Since there is no other source of contamination of wells, it is concluded that due to leachate from dump site groundwater has been contaminated. Proper management of dump site should be done to minimize the effect of leachate on groundwater. Engineered landfill sites should be provided with impermeable liner and drainage.

References

Abd El-Salam MM, Abu-Zuid GI (2015) Impact of landfill leachate on the groundwater quality: a case study in Egypt. J Adv Res 6(4):579–586

Aderemi A, Oriaku Ada V, Adewumi Gbenga A, Otitoloju Adebayo A (2011) Assessment of groundwater contamination by leachate near a municipal solid waste landfill. Afr J Environ Sci Technol 5:933–940

Agrawal A, Pandey R, Agrawal ML (2011) Impact of solid waste leachate on ground water. Int J chem Environ Eng 2(2):113–118

Akinbile CO (2012) Environmental impact of landfill on groundwater quality and agricultural soils in Nigeria. Soil Water Res 7(1):18–26

Al Sabahi E, Rahim SA, Zuhairi WYM, Selangor B (2009) The characteristics of leachate and groundwater pollution at municipal solid waste landfill of Ibb City, Yemen. Water and Environ 5(3):256–266

Albores P, Petridis K, Dey PK (2016) Analysing efficiency of waste to energy systems: using data envelopment analysis in municipal solid waste management. Proc Environ Sci 35:265–278

Alvarez-Vazquez E, Jefferson B, Judd SJ (2004) Membrane bioreactors vs conventional biological treatment of landfill leachate: a brief review. J Chem Technol Biotechnol 79(March):1043–1049

Anilkumar A, Sukumaran D, Vincent SGT (2015) Effect of municipal solid waste leachate on ground water quality of Thiruvananthapuram District, Kerala. Appl Ecol Environ Sci 3(5):151–157

Annual Status Report on Municipal Solid Waste Management 2014–2015, Central Pollution Control Board, India

Archana, Dutta V (2014) Seasonal variation on physico-chemical characteristics of leachate in active and closed municipal solid waste landfill site in Lucknow, India. J Environ Sci Technol 1(4):90–95

Bali L, Devi KS (2013) Leachate characterization and assessment of water pollution near municipal solid waste landfill site. Ind J Appl Res 3:62–64

Banar M, Ozkan A, Kurkcuoglu M (2006) Characterization of the leachate in an urban landfill by physicochemical analysis and solid phase microextraction-GC/MS. Environ Monit Assess 121 (1–3):437–457

Bhalla B, Saini MS, Jha MK (2012) Characterization of leachate from municipal solid waste (MSW) landfilling sites of Ludhiana, India: a comparative study. Int J Eng Res Appl 2(6):732–745

Bhalla B, Saini M, Jha M (2014) Assessment of municipal solid waste landfill leachate treatment efficiency by leachate pollution index. Assessment 3(1):8447–8454

Boeckx P, Van Cleemput O, Villaralvo I (1996) Methane emission from a landfill and the methane oxidising capacity of its covering soil. Soil Biol Biochem 28(10–11):1397–1405

Census 2001 and 2011, Ministry of Home affairs, Government of India

Chandrappa R, Das DB (2012) Solid waste management: principles and practice. Environ Sci Eng 3:413

Christensen TH et al (2001) Biogeochemistry of landfill leachate plumes. Appl Geochem 16(7–8):659–718

Das D, Majhi BK, Pal S, Jash T (2016) Estimation of land-fill gas generation from municipal solid waste in Indian Cities. Energy Proc 90:50–56

Fatta D, Papadopoulos A, Loizidou M (1999) A study on the landfill leachate and its impact on the groundwater quality of the greater area. Environ Geochem Health 21:175–190

Giusti L (2009) A review of waste management practices and their impact on human health. Waste Manage 29(8):2227–2239

Gupta B, Arora SK (2016) Assessment of impact of leachate on groundwater, in the vicinity of the first engineered landfill site in Delhi, India. Asian Rev Civ Eng 5(1):13–20

Gupta N, Yadav KK, Kumar V (2015) A review on current status of municipal solid waste management in India. J Environ Sci (China) 37:206–217

Han Z, Ma H, Shi G, He L, Wei L, Shi Q (2016) A review of groundwater contamination near municipal solid waste landfill sites in China. Sci Total Environ 569–570(1):1255–1264

Harmsen J (1983) Identification of organic compounds in leachate from a waste tip. Water Res 17 (6):699–705

Hoornweg D, Bhada-Tata P (2012) What a waste: a global review of solid waste management. Urban Development Series Knowledge. Paper no.15, World Bank, p 116

Jha AK, Singh SK, Singh GP, Gupta PK (2011) Sustainable municipal solid waste management in low income group of cities: a review. Trop Ecol 52(1):123–131

Jorstad LB, Jankowski J, Acworth RI (2004) Analysis of the distribution of inorganic constituents in a landfill leachate-contaminated aquifer: Astrolabe Park, Sydney, Australia. Environ Geol 46 (2):263–272

Kalyani KA, Pandey KK (2014) Waste to energy status in India: a short review. Renew Sustain Energy Rev 31:113–120

Karak T, Bhagat RM, Bhattacharyya P (2012) Municipal solid waste generation, composition, and management: the world scenario. Crit Rev Environ Sci Technol 42(15):1509–1630

Kjeldsen P, Barlaz MA, Rooker AP, Baun A, Ledin A, Christensen TH (2002) Present and long-term composition of MSW landfill leachate: a review. Crit Rev Environ Sci Technol 32 (4):297–336

Kulikowska D, Klimiuk E (2008) The effect of landfill age on municipal leachate composition. Bioresour Technol 99(13):5981–5985

Kumar PJS (2013) Interpretation of groundwater chemistry using piper and chadha's diagrams: a comparative study from Perambalur taluk. Elixir Geosci 54:12208–12211

Kumar D, Alappat BJ (2005) Evaluating leachate contamination potential of landfill sites using leachate pollution index. Clean Technol Environ Policy 7(3):190–197

Kurakalva RM, Aradhi KK, Mallela KY, Venkatayogi S (2016) Assessment of groundwater quality in and around the Jawaharnagar Municipal Solid Waste Dumping Site at Greater Hyderabad, Southern India. Proc Environ Sci 35:328–336

Lavee D (2007) Is municipal solid waste recycling economically efficient? Environ Manage 40 (6):926–943

Lee AH, Nikraz H, Hung YT (2010) Characterization of acetogenic and methanogenic leachates generated from a sanitary landfill site. Int J Civil Environ Eng 4(7):595–600

Loizidou M, Kapetanios EG (1993) Effect of leachate from landfills on underground water quality. Sci Total Environ 128(1):69–81

Mahapatra DM, Chanakya HN, Ramachandra TV (2011) Assessment of treatment capabilities of Varthur Lake, Bangalore, India. Int J Environ Technol Manage 14(1/2/3/4):84

Maiti SK, De S, Hazra T, Debsarkar A, Dutta A (2016) Characterization of leachate and its impact on surface and groundwater quality of a closed dumpsite—a case study at Dhapa, Kolkata, India. Procedia Environ Sci 35:391–399

Management of Municipal Solid Waste (2010) Central Pollution Control Board (Ministry of Environemntal and Forests), Government of India

Mani S, Singh S (2016) Sustainable municipal solid waste management in India: a policy agenda. Proc Environ Sci 35:150–157

Marzougui A, Ben Mammou A (2006) Impacts of the dumping site on the environment: case of the Henchir El Yahoudia Site, Tunis, Tunisia. Comptes Rendus Geosci 338(16):1176–1183

McKay G (2002) Dioxin characterisation, formation and minimisation during municipal solid waste (MSW) incineration: review. Chem Eng J 86(3):343–368

Moody CM, Townsend TG (2017) A comparison of landfill leachates based on waste composition. Waste Manage 63:267–274

Mor S, Ravindra K, Dahiya RP, Chandra A (2006) Leachate characterization and assessment of groundwater pollution near municipal solid waste landfill site. Environ Monit Assess 118(1–3):435–456

Mor S, Ravindra K, Dahiya RP, Chandra A (2006b) Leachate characterization and assessment of groundwater pollution near municipal solid waste landfill site. Environ Monit Assess 118(1–3):435–456

Nagarajan R, Thirumalaisamy S, Lakshumanan E (2012a) Impact of leachate on groundwater pollution due to non-engineered municipal solid waste landfill sites of erode city, Tamil Nadu, India. Iran J Environ Health Sci Eng 9(1):35

Nagarajan R, Thirumalaisamy S, Lakshumanan E (2012) Impact of leachate on groundwater pollution due to non-engineered municipal solid waste landfill sites of erode city, Tamil Nadu, India. Iran J Environ Health Sci Eng 9(1):35

Naveen BP, Mahapatra DM, Sitharam TG, Sivapullaiah PV, Ramachandra TV (2016) Physico-chemical and biological characterization of urban municipal landfill leachate. Environ Pollut 220:1–12

Naveen BP, Mahapatra DM, Sitharam TG, Sivapullaiah PV, Ramachandra TV (2017) Physico-chemical and biological characterization of urban municipal landfill leachate. Environ Pollut 220:1–12

Omar D, Karuppanan S, AyuniShafiea F (2012) Environmental health impact assessment of a sanitary landfill in an urban setting. Proc Soc Behav Sci 68:146–155

Pappu A, Saxena M, Asolekar SR (2007) Solid wastes generation in India and their recycling potential in building materials. Build Environ 42(6):2311–2320

Planning Commission's Report of the Task Force on Waste to Energy (2014) In the context of integrated municipal solid waste management, vol I

Prechthai T, Parkpian P, Visvanathan C (2008) Assessment of heavy metal contamination and its mobilization from municipal solid waste open dumping site. J Hazard Mater 156(1–3):86–94

Raju MVS (2012) Contamination of ground water due to landfill leachate. Int J Eng Res 53(1):48–53

Raman N, Narayanan DS (2008) Impact of solid waste effect on ground water and soil quality nearer to Pallavaram Solid Waste Landfill Site in Chennai. Rasayan J Chem 1(4):828–836

Rathod M, Mishra H, Karmakar S (2013) Leachate characterization and assessment of water pollution near municipal solid waste landfill site. Int J Chem Phys Sci 2(March):186–199

Reddy PJ (2011) Municipal solid waste management: processing—energy recovery—global examples. BS Publications, CRC Press

Regadio M et al (2012) Pollution profiles and physicochemical parameters in old uncontrolled landfills. Waste Manage 32(3):482–497

Reyes-López JA, Ramírez-Hernández J, Lázaro-Mancilla O, Carreón-Diazconti C, Garrido MML (2008) Assessment of groundwater contamination by landfill leachate: a case in Mexico. Waste Manage 28(Suppl 1):33–39

Rovira J, Vilavert L, Nadal M, Schuhmacher M, Domingo JL (2015) Temporal trends in the levels of metals, PCDD/Fs and PCBs in the vicinity of a municipal solid waste incinerator. Preliminary assessment of human health risks. Waste Manage 43:168–175

Saravanan P, Bhagavanulu DVS (2004) Analysis of municipal solid waste for energy generation potential—a case study of Vellore City. Nat Environ Pollut Technol 3:137–139

Shaikh PR, Bhosle AB, Yannawar VB (2012) The impact of landfill on soil and groundwater quality of the Nanded city, Maharashtra. Researcher 4(7):56–63

Sharholy M, Ahmad K, Mahmood G, Trivedi RC (2008) Municipal solid waste management in Indian cities—a review. Waste Manage 28(2):459–467

Shekdar AV (2009) Sustainable solid waste management: an integrated approach for Asian countries. Waste Manage 29(4):1438–1448

Shenbagarani S (2013) Analysis of groundwater quality near the solid waste dumping site. J Environ Sci Toxicol Food Technol 4(2):1–5

Singh UK, Kumar M, Chauhan R, Jha PK, Ramanathan AL, Subramanian V (2008) Assessment of the impact of landfill on groundwater quality: a case study of the Pirana site in western India. Eviron Monit Assess 141(1–3)

Sizirici B, Tansel B (2015) Parametric fate and transport profiling for selective groundwater monitoring at closed landfills: a case study. Waste Manage 38(1):263–270

Smahi D, El Hammoumi O, Fekri A (2013) Assessment of the impact of the landfill on groundwater quality: a case study of the Mediouna Site, Casablanca, Morocco. J Water Resour Prot 5(April):440–445

Solid Waste Management Rules, 2016, Ministry of Environment, Forest and Climate Change, India

Umar M, Aziz HA, Yusoff MS (2010) Variability of parameters involved in leachate pollution index and determination of LPI from four landfills in Malaysia. Int J Chem Eng 2010:1–6

Vij D (2012) Urbanization and solid waste management in India: present practices and future challenges. Proc Soc Behav Sci 37:437–447

Xiaoli C, Shimaoka T, Xianyan C, Qiang G, Youcai Z (2007) Characteristics and mobility of heavy metals in an MSW landfill: implications in risk assessment and reclamation. J Hazard Mater 144(1–2):485–491

Zhu D, Asnani PU, Zurbrügg C, Anapolsky S, Mani S (2008) Improving municipal solid waste management in India: a sourcebook for policy makers practitioners. World Bank Institute Development Studies, The World Bank

Author Biographies

Mr. N Manoj Kumar is currently working as Assistant Professor in the department of Civil Engineering at Builders Engineering College, Tiruppur, Tamil Nadu. He have almost 2 years of working experience in teaching field. He obtained his bachelor degree in Civil Engineering from EBET- Group of Institutions affiliated to Anna University, Chennai. Later he completed his in the field of Energy & Environmental Engineering at VIT University, Vellore. His research interests are inclined towards Air pollution and Municipal solid waste management.

Dr. M Chaithanya Sudha is an environmental expert working in the fields of Pollution Control and Water quality monitoring. Her major research areas include Lake Ecosystem Monitoring, Wastewater treatment, Solid waste management and Bioremediation.

Chapter 16
Effectiveness of Plant Growth-Promoting Rhizobacteria in Phytoremediation of Chromium Stressed Soils

Pratishtha Gupta, Rupa Rani, Avantika Chandra, Sunita J. Varjani and Vipin Kumar

Abstract Chromium pollution is increasing ceaselessly due to unending industrialization. Of the various oxidation states, Cr^{6+} is highly detrimental due to its mutagenic and carcinogenic nature. Lack of effectiveness of various conventional methods due to economic and technical constraints resulted in a search for an eco-friendly and cost-effective biological techniques for Cr^{6+} removal from the soil. Phytoremediation came up as an innovative technique to address the problem. However, the effectiveness of phytoremediation process is greatly hindered in high metal contamination environments. Recently, microbial-mediated plant stress improvement has occurred as a major element of metal stress management in plants, and their role in enhancing plant growth and improving phytoremediation process has been well studied. The inoculation of plants with metal-resistant plant growth-promoting rhizobacteria (PGPR) plays an important role in enhancing the efficiency of heavy metal phytoremediation. PGPR improves the plant growth through innumerable mechanisms, such as production of siderophores, solubilization of mineral nutrients, 1-aminocyclopropane-1-carboxylic acid (ACC) deaminase, and phytohormone production. These microbes transform heavy metals into soluble and bioavailable forms, hence facilitates metal removal through phytoremediation. Correspondingly, application of plant growth-promoting bacteria exhibiting Cr^{6+} reduction potential when used as an inoculant with phytoremediation plant may result in improved plant growth and chromium remediation efficiency. This chapter focuses on the Cr^{6+} remediation potential of PGPR through metal–microbe–plant interactions.

P. Gupta · R. Rani · V. Kumar (✉)
Laboratory of Applied Microbiology, Department of Environmental Science & Engineering, Indian Institute of Technology (ISM), Dhanbad, Jharkhand 826004, India
e-mail: vipinmicro1@gmail.com

A. Chandra
Department of Environmental Science & Engineering, Indian Institute of Technology (ISM), Dhanbad, Jharkhand 826004, India

S. J. Varjani
Gujarat Pollution Control Board, Gandhinagar, Gujarat 382010, India

S. J. Varjani et al. (eds.), *Waste Bioremediation*, Energy, Environment, and Sustainability, https://doi.org/10.1007/978-981-10-7413-4_16

Keywords Chromium · Phytoremediation · PGPR · Phytohormones Soil

1 Chromium: Sources and Speciation

The natural atmosphere is extremely diverse and exhibits numerous microbes which reflect the quality of the habitat and ability of the individuals to exist together successfully by maintaining the dynamics of the ecosystem (Ahemad et al. 2009). Different types of pollutants, such as pesticides, heavy metals, insecticides affect the stability and sustainability of the environment differently. These cause long-term negative impacts on the functional abilities of the microorganisms (Ahemad and Malik 2011). Among them, chromium (Cr) pollution is increasing hastily and incessantly worldwide. The major sources of Cr pollution are mine tailing, effluents from nonferrous metal industries, electronics, corrosion control, electroplating industries, manufacturing of nuclear reactor vessels, leather tanning, negative and film making, dye productions, and cement-producing plants (Baral et al. 2006; Alam and Ahmad 2012; Chirwa and Molokwane 2011).

Cr is a redox active 3d transition element which belongs to group VI-B of the periodic table and has various oxidation forms (Table 1). Among them, hexavalent Cr (Cr^{6+}) and trivalent Cr (Cr^{3+}) are the most significant and stable species found in the natural environment. Cr^{3+} forms insoluble complexes with organics and precipitates in the form of oxides, hydroxides, and sulfates (Chirwa and Molokwane 2011). In contrast, Cr^{6+} is persistent in the environment due to its high solubility and it is mainly the result of anthropogenic activities (Dixit et al. 2015). Moreover, Cr^{6+} is a known human carcinogen that generally exists as an oxyanion (CrO_4^-) in aqueous systems (Yang et al. 2009). CrO_4^- is a strong oxidizing agent which can cause impairment to nucleic acids, proteins, and cell structures (Ryan et al. 2002).

Table 1 Different chemical species of Cr and their characteristics (adopted and modified from Ahemad 2014)

Species	Characteristics
Cr^{o}	Naturally absent
Cr^{2+}	Highly unstable; readily oxidizes to Cr^{3+}
Cr^{3+}	Naturally occurs in ores; forms unstable compounds
Cr^{4+}	Naturally absent; short half-life
Cr^{5+}	Naturally absent; long half-life; derived from CrO_4
Cr^{6+}	Most stable; anthropogenic source

2 Cr Toxicity to Plants

2.1 Cr Uptake and Translocation in Plants

The mechanism of Cr uptake by plants is yet not understood clearly. There is no specific mechanism for Cr uptake by plants, while the uptake mechanism is widely dependent on Cr speciation. The uptake of Cr^{3+} by plants is a passive process, that is, no energy expenditure is required by the plants (Zayed and Terry 2003; Skeffington et al. 1976). The uptake of Cr^{6+} is an active mechanism which is carried out by the carriers required for the uptake of vital elements (Cervantes et al. 2001). Also, Cr competes with Fe, P, and S for carrier binding (Shanker et al. 2005). Cr accumulates principally in shoots and roots; however, roots accumulate more Cr followed by stem and leaves (Paiva et al. 2009; Cary 1982).

2.2 Effects of Cr on Plants

Cr accumulation can reduce plant growth and pigment content, decrease nutrient uptake, induce chlorosis in young leaves, alter enzyme activities, disrupts ultra-structure of chloroplasts and cell membranes, and damage root cells (Panda 2003; Panda and Choudhary 2005). The reduction in seed germination and radical growth in plants is also a symptom of Cr toxicity. The inhibition of growth in plants can be due to interference in cell division by inducing chromosomal aberrations (Liu et al. 1993). Thus, the toxicity of Cr is dependent on metal speciation, which determines its uptake, translocation, and accumulation in plants. Cr is toxic for plants at about 0.5–5.0 mg mL^{-1} concentration in the nutrient solution and 5–100 mg g^{-1} in soil (Hossner 1996). Under normal conditions, Cr concentration in plants is less than 1 μg g^{-1}.

3 Phytoremediation: An Overview

Currently, several physicochemical techniques such as soil washing, soil vapor extraction, electrokinetic remediation, solidification, and stabilization are in practice to remediate the metal-contaminated soils. The remediation based on physico-chemical parameters is very costly and adversely affects the soil quality.

Phytoremediation is an emerging, promising, eco-friendly, and cost-effective technique that uses various plants to degrade, extract, and immobilize contaminants from surface and ground water, sediment, and soils (Oh et al. 2014). The term 'phytoremediation' is derived from Greek word 'phyto' meaning 'plant' and Latin word 'remedium' meaning 'able to cure' or 'restore' (Vamerali et al. 2010). The concept of phytoremediation was first proposed by Chaney (1983) and then developed through the study of plant species ability to remove pollutants from the environment (Laghlimi et al. 2015). Different processes involved in phytoremediation are as follows:

(a) ***Phytoextraction***—Phytoextraction is the process of uptake of contaminants from soil or water through the roots and its translocation and accumulation in the above ground parts, shoots. This technique is mainly used for the remediation of diffusely polluted soils where pollutants occur at relatively low concentration and superficially (Rulkens et al. 1998). Phytoextraction aids in removal of inorganic form of contaminants (McGrath and Zhao 2003).
(b) ***Phytovolatilization***—This process makes use of such plants that can absorb contaminants from water or soil, convert them into volatile forms, and transpire them back into the atmosphere, e.g., removal of mercury from soil (Ali et al. 2013).
(c) ***Phytostabilization***—This method implicates the ability of plant roots to remove contaminants in soil and reduce their bioavailability to prevent the movement of metal contaminants from soil and their entry into food chain. Phytostabilization can occur through complex action, sorption, metal valence reduction, and precipitation (Ghosh and Singh 2005).
(d) ***Rhizofilteration***—Rhizofilteration involves the use of plants (terrestrial and aquatic) to absorb, precipitate, and concentrate contaminants from water in their roots. This method can partially treat acid mine drainage, agricultural runoff, and industrial discharge and can be used for cadmium, copper, chromium, nickel, lead, and zinc, which are primarily retained within the roots (Chaudhry et al. 1998).
(e) ***Phytodegradation***—Phytodegradation employs the breakdown of organic pollutants into the simpler molecules that are incorporated into the plant tissues. Plants contain enzymes, such as dehalogenases, oxygenases, and reductases, that can efficiently break down and convert chlorinated solvents, herbicides, and ammunition wastes and helps in remediation of such wastes (Ghosh and Singh 2005).

4 Plant Growth-Promoting Bacteria (PGP) and Their Beneficial Traits

Plant growth-promoting rhizobacteria (PGPR) is the group of bacteria that colonize the rhizosphere (root of the plants) and promotes the plant growth. 'Rhizosphere' is the zone of maximum microbial activity due to the presence of high nutrient levels from which essential macronutrients and micronutrients can be extracted. Based on their interactions with the host plants, PGPR can be classified into two groups: (1) symbiotic PGPR (also known as intracellular PGPR, e.g., nodule bacteria): These may invade the interior of the cell and live inside the cell, (2) free-living rhizobacteria (extracellular PGPR, e.g., *Bacillus* sp., *Burkholderia* sp., *Pseudomonas* sp.): They exist outside the root cells. The population of bacteria in the rhizosphere is higher than the bulk soil which may be due to the presence of root exudates that function as a source of nutrients for microbial growth. The beneficial traits of PGPR for plant growth are as follows:

(a) ***Siderophore***—Siderophores are low molar mass organic compounds, produced by plants and animals, under Fe-deficient condition to sequester iron (Ahmed and Holmstorm 2014; Banik et al 2014; Bharti et al. 2014). Siderophores can also bind to other metals, such as Al, Cd, Cu, Ga, and form stable compounds (Rajkumar et al. 2010). Thus, the ubiquitous effect of siderophores is not only limited to iron, but can be extended to other heavy metals also (Mosa et al. 2016). Siderophores could contribute to the mobilization of Cr in sediments and soils where it is abundant (Duckworth et al. 2014).
(b) ***Indole Acetic Acid*** (IAA)—IAA belongs to the class of auxins and is the main plant hormone that regulates plant growth and development (Fu et al. 2015; Leveau and Lindow 2005). The existence of various PSBs capable of producing IAA belonging to different genera had been reported in several researches (Glick et al. 2007). The major pathway for IAA biosynthesis in bacteria is the removal of amino and carboxyl group from the tryptophan leading to oxidized end product IAA (Patten et al. 2013). Auxin plays an important role in regulation and protection of plants' metabolism under stress conditions.
(c) ***ACC deaminase***—ACC deaminase is the enzyme that causes the breakdown of ACC which is a plant ethylene precursor into ammonia and α-ketobutyrate. The decrease in amount of ACC in plants concurrently decreases the amount of ethylene level, hence reduces ethylene stress (Glick 2014). Besides, some PGP bacteria contain ACC deaminase that regulates the production of ethylene in the plant roots and hence results in enhanced root growth. Thus, the ACC deaminase mediated plant growth promotes the efficacy of phytoremediation processes in contaminated soils (Arshad et al. 2007).
(d) ***Organic acids***—The major mechanisms by which microbes produce organic acids to solubilize inorganic phosphates are lowering of pH, chelation of cations, and competing with phosphate for adsorption (Rathi and Gaur 2016). Among the various organic acids produced, gluconic acids and keto-gluconic acids are more common while citric, lactic, isovaleric, isobutyric, malonic, oxalic, glycolic, tartaric and pyruvic and succinic acids are also reported (Kumar 2016). Park et al. (2009) affirmed the production of these acids from the strains of *Pseudomonas*, while Lin et al. (2006) from *Burkholderia* species. Besides organic acids, inorganic acids such as nitric and sulfuric acids are also produced by *Nitrosomonas* and *Thiobacillus*. PGP activities of different strains are described in Table 2.

Table 2 Plant growth-promoting activities by Cr resistant PGPB

PGPB	Activities	References
Pseudomonas sp. VRK3 *Bacillus* spp. *Pseudomonas* sp. NBRI 4014 *Pseudomonas* sp. RNP4 *Pseudomonas* sp. PsA4, *Bacillus* sp. Ba32 *Bacillus* spp. *Mesorhizobium* sp. RC3 *Rhizobacterial* strains A3, S32 *Pseudomonas aeruginosa* OSG41 *Enterobacter aerogenes*, *Rahnella aquatilis*	IAA, P solubilization, siderophores IAA, siderophores, P solubilization, antifungal activity IAA, P solubilization, siderophores IAA, P solubilization, siderophores IAA, P solubilization IAA, siderophores, HCN, ammonia, P solubilization IAA IAA, siderophores P solubilization, EPS, siderophores, HCN, ammonia Siderophores, ACC deaminase, IAA, P solubilization	Hemambika et al. (2013) Karuppiah and Rajaram (2011) Rajkumar et al. (2005) Rajkumar et al. (2006) Wani et al. (2007) Wani et al. (2008) Rajkumar et al. (2005) Oves et al. (2013) Kumar et al. (2009)

PGPB Plant growth-promoting bacteria; *IAA* Indole-3-acetic acid; *ACC* 1-aminocyclopropane-1-carboxylic acid

5 Mechanism of PGP Enhanced Cr Phytoremediation

Bioremediation, an in situ approach to clean up metal-contaminated soils, has been proposed and practices delineating encouraging results. Among bioremediation, phytoremediation is an emerging, promising, environmentally friendly, and profitable technique that uses various plants and their associated rhizospheric microorganisms to concentrate, immobilize, and degrade contaminants from soil, sediment, surface, and ground water (Oh et al. 2014). However, this approach is very slow and time-consuming and requires increased plant biomass, root growth, and metal mobility in soils to decontaminate heavily metal stressed soils. Thus, the efficiency of phytoremediation process is greatly hindered under the high soil metal concentration conditions. In this regard, application of PGPB in phytoremediation (either in phytoextraction or phytostabilization) has gained wider suitability due to excellent performance in enhancing the remediation efficiency as well as plant growth. The application of PGPB in enhancing phytoextraction and phytostabilization of pollutants from soil is shown in Fig. 1.

The schematic representation of phytoremediation with and without PGPR inoculation is presented in Fig. 2.

In this regard, the potential of Cr^{6+} resistant PGPB in reducing Cr toxicity and enhancing plant growth in Cr stressed soils has been studied (Maqbool et al. 2014). Many bacterial genera inhabiting Cr^{6+} reduction ability and PGPB potential have been isolated from soils. These are *Bacillus* (Wani et al. 2007), *Pseudomonas* (Rajkumar et al. 2005), *Mesorhizobium* (Wani et al. 2008), *Cellulosimicrobium* (Chatterjee et al. 2009), *Ochrobactrum* (Faisal and Hasnain 2005), and *Defltia* (Morel et al. 2011). The mechanism of Cr^{6+} reduction through bacterial traits of reductive immobilization

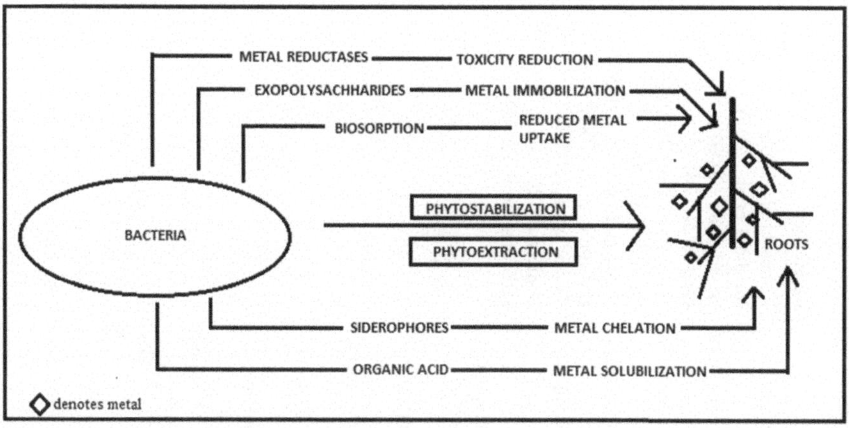

Fig. 1 Schematic representation of PGPR-assisted metal phytoremediation (adopted and modified from Ahemad 2015)

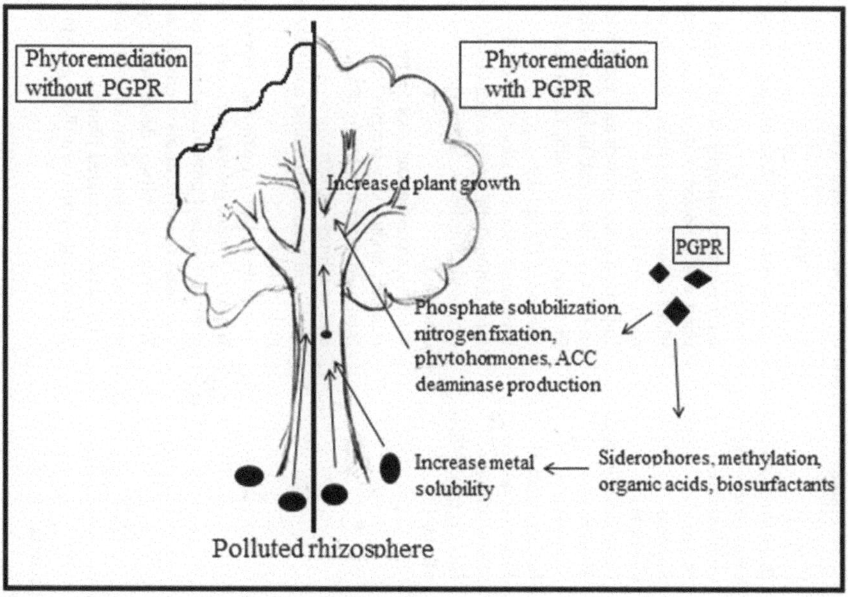

Fig. 2 Schematic representation of phytoremediation with and without PGPR inoculation (adopted and modified from Ullah et al. 2015)

renders the toxic Cr derivatives into environmentally less harmful products (Chatterjee et al. 2009). Inoculation of plants with PGPB exhibiting Cr^{6+} reducing trait has shown better adaptation while growing under Cr stress as these beneficial bacteria induce changes in plant metabolism, thereby they become tolerant to Cr stress. Recently, various studies using different plants inoculated with different genera of Cr^{6+} reducing

Table 3 Case study of PGPB-mediated phytoremediation of Cr contaminated soils

Strains	Plants	Role of PGPB	References
Enterobacter aerogenes, Rahnella aquatilis	*Brassica Juncea*	Strains enhanced plant biomass, protein, and chlorophyll content	Kumar et al. (2009)
Brucella sp. K12	Okra (*Hibiscus esculentus* L.)	Increase in root length, shoot length, fresh and dry weight of plant was reported with various treatments	Maqbool et al. (2014)
Pseudomonas aeruginosa, Pseudomonas fluorescens, Ralstonia Metallidurans	Maize	Plant showed a significant amount of metal accumulation in roots than in shoots	Braud et al. (2009)
Pseudomonas sp. NBRI 4014	Soybean	Improved plant growth and yield with a significant reduction in Cr(VI) concentration both in soils and plant parts	Oves et al. (2013)
Pseudomonas aeruginosa OSG41	Chickpea	Promoted plant growth, facilitated soil metal mobilization, enhanced Cr uptake	Wani and Khan (2010)
Bacillus sp.	*Cicer arietinum* L.	Promoted root and shoot growth by plants	Faisal and Hasnain (2006)
Brevibacterium sp.	Sunflower	Inoculation with strain enhanced dry matter accumulation, symbiotic attributes, grain yield, and protein content of chickpea	Faisal and Hasnain (2005)
Ochrobactrum intermedium	Sunflower	*Bacillus* sp. enhanced root length, shoot length, nodule number, nodule dry weight, and total dry weight	Rajkumar et al. (2006)
Pseudomonas sp. PsA4, *Bacillus* sp. Ba32	Indian mustard	Increased plant height, fresh and dry weight, auxin content, and seedlings growth	Kumar et al. (2008)
Enterobacter sp. NBRI K28, mutant NBRI K28 SD1	*Brassica juncea*	Increased seed germination and plant height and decreased Cr (VI) uptake	Soni et al. (2014)
Microbacterium sp. SUCR140	*Pisum sativum*	Stimulated plant growth and decreased Cr (VI) content	Rajkumar et al. (2005)
Pseudomonas sp. RNP4	Black gram, Indian mustard	Improved plant growth parameters such as biomass, chlorophyll, and protein and increased Ni, Cr, and Zn uptake	
		Increased the overall plant growth and *Pisum sativum*–Rhizobium symbiosis; decreased Cr (VI) toxicity to plants by minimizing its soil bioavailability and uptake in SUCR140-inoculated plants	
		Significantly promoted plant growth	

PGPB Plant growth-promoting bacteria

PGPB have been conducted by many authors to phytoremediate the Cr stresses soils with concurrent plant growth promotion (Table 3). In PGPB-mediated Cr phytoextraction process, PGPB enhance the bioavailability of Cr in soils for an efficient phytoextraction by producing various metabolites, like siderophores, organic acids, IAA, ACC, and biosurfactants.

6 Conclusion

This chapter discloses the potential of PGP bacteria in phytoremediation of chromium polluted soils. Since the efficiency of specific PGP strain varies according to the soil type and environmental conditions, an extensive research is required to enhance the root colonization and chromium phytoextraction ability of PGP bacteria under soil stressed conditions. In addition, the use of consortium instead of monoculture can be undertaken for research in order to enhance the PGP activities and chromium phytoextraction process. Furthermore, most of the studies undertaken are under limited conditions (green house, pot); in this respect, field trails are also needed to reveal the exact working potential of the PGP strain in accelerating the pace of remediation.

References

Ahemad E, Holmstrom SJM (2014) Siderophores in environmental research: roles and applications. Microbiol Biotechnol 7:196–208

Ahemad M (2014) Bacterial mechanisms for Cr (VI) resistance and reduction: an overview and recent advances. Folia Microbiol 59:321–332

Ahemad M (2015) Phosphate-solubilizing bacteria-assisted phytoremediation of metalliferous soils: a review. Biotech 5:111–121. https://doi.org/10.1007/s13205-014-0206-0

Ahemad M, Malik A (2011) Bioaccumulation of heavy metals by zinc resistant bacteria isolated from agricultural soils irrigated with wastewater. Bacteriol J 2:12–21

Ahemad M, Zaidi A, Khan MS, Oves M (2009) Factors affecting the variation of microbial communities in different agro-ecosystems. In: Khan MS, Zaidi A, Musarrat J (eds) Microbial strategies for crop improvement. Springer, Berlin, pp 301–324

Alam MZ, Ahmad S (2012) Toxic chromate reduction by resistant and sensitive bacteria isolated from tannery effluent contaminated soil. Ann Microbiol 62:113–121

Ali H, Khan E, Sajad MA (2013) Phytoremediation of heavy metals-Concepts and applications. Chemosphere. https://doi.org/10.1016/j.chemosphere.2013.01.075

Arshad M, Saleem M, Hussain S (2007) Perspectives of bacterial ACC deaminase in phytoremediation. Trends Biotechnol 25:356–362. https://doi.org/10.1016/j.tibtech.2007.05.005

Banik S, Das K, Islam M, Salimullah M (2014) Recent advancements and challenges in microbial bioremediation of heavy metals contamination. JSM Biotechnol Biomed Eng 2:1035–1043

Baral A, Engelken R, Stephens W, Farris J, Hannigan R (2006) Evaluation of aquatic toxicities of chromium and chromium containing effluents in reference to chromium electroplating industries. Arch Environ Contam Toxicol 50:496–502

Bharti RP, Shrivastava A, Soni N, Tiwari A et al (2014) Phytoremediation of heavy metal toxicity and role of soil in rhizobacteria. Int J Sci Res Publ 4:1–5

Braud A, Jezequel K, Bazot S, Lebeau T (2009) Enhanced phytoextraction of an agricultural Cr and Pb contaminated soils by bioaugmentation with siderophore producing bacteria. Chemosphere 74:280–286

Cary EE (1982) Chromium in air, soils, and natural waters. In: Langard S (ed) Biological and environmental aspects of chromium, pp 49–63. Elsevier Biomedical, New York, NY, USA

Cervantes C, García JC, Devars S et al (2001) Interactions of chromium with microorganisms and plants. FEMS Microbiol Rev 25:335–347

Chaney RL (1983) Plant uptake of inorganic waste constituents. In: Parr JFEA (ed) Land Treatment of Hazardous Wastes. Noyes Data Corp., Park Ridge, NJ, pp 50–76

Chatterjee S, Sau GB, Mukherjee SK (2009) Plant growth promotion by hexavalent chromium reducing bacterial strain, *Cellulosimicrobiumcellulans* $KUCr_3$. World J Microbiol Biotechnol 25:1829–1836

Chaudhry TM, Hayes WJ, Khan AG, Khoo CS (1998) Phytoremediation—focusing on accumulator plants that remediate metal contaminated soils. Aus J Ecotoxicol 4:37–51

Chirwa EMN, Molokwane PE (2011) Biological Cr(VI) reduction: microbial diversity, kinetics and biotechnological solutions to pollution. In: Sofo DA (ed) Biodiversity. InTech. Available from: http://www.intechopen.com/books/biodiversity/biological-cr-vireductionmicrobialdiversity-kinetics-andbiotechnologicalsolutions-to-pollution

Dixit R, Wasiullah, Malaviya D, Pandiyan K, Singh UB et al (2015) Bioremediation of heavy metals from soil and aquatic environment: an overview of principles and criteria of fundamental processes. Sustainability 7:2189–2212. https://doi.org/10.3390/su7022189, https://doi.org/10.1016/j.jhazmat.2009.04.132

Duckworth OW, Akafia MM, Andrews MY, Bargar JR (2014) Siderophore-promoted dissolution of chromium from hydroxide minerals. Environ Sci Processes Impacts 16:1348–1359

Faisal M, Hasnain S (2005) Growth improvement of sunflower seedlings by Cr (VI) resistant bacteria. Iranian J Biotechnol 3:114–120

Faisal M, Hasnain S (2006) Growth stimulatory effect of *Ochrobactrum intermedium* and *Bacillus cereus* on *Vigna radiate* plants. Lett Appl Microbiol 43:461–466

Fu S, Wei J, Chen H, Liu Y, Lu H, Chou J (2015) Indole-3-acetic acid: a widespread physiological code in interactions of fungi with other organisms. Plant Signal Behav. https://doi.org/10.1080/15592324.2015.1048052

Ghosh M, Singh SP (2005) A review on phytoremediation of heavy metals and utilization of its byproducts. Appl Ecol Environ Res 3:1–18

Glick BR (2014) Bacteria with ACC deaminase can promote plant growth and help to feed the world. Microbiol Res 169:30–39. https://doi.org/10.1016/j.micres.2013.09.009

Glick BR, Todorovic B, Czarny J, Cheng Z, Duan J, McConkey B (2007) Promotion of plant growth by bacterial ACC deaminase. Crit Rev Plant Sci 26:227–242. https://doi.org/10.1016/j.micres.2013.09.009

Hemambika B, Balasubramanian V, Kannan VR, James RA (2013) Screening of chromium-resistant bacteria for plant growth-promoting activities. Soil Sediment Contamination Int J 22:717–736

Hossner LR (1996) Phytoaccumulation of selected heavy metals, uranium, and plutonium in plant systems. Quarterly Progress Report, Texas A&M University: College Station, TX, Project UTA96–0043

Karuppiah P, Rajaram S (2011) Exploring the potential of chromium reducing *Bacillus* sp. and there plant growth promoting activities. J Microbiol Res 1:17–23

Kumar A (2016) Phosphate solubilizing bacteria in agriculture biotechnology: diversity, mechanism and their role in plant growth and crop yield. Int J Adv Res 4:116–124. https://doi.org/10.21474/IJAR01

Kumar KV, Singh N, Behl HM, Srivastava S (2008) Influence of plant growth promoting bacteria and its mutant on heavy metal toxicity in *Brassica juncea* grown in fly ash amended soil. Chemosphere 72:678–683

Kumar KV, Srivastava S, Singh N, Behl HM (2009) Role of metal resistant plat growth promoting bacteria in ameliorating fly ash to the growth of *Brassica Juncea*. J Hazard Mater 170:51–57

Laghlimi M, Baghdad B, El Hadi H, Bouabdli A (2015) Phytoremediation mechanisms of heavy metal contaminated soils: a review. Open J Ecol 5:375–388. https://doi.org/10.4236/oje.2015.58031

Leveau JHJ, Lindow SE (2005) Utilization of the plant hormone indole-3-acetic acid for growth by *Pseudomonas putida* strain 1290. Appl Environ Microbiol 71:2365–2371. https://doi.org/10.1128/AEM.711.5.23652371.2005

Lin TF, Huang HI, Shen FT, Young CC (2006) The protons of gluconic acid are the major factor responsible for the dissolution of tricalcium phosphate by *Burkholderia cepacia* CC-A174. Bioresour Technol 97:957–960. https://doi.org/10.1016/j.biortech.2005.02.017

Liu DH, Jaing WS, Li MX (1993) Effect of chromium on root growth and cell division of *Allium cepa*. Israel J Plant Sci 42:235–243

Maqbool Z, Asghar HN, Shahzad T, Hussain S, Riaz W, Ali S, Arif MS, Maqsood M (2014) Isolating, screening and applying chromium reducing bacteria to promote growth and yield of okra (*Hibiscus esculentus* L.) in chromium contaminated soils. Ecotoxicol Environ Safety. https://doi.org/10.1016/j.ecoenv.2014.07.007

McGrath SP, Zhao FJ (2003) Phytoextraction of metals and metalloids from contaminated soils. Curr Opin Biotechnol 14:277–282

Morel MA, Ubalde MC, Brana V, Castro-Sowinski (2011) Delftia sp. JD2: a potential Cr (VI) reducing agent with plant growth promoting activity. Arch Microbiol 193:63–68

Mosa KA, Saadoun I, Kumar K, Helmy M, Dhankher OP (2016) Potential biotechnological strategies for the cleanup of heavy metals and metalloids. Front Plant Sci 7:303. https://doi.org/10.3389/flps.2016.00303

Oh K, Cao T, Li T, Cheng H (2014) Study on application of phytoremediation technology in management and remediation of contaminated soils. J Clean Energy Technol 2:216–220

Oves M, Khan MS, Zaidi A (2013) Chromium reducing and plant growth promoting novel strain *Pseudomonas aeruginosa* OSG41 enhances chickpea growth in chromium amended soils. Eur J Soil Biol 56:72–83. https://doi.org/10.1007/s10529-007-9515-2

Paiva LB, de Oliveira JG, Azevedo RA, Ribeiro DR, da Silva MG, Vitória AP (2009) Ecophysiological responses of water hyacinth exposed to Cr^{3+} and Cr^{6+}. Environ Exper Bot 65:403–409

Panda SK (2003) Heavy metal phytotoxicity induces oxidative stress in *Taxithelium* sp. Curr Sci 84:631–633

Panda SK, Choudhary S (2005) Chromium stress in plants. Brazilian J Plant Physiol. https://doi.org/10.1590/S1677-04202005000100008

Park KH, Lee CY, Son HJ (2009) Mechanism of insoluble phosphate solubilization by *Pseudomonas fluorescens* RAF15 isolated from ginseng rhizosphere and its plant growth-promoting activities. Lett Appl Microbiol. https://doi.org/10.1111/j.1472-765X.2009.02642.x

Patten CL, Blakney AJC, Coulson TJD (2013) Activity, distribution and functions of indole-3-acetic acid biosynthesis pathways in bacteria. Crit Rev Microbiol 39:395–415. https://doi.org/10.3109/1040841X.2012.716819

Rajkumar M, Ae N, Prasad MNV, Freitas H (2010) Potential of siderophore-producing bacteria for improving heavy metal phytoextraction. Trends Biotechnol. https://doi.org/10.1016/j.tibtech.2009.12.002

Rajkumar M, Nagendran R, Lee KJ, Kim SZ (2006) Influence of plant growth promoting bacteria and Cr^{6+} on the growth of Indian mustard. Chemosphere 62:741–748

Rajkumar M, Nagendran R, Lee KJ, Lee WH (2005) Characterization of a novel Cr^{6+} reducing *Pseudomonas* sp. with plant growth-promoting potential. Curr Microbiol 50:266–271

Rathi M, Gaur N (2016) Phosphate solubilizing bacteria as biofertilizer and its applications. J Pharm Res 10:146–148

Rulkens WH, Tichy R, Grotenhuis JTC (1998) Remediation of polluted soil and sediment: perspectives and failures. Water Sci Technol 37:27–35

Ryan MP, Williams DE, Chater RJ, Hutton BM, McPhail DS (2002) Why stainless steel corrodes. Nature 415:770–774

Shanker AK, Cervantes C, Loza-Tavera H, Avudainayagam S (2005) Chromium toxicity in plants. Environ Int 31:739–753

Skeffington RA, Shewry PR, Peterson PJ (1976) Chromium uptake and transport in barley seedlings (*Hordeum vulgare* L.). Planta 132:209–214

Soni SK, Singh R, Awasthi A, Kalra A (2014) In vitro Cr (VI) reduction by cell-free extracts of chromate-reducing bacteria isolated from tannery effluent irrigated soil. Environ Sci Pollut Res Int 21:1971–1979

Ullah A, Heng S, Munis MFH, Fahad S, Yang X (2015) Phytoremediation of heavy metals assisted by plant growth promoting (PGP) bacteria: a review. Environ Exper Bot 117:28–40

Vamerali T, Bandiera M, Mosca G (2010) Field crops for phytoremediation of metal-contaminated land: a review. Enviro Chem Letters 8:1–17

Wani PA, Khan MS (2010) Bacillus species enhance growth parameters of chickpea (*Cicer arietinum* L.) in chromium stressed soils. Food Chem Toxicol 48:3262–3267. https://doi.org/10.1016/j.fct.2010.08.035

Wani PA, Khan MS, Zaidi A (2007) Chromium reduction plant-growth promoting potentials and metal solubilization by *Bacillus* sp. Isolated from Alluvial Soil Curr Microbiol 54:237–243

Wani PA, Khan MS, Zaidi A (2008) Chromium reducing and plant-growth promoting *Mesorhizobium* improves chickpea growth in chromium amended soil. Biotechnol Lett 30:159–163

Yang J, He M, Wang G (2009) Removal of toxic chromate using free and immobilized Cr (VI)-reducing bacterial cells of *Intrasporangium* sp. Q5-1. World J Microbiol Biotechnol 25:1579–1587

Zayed AM, Terry M (2003) Chromium in the environment: factors affecting biological remediation. Plant Soil 249:139–156

Author Biographies

Pratishtha Gupta graduated from the CSJM University, Kanpur in 2012. Subsequently she completed her Master's degree in Environmental Science from BBA University, Lucknow in 2014. She started her Ph.D. in 2015 at Indian Institute of Technology (Indian School of Mines), Dhanbad, India. Her research work focuses on microbial assisted phytoremediation of toxic metals.

Rupa Rani has completed her B. Sc. in Biotechnology from VB University, Hazaribag and M. Sc. in Biotechnology from DAV, Indore. She is currently Senior research fellow at the Indian Institute of Technology (Indian School of Mines), Dhanbad, India. Her research work focuses on microbial degradation of pesticides.

Avantika Chandra has completed her master's degree in Environmental Science and Management from Patna University in 2010. She has completed her Ph.D. in Environmental Science from the Indian Institute of Technology (Indian School of Mines), Dhanbad, India. Her research work focuses on remediation of pollutants on aquatic and terrestrial ecosystems.

Dr. Sunita Varjani is Scientific Officer at Gujarat Pollution Control Board, Gandhinagar, Gujarat, India. Her major areas of research are Industrial and Environmental Microbiology/Biotechnology and Molecular biology. She has delivered talks as invited speaker in many international conferences including the countries India and China. She is recipient of visiting scientist fellowship from EPFL, Lausanne, Switzerland (2017).

Vipin Kumar is an Assistant Professor in the Department of Environmental Science and Engineering at Indian Institute of Technology (Indian School of Mines), Dhanbad, India. He completed his Post Graduate in Microbiology and Doctorate in Environmental Biology. Dr. Kumar's research interest includes microbial remediation of pollutants, biological treatment of solid waste, etc. Earlier, he served as a Research Associate and Senior Research Fellow under the Department of Biotechnology (DBT) Government of India, New Delhi.

Chapter 17
Biomining of Natural Resources

P. Senthil Kumar, P. R. Yaashikaa and G. Baskar

Abstract Biomining is the way towards removing significant metals from minerals and mine tailings with the help of micro-organisms. This process has emerged as an innovative biotechnological approach for extracting the essential metals from low-grade ores. Micro-organisms are utilized to filter out the minerals, instead of the ancient strategies including utilization of outrageous temperature or toxic chemicals, for example simmering and purifying, which adversely affect the earth and furthermore expensive. Many industries have now turned to biomining due to low production cost and high yield maximum to 90%. The need emerges from recent patterns in the industries: progresses with the consumption of high-review mineral assets, the subsequent inclination for mining to be amplified further underground, the developing familiarity with ecological issues related with the purifying of sulphide minerals and the consuming of sulphur-rich fossil energizes and the increasing expense of the huge measures of vitality required in the routine recuperation techniques. Natural resources are assets that exist without activities of mankind. Natural resources can be segregated as biotic and abiotic natural resources. Biotic natural resources include all living creatures on earth including flora, fauna, vertebrates and invertebrates. Abiotic natural resources include all minerals such as gold, copper, iron. Biomining will turn out to be more vital as high-review surface mineral stores are worked out and turned out to be less feasible and mining organizations will be compelled to discover other mineral assets. Biomining can without a doubt give such a green innovation to adventure natural mineral assets.

Keywords Biomining · Natural resources · Sulphide minerals
Microbial consortium · Green innovation

P. Senthil Kumar (✉) · P. R. Yaashikaa
Department of Chemical Engineering, SSN College of Engineering, Chennai 603110, India
e-mail: senthilchem8582@gmail.com

G. Baskar
Department of Biotechnology, St. Joseph's College of Engineering, Chennai 600119, India

S. J. Varjani et al. (eds.), *Waste Bioremediation*, Energy, Environment,
and Sustainability, https://doi.org/10.1007/978-981-10-7413-4_17

1 Introduction

Natural resources are boon to mankind without which life is impossible on earth. All living beings in one way or the other depend on natural resources for their existence. A natural resource may exist as a different element organic or inorganic, for example water, air and additionally a living being or it might exist in another frame that must be handled to acquire the resource, metallic or non-metallic, for example, metal minerals, oil. Natural resources can also be utilized as crude materials for extraction of product (Bridge 2009). The distribution and quantity of natural resources greatly vary from one place to another. For example, India is the fourth largest reserve in the world for coal, whereas Syria is rich in crude oil. Natural resources contribute major part to a country's economy. The wealth of a nation greatly depends on the type and availability of natural resources present. Russia stands first among other countries in the world for its huge availability of natural resources (Table 1). India is also rich in mineral resources like coal (4th largest reserve in the world), bauxite (5th largest reserve), manganese (7th largest reserve), natural gas, limestone, mica and precious metal diamonds. India's major exports include crude oil, gold, petroleum gas, diamonds and coal. In spite of vast availability of natural resources, equal proportion has been exploited due to mankind activities. The process of mining has been employed by many industries for extraction of metals from their ores through chemical methods. The drawbacks of chemical methods mainly involve the release of toxic and reagents pollution that are hazardous to environment. To overcome these drawbacks, new method of extraction using micro-organisms has been introduced. Biomining process replaces chemical methods owing to its increased advantages and applications.

Table 1 List of top 10 countries with natural resources

S. no.	Country	Estimated quantity (trillion)	Natural resources present
1.	Russia	\$75	Gold, timber, natural gas, coal and oil
2.	USA	\$45	Coal, timber, natural gas, copper, gold and oil
3.	Saudi Arabia	\$34.4	Natural gas, oil and timber
4.	Canada	\$33.2	Oil, timber, uranium, natural gas and phosphate
5.	Iran	\$27.3	Oil and natural gas
6.	China	\$23	Coal, rare earth metals and timber
7.	Brazil	\$21.8	Gold, uranium, timber and iron
8.	Australia	\$19.9	Gold, copper, uranium, timber, coal and iron
9.	Iraq	\$15.9	Oil and phosphate rock
10.	Venezuela	\$14.3	Natural gas, iron and oil

2 Natural Resources

Natural resources are resources that exist without activities of mankind. This incorporates every esteemed trademark, for example magnetic, gravitational and electrical properties. On earth, it incorporates daylight, climate, water, flora and fauna.

3 Classification of Natural Resources

Natural resources can be classified into different categories based on the nature of their existence as shown in Fig. 1.

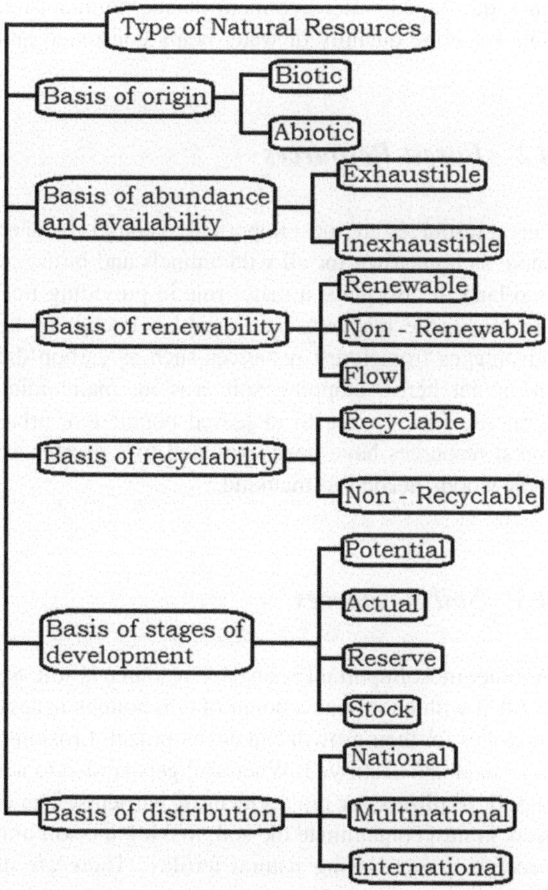

Fig. 1 Classification of natural resources

4 Types of Natural Resources

4.1 Water Resources

Water is the basic necessity of all living beings and is one of the biggest resources gifted by nature. No chemical process will be complete without water. It is estimated that nearly 71% of the world is covered with water resources. Water plays a major role in different fields such as agriculture, irrigation, industries and domestic purposes. Of the total water content in the world, only 2.5% are portable. Due to industrialization and population expansion, water resources are polluted to a greater extent. Groundwater level has been decreased and is contaminated with toxic pollutants such as heavy metals, dyes, hazardous chemicals. Huge quantity of water is also required for recovering of other resources. Industrial wastewater with toxic substances is let into lakes and rivers thus polluting and making it unfit for agricultural or domestic purposes. Biotic natural resources consuming this contaminated water get affected with various infectious diseases. Though many countries have developed water treatment plants, pollutant-free water still remains unavailable since the quantity of water being consumed and replaced is not equal.

4.2 Forest Resources

Forest resources are most important to conserve the ecology of the universe. Forests serve as hometown for all wild animals and birds; cease them from being entering into land. It also plays a major role in providing fresh air and clean water. Forests are renewable resources and are also known as "green gold". There are many advantages from forest resources such as carbon dioxide consumption, providing medicinal herbs, stopping soil erosion, maintaining soil fertility and providing domestic fuel. Owing to increased population, urbanization and industrialization, forest resources have been exploited to a greater extent. Trees are destructed for their woods needed to mankind.

4.3 Soil Resources

Another most important renewable resource is soil. Soil covers the earth surface and is filled with enormous amount of compounds necessary for the plants and also for microbes for their growth and development. Erosion is the only factor by which soil is formed and destroyed. When soil gets eroded to a larger extent, there is a need to supply fertilizers for replacement of nutrients. The chemicals present in the fertilizers in turn contaminate the soil, making the soil unfit for agriculture, and the land becomes barren losing natural fertility. There are different types of soil, namely alluvial soil, red soil, desert soil, black soil and mountain soil. These types of soil are suited for different types of crops to grow but desert soil is not fit for agriculture.

4.4 Mineral Resources

Mineral belong to non-renewable resources. Minerals are also present in microbes as organic and inorganic compounds. Mineral resources in plants and animals can be classified as major and minor minerals listed in Table 2.

The distribution of mineral resources is not uniform in all countries. The minerals must be preserved and ought to be reused routinely. They should be utilized as a crude material where there is a need. They should be investigated consistently. They should be substituted, and new methods must be utilized to prevent exploitation. Few mineral resources and their uses are listed in Table 3.

Increased necessity for these minerals has resulted in increased extraction of minerals by conventional methods which cause serious threats to human life. High energy is needed for extraction of minerals with outcome of enormous amount of

Table 2 List of minerals present in plants and animals

S. no.	Source	Major minerals	Minor minerals
1.	Plants	Calcium, sulphur, iron and magnesium	Zinc, Cobalt, Chlorine and Manganese
2.	Animals	Calcium, phosphorous, chlorine, sulphur and sodium	Cobalt, zinc, iron, copper, selenium and fluorine

Table 3 Minerals and their uses

S. no.	Minerals	Leading producer	Uses
1.	Coal	China	Electricity generation, steel manufacturing, gasification for production of syngas
2.	Natural gas	USA	Fuel for vehicles, manufacturing of steel, domestic purposes and electricity generation
3.	Petroleum	Russia	Fuel for transportation, electricity generation, fuel oils for heating and chemical feedstock for preparing chemicals
4.	Gold	China	Ornament and field of dentistry
5.	Copper	Chile	Metallic mineral in alloys and electric products
6.	Uranium	Kazakhstan	Electricity, nuclear bombs and in tinting glass
7.	Aluminium	China	Wiring, utensils, building materials and aircrafts
8.	Iron	Chile	Manufacture of steel
9.	Manganese	South Africa	Alloys of steels and disinfectants
10.	Zinc	China	Soldering, manufacture of plastics, rubber, batteries, etc.

waste products. Disposal and transportation of these waste materials released require huge quantities of energy which come from other sources such as natural gas, petroleum, firewood. Top layer of soil have been contaminated completely causing land, water and soil pollution. Chemical reagents and high temperature are being employed during the extraction process which affects the mankind working with the process causing lung cancer. Mineral resources determine the wealth of a country, and each nation in the world is gifted with different minerals.

5 Significance of Natural Resources

Natural resources were truly a critical state of the fruitful advancement of nations. Indeed, the financial and innovative improvement of nations and the distinction in their advancement was driven by the accessibility of natural resources. It fortified the improvement of innovation and, in this way, financial advance of nations, which profited from the accessible natural resources and the backwardness of those nations and groups, which needed natural resources. Despite the fact that today the procedure of globalization makes natural resources accessible around the world, however the world economy still relies upon natural resources and those nations, which acquire natural resources, particularly non-renewable energy sources, for example oil and gas. For instance, coal mining and the accessibility of iron to Great Britain permitted the nation to end up noticeably one of the pioneers of the world economy and direct its industrial transformation effectively. The three structures of natural resources required by mankind are food and drink, housing and framework, and mobility. These three make up over 60% of total resource utilized. Natural resources are consumed at a faster rate than being replaced. The major significance of natural resources is that it promotes employment. The natural resources for economy, for example gold, manganese, timber, should be saddled for exports or commercially used. A great deal of work is created in such conditions. Foreign exchange is greatly enhanced by natural resources by exporting the raw or crude mined resources to other nations. Natural resources also provide crude materials to industries for production of required products. For instance, cocoa is utilized as a crude material in the generation of chocolates and different drinks. The timber is utilized as a part of the generation of furniture. Natural resources also provide eco-friendly environment to living creatures without causing any negative impacts or hazardous effects. Man is abusing the natural resources unnecessarily as a major aspect of his reality. If the present situation continues, then the availability of the resources decreases greatly. So it is necessary to conserve the natural resources. Thus, human life in earth is possible because of natural resources.

6 Exploitation of Natural Resources

For quite a while, mankind trusted that resources provided by nature were boundless. An ever-increasing number of nations have turned out to be industrialized, and individuals have started to understand that there is a danger of overexploitation of natural resources. All of a sudden, the risk of overexploitation of natural resources has pulled in broad consideration. Natural resources misuse, investigation, mining and preparing have brought about various sorts of ecological harms which incorporate environmental unsettling influences, contamination of air, water and land, desertification and weather change. The ecological harm has thus brought about misuse of arable land and in addition financial yields and trees. The oil and gas ventures, mining organizations and other natural resources abuse bodies are required to complete safeguards, cures or pay for harm done. An unmonitored and uncontrolled worldwide increment of mining process will be exceptionally inconvenient from a natural viewpoint. Environmental degradation occurs. Today, extraction infiltrates profoundly into the earth and discharges lethal chemicals that are toxic to soil and water away from the site of operation. Mining process requires tremendous measures of water making issues to both upstream and downstream operations. It likewise produces corrosive waste into the land, water bodies, contaminating the air by discharging dust and poisons. It causes deforestation; loss of biodiversity and loss of topsoil through mining and erosion. Since a great part of the harm is unavoidable if the resources must be produced, both the administration and the natural resource industry must be included in taking preparatory and therapeutic measures that can limit the evil impacts of natural resources abuse. Change of condition from waste transfer to waste minimization through reusing, bioremediation, afforestation, sewage treatment and contamination control must prevail.

7 Methods of Extraction of Natural Resources

Many commercially used chemical and mechanical methods are available for extraction of natural resources. These common methodologies involve the application of chemicals for the extraction process. Some of the techniques used by industries are listed below:

1. Mechanical separation (crushing, pulverization and froth flotation)
2. Chemical extraction (by means of chemical reagents)
3. Biological approach
4. Displacement of one element by other
5. Thermal decomposition
6. Electrolytic reduction.

7.1 Biological Approach

This method of biological approach utilizes microbes for recovering of natural resources. Microbes are an integral part of soil and play a tremendous role in recovering of resources without causing any threats to the environment. Micro-organisms are proven to be less expensive, more proficient and cleaner at mineral extraction than traditional mining procedures (Rawlings 1997).

8 Biomining

Mining organizations have turned out to be progressively aware of the capability of microbiological methodologies for recovery of valuable metals from poor-quality minerals and for bioremediating corrosive mine waste and mine tailings (Jerez 2009). Biomining is the biotechnological approach of utilizing micro-organisms for recovering of metals from low-grade ores (Fig. 2). Biomining is an interdisciplinary field including metallurgy, molecular biology, microbiology and chemical engineering (Thosar et al. 2014). In a nation like India, biomining has extraordinary national importance where there is limitless unexploited mineral potential. Biomining will turn out to be more significant as high-review surface mineral stores are worked out and turned out to be less reasonable, and mining organizations will be compelled to discover other mineral sources. The mechanical and chemical techniques of extraction metals from ores are difficult and costly; however, biological strategies are of low cost, low consumption of energy, work well at low concentration of metals, do not release any toxic products and decrease the contamination of metal-containing outlets. Biomining is a natural and safe technique for recovering of resources. Biomining can be used for the extraction of natural resources without contaminating environment.

Fig. 2 Biomining process

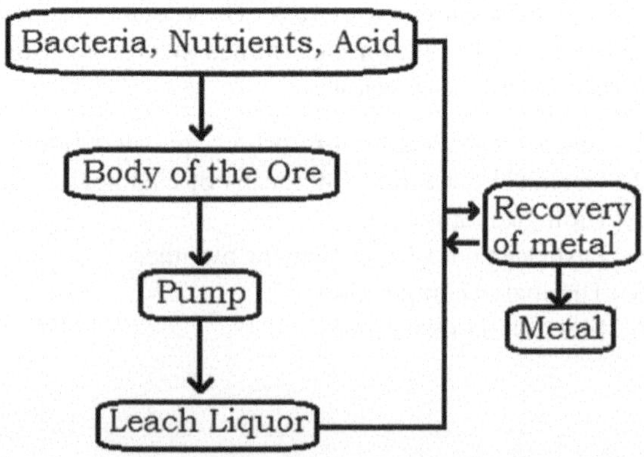

9 Mechanism of Biomining

The minerals are extracted from crude ores with the application of microbes by the following two mechanisms:

9.1 *Oxidation*

Iron and sulphur can be extracted through oxidation process. The two bacteria, namely *Thiobacillus thiooxidans* and *Thiobacillus ferrooxidans,* are effective for this oxidation process. *T. thiooxidans* can effectively oxidize ferrous ion to ferric ion, and the reaction occurs as shown in Eq. (1).

$$4Fe^{++} + O_2 + 4H^+ \rightarrow Fe^{+++} + 2H_2O \quad (1)$$

The bacteria are attached to the surface of the ore and oxidize either directly or indirectly.

9.1.1 Direct Oxidation

The organism oxidizes the ore or metal sulphides by direct contact on the surface of the compounds by obtaining electrons from the reduced minerals (Chan Jang and Valix 2017). Bacterial cells get attached to the surface of the ore, and this interaction may occur within minutes to several hours. Due to this, close contact between the organism and the metal surface occurs. Direct method of oxidation can be described by radioactively labelling of organism used. Physical interaction occurs, and oxidation process is enhanced by several enzyme-catalysed reactions as in Eq. 2.

$$2FeS_2 + 7O_2 + 2H_2O \overset{T.ferrooxidans}{\longrightarrow} 2FeSO_4 + 2H_2SO_4 \quad (2)$$

9.1.2 Indirect Oxidation

In this method, mineral is extracted indirectly from the ore using an oxidizing agent produced by direct method of oxidation. The biomining bacteria are not in direct contact with the crude ore instead it produces strong oxidation agents for extraction of minerals (Pakostova et al. 2017). For example, ferric ion produced during direct oxidation method is capable of extracting sulphur from metal sulphides (Eq. 3). Thus, the product from direct oxidation process breaks down the crystal lattice energy of the heavy metal sulphide, thus separating the ore and mineral. Acidic environment is required for ferric ions to maintain in solution (Eq. 4). The reaction occurs as follows:

$$FeS_2 + Fe_2(SO_4) \rightarrow 3FeSO_4 + 2S^{\circ} \quad (3)$$

$$2S^{\circ} + 3O_2 + 2H_2 \rightarrow 2H_2SO_4 \quad (4)$$

9.1.3 Thiosulphate Mechanism

Solubilization occurs when ferric ion acts on acid-insoluble metal sulphides such as molybdenite (MoS) and pyrite (FeS) through an intermediate thiosulphate and releases sulphur as the end product. The bond between the metal compound and sulphur breaks resulting in release of thiosulphate as shown in Eqs. 5 and 6. The reaction occurs as follows:

$$FeS_2 + 6Fe^{3+} + 3H_2O \rightarrow S_2O_3{}^{2-} + 7Fe^{2+} + 6H^+ \quad (5)$$

$$S_2O_3^2 + 8Fe^{3+} + 5H_2O \rightarrow 2SO_4{}^{2-} + 8Fe^{2+} + 10H^+ \quad (6)$$

9.1.4 Polysulphate Pathway

In this mechanism, solubilization occurs due to combined reaction of ferric ion and protons of acid-insoluble metal sulphides with elemental form of sulphur as an intermediate. The sulphur-oxidizing microbes convert the elemental form of sulphur to sulphate. This reaction can be represented in Eqs. 7 and 8.

$$MS + Fe^{3+} + H^+ \rightarrow M^{2+} + 0.5H_2Sn + Fe^{2+} \quad (7)$$

$$0.5H_2Sn + Fe^{3+} \rightarrow 0.125S^8 + Fe^{2+} + H^+ \quad (8)$$

9.2 *Reduction*

In this process, sulphate reduction occurs resulting in formation of hydrogen sulphides as shown in Eq. 9 (Johnson et al. 2013). *Desulfovibrio desulfuricans* is the commonly used bacterium for this reduction process.

$$4H_2 + H_2SO_4 \rightarrow H_2S + 4H_2O \quad (9)$$

10 Micro-organisms Involved in Biomining

Many species of bacteria, fungi, algae and moulds are employed in biomining process (Table 4). Microbes can be employed for recovery of resources by two biomining processes: bioleaching and biosorption (Jerez 2011). Bioleaching is the process of extensive application of microbes for resource recovery through solubilization. Biosorption involves the adsorption of minerals along the surface of microbes from dilute solutions or mineral waste. Time-consuming is the only major drawback of using microbes for resource recovery since microbes are microscopic and produce products at low quantities. Research activities are being done to increase the efficiency of the organism and also reaction time. The microbes incorporated in this mechanism are resistant to the metals (Orell et al. 2010). Then only the microbes can be effectively used in biomining process.

10.1 Bacteria

Recent advance in the genetic manipulation of bacteria for industrial processes enables to rejuvenate bacterial mining of metal-bearing minerals and also the

Table 4 List of micro-organisms involved in biomining

S. no.	Micro-organism	Name
1.	Bacteria	*Thiobacillus thiooxidans* *Thiobacillus ferrooxidans* *Thiobacillus thermophilica* *Sulfolobus acidocaldarius* *Bacillus licheniformis* *Bacillus luteus* *Bacillus megaterium* *Bacillus polymyxa* *Rhodospirillum* sp. *Leptospirillum ferrooxidans* *Pseudomonas fluorescens* *Thermothrix thioparus*
2.	Fungi	*Aspergillus niger* *Penicillium simplicissimum* *Penicillium lapidorum* *Penicillium spimuiosum* *Rhizopus arrhizus* *Mucor haemalis* *Penicillium chrysogenum*
3.	Algae	*Hydrodictyon reticulatum* *Chlorella vulgaris* *Chlorella regularis* *Luminaria* sp. *Codium* sp.

microbiological treatment of metal-polluted wastewater. The idea of using bacteria for mining technique was first proposed by Rio Tinto mines in eighteenth century. Bacteria play tremendous role in extracting metals from their ores (Johnson 1999). Compared to other micro-organisms, bacteria produce more efficient product. Enzymes produced by bacteria on the crude ores solubilize the ore and retain the pure metals. Most commonly used bacterium for biomining is *T. thiooxidans* (Campodonico et al. 2016). It is a spore-forming rod-shaped gram-negative bacterium which is motile. This bacterium helps in the oxidation of ferrous iron and sulphur to ferric iron and sulphate, respectively. Another bacterium *Thiobacillus ferrooxidans* possesses the same characteristic features of *T. thiooxidans* feeding mostly on sulphur compounds. The main advantage of using these two organisms is that they can tolerate maximum concentration of metals. A few reviews state that combined effect of these two bacteria *T. thiooxidans* and *T. ferrooxidans* results in high efficiency with enhanced metal extraction. Survey results depict that in USA, more than 10% of copper obtained is through bacterial mining. Table 5 shows the list of other bacteria involved in biomining process. Optimizing parameters such as pH, temperature, aeration have a major impact on the microbial populations and metal recovery efficiency. A variety of mineral-oxidizing bacteria readily found can easily oxidize iron- and sulphur-containing minerals. Several species of fungi can be used for biomining. Physical conditions like pH and temperature are although the inherent property of micro-organism and so vary with micro-organism. Still it is mainly decided by the dominant bacteria in the consortium like *T. ferrooxidans* and *Leptospirillum ferrooxidans*. The significant explanation behind the predominance of *Leptospirillum* and *T. ferrooxidans* in industrial procedures is more likely than not the Fe/Fe proportion, i.e. redox potential. In any case, there might be different reasons which add to the strength. The ideal pH for the development of *T. ferrooxidans* is inside the range pH 1.8–2.5, though *L. ferrooxidans* is more corrosive safer than *T. ferrooxidans* and will develop at a pH of 1–2. Concerning temperature, *T. ferrooxidans* is more tolerant at low temperatures and less tolerant at high temperatures than *L. ferrooxidans*. Some strains of *T. ferrooxidans* can oxidize pyrite at temperatures as low as 10 °C but 30–35 °C is thought to be ideal. *Leptospirillum*-like microbes have been accounted for to have a farthest point of around 45 °C with a lower range of around 20 °C. The greater part of continuous

Table 5 Classification of bacteria for biomining based on their shape and growth at different temperature (Campodonico et al. 2016)

S. no.	Bacterial type	Temperature (°c)	Shape	Examples
1.	Mesophilic	10–40	Cylindrical	*Thiobacillus ferrooxidans*
2.	Moderately thermophilic	40–60	Larger than mesophiles	*Thiobacillus caldus*
3.	Extremely thermophilic	60–85	Spherical	*Sulfolobus metallicus*

stream bio-oxidation forms which are utilized to treat gold-bearing arsenopyrite minerals or concentres work at 40 °C and pH 1–6. The oxidative conditions are required for the most metal dissolution reaction, and these engineered heaps are aerated with external blowers. Forced air is a fairly common practice and is also important for the cultivation of microbes within the heap. The heaps are aerated to provide both oxygen and carbon dioxide for the bioleaching micro-organisms. The micro-organisms used in biomining are aerobic thus require an abundant supply of oxygen for survival and growth. Oxygen can be provided by aerators and pipes. Mechanical agitation is also an effective method to provide continuous air supply uniformly and also to mix the contents. Mineral concentrates are coated onto rock particle supports and then stacked as thin-layer heaps and have been demonstrated to promote more rapid recovery of base and precious metals than conventional bioheap leaching. The composition of the microbial community can change radically in stirred tanks configured sequentially, with heterotrophic iron-oxidizing eukaryotes being more prominent in downstream tanks. These are characterized by spatial and temporal heterogeneity of key physicochemical parameters such as pH, temperature and leachate chemistry. Consequently, the microbial biodiversity of biomining heaps (which have been far more intensively studied than dumps) are known to be far greater than that of stirred tanks and can also change greatly as heaps evolve. Indigenous prokaryotes may, for example, become increasingly supplanted by introduced microflora in situations where heaps are inoculated. Also in contrast to stirred tanks, the ability of micro-organisms to attach to minerals is an important attribute in irrigated systems, to prevent their washout in percolating leach liquors. Leaching of these ores using heterotrophic micro-organisms requires an organic carbon source for their growth and as a source of energy (Leduc and Ferroni 1994).

10.1.1 Bacterial Characteristics for Biomining Process

Certain features of bacterial species make it fit for biomining process. This characteristic differs among the bacterial group based on their tolerance to temperature and metal concentration, ability to solubilize and absorb metals, etc. Advanced techniques such as genetic engineering and conjugation have been developed for engineering bacteria with necessary features for increasing the yield and efficiency of biomining.

The bacteria employed for biomining process should possess the following features:

- Thermophilic bacteria—Ability to withstand high temperatures since metal extraction high temperature is required.
- Chemophilic bacteria—Bacterial population growing at high acidic conditions.
- Autotrophic bacteria—Ability of bacteria to produce energy from inorganic compounds through photosynthesis or chemosynthesis.
- Ability of bacteria to form biofilms.

10.1.2 Techniques for Bacterial Identification

Immunofluorescence

This technique is used to identify specific antigen or antibodies in the biological fluid. Bacteria suitable for biomining process are identified using fluorescent antibodies. Immunofluorescence works on the basic principle of specificity of antibodies to their antigens. Through this technique, antibodies are used to identify the desired bacteria for biomining. Immunofluorescence assay can be performed either directly or indirectly.

Dot Immunoassay

This technique is employed for identification of ore-adhering bacteria. The two commonly used biomining bacteria, namely *T. thiooxidans* and *T. ferrooxidans,* can be identified using dot immunoassay method. Nitrocellulose films are used on which the bacterial colonies are applied as dots. The interaction between antigen and antibody takes place in the film during which binding of specific antibody to antigen occurs. Then, the nitrocellulose film is treated with secondary antibody to make the response visible by emitting colour. Comparison between the test sample and known sample reveals the bacterial identification.

Dot-Plot Identification

Dot-plot technique is used for identification of bacteria from ores and soil samples. *Thiobacillus ferrooxidans* can be identified using dot-plot method. It is a DNA-based technique. Bacterial colonies are treated using chemicals such as sodium dodecyl sulphate (SDS). Bacterial cells are disrupted to extract the DNA and then followed by DNA purification. Using southern blotting technique, the isolated DNA is fixed on nitrocellulose membrane. Specific genetic probes are used for identification of bacteria suitable for biomining. DNA strands embedded on the membrane are treated using DNA probes. Hybridization occurs between the DNA strands and the probes followed by washing of the unhybridized probes. The fluorescence emitted by the hybridized strand and probes depicts the specific bacterial strain for biomining process.

10.2 Fungi

Apart from bacteria, fungal species also plays a significant role in conversion of metal compounds by acting as biocatalysts. Many fungal species are involved in solubilization process for metal extraction. Research work shows that the fungal strains, namely *Aspergillus niger* and *Penicillium simplicissimum,* possess high capability of mobilizing aluminium, lead, zinc, uranium and nickel. Edible mushrooms, namely *Agaricus bisporus,* can absorb metals like mercury. Yeasts like *Saccharomyces cerevisiae* and *Sporobolomyces salmonicolor* are also capable of absorbing metals (Das et al. 2009).

10.2.1 Mechanism

A few types of growths can be utilized for biomining. Parasitic strains, *A. niger* and *P. simplicissimum,* could prepare Cu, Sn, Al, Ni, Pb and Zn. Based on this utilization, the microbes create metabolites, for example natural acids, extracellular polymeric substances (EPS) or complexing compounds, for example siderophores, that may collaborate with the mineral surfaces and can solubilize the metals. In addition, microbial-mediated reduction processes, acidification or alkanization can contribute to the dissolution of minerals. This form of mining does not rely on microbial oxidation of metal, but rather uses microbial metabolism as source of acids which directly dissolve the metal. Microfungi are heterotrophic organisms. They exist in all ecological niches, e.g. supporting the weathering of rocks as well as the mineralization of materials containing metals. Their development is encouraged by the acidic reaction, the presence of sugars and the appropriate humidity. These micro-organisms can produce large amounts of organic acids, such as citric, glycolic, oxalic, and other acids, which work as chemical solvents, can be used on an industrial scale in bioleaching processes and impact the change of the environment's reaction. The microfungi, due to their biochemistry and relatively high immunity to hostile factors (pH, temperature, etc.), provide an excellent alternative in the bioleaching of metals, since the classical chemical methods of acidic bioleaching cannot be used for environmental reasons. The extraction through microfungi consists mainly of producing metabolites such as organic acids, amino acids and peptides that serve as leaching agents for the dissolution of metals. Micro-organisms involved in biomining actually gain energy by breaking down minerals into their constituent elements. The mineral dissolution reaction is not identical for all metal sulphides. A two-step process seems appropriate to increase leaching efficiencies (e.g. for an industrial application): in the first step, organisms are grown in the absence of electronic scrap followed by a second step where the formed metabolites are used for metal solubilization (Bindschedler et al. 2017).

10.3 Algae

Fresh water and marine algae act as metal bioaccumulators. Algal species such as *Chlorella vulgaris, Scenedesmus quadricauda, Nostoc linckia* and *Hydrodictyon reticulatum* adsorb high amount of metals like mercury, uranium and lead (Calmoi and Cecal 2007). Research activities are trying to treat rivers polluted with metals using algal species.

10.4 Microbe–Mineral Interactions

Research activities have demonstrated that connection of the bacterial cells to metal sulphides does not occur arbitrarily. For instance, *A. ferrooxidans* cells specially stick to places with noticeable surface scratches. Microbes, for example *L. ferrooxidans* and *A. thiooxidans,* that are exceedingly motile have been appeared to have a chemosensory framework that permits them to have chemotaxis, that is, the ability to identify inclinations of oxidizable substrates being removed from metals, for example Fe^{+2}/Fe^{+3} particles, thiosulphate and others. A chemotactic receptor has been portrayed in *L. ferrooxidans*. Bacterial cells used in biomining always grow and attach on the surface of the elemental form of sulphur and metal sulphides (Mapelli et al. 2012). Extracellular polymeric substances (EPS) found outside the cell are responsible for this attachment of bacterial cells to the ores. The composition of this EPS varies according to the growth of the bacterial cell. Bacteria containing ferric ions are bound to the exopolysaccharides get attach to the surface of the mineral, and biofilm formation occurs here. This ferric ion reacts with the metal sulphides, for example pyrite ferrous ion is reoxidized to ferric ion and in case of thiosulphate, sulphuric acid is formed.

11 Techniques Involved in Biomining

The success of biomining process depends on various parameters mainly on choice of bacterial selection and mineral composition of the ore. The efficiency is also related to the experimental set-up used for biomining process. Optimum conditions have to be maintained for high recovery of minerals. These conditions have to be optimized at laboratory scale then applied on large-scale process in industries. The yield efficiency, process time and conditions can be altered easily at laboratory condition than at large-scale operations. The major techniques involved in biomining process are given in Fig. 3.

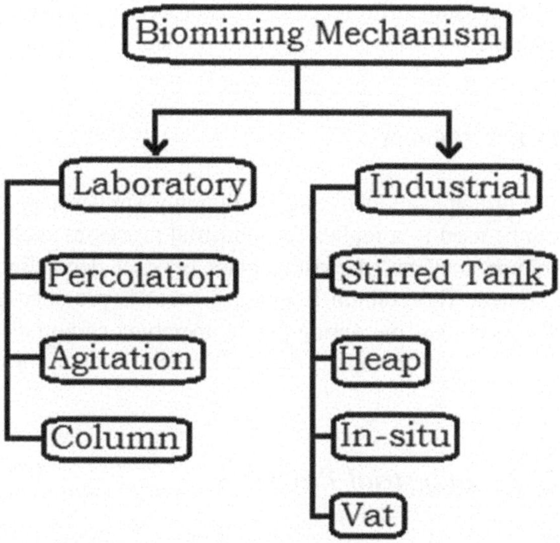

Fig. 3 Techniques in biomining process

11.1 Laboratory Methods

11.1.1 Percolation

This is the first experimental method used in laboratory for extraction of metals. Percolation technique involves the movement of liquid medium on a static bed. The movement of liquid medium is done using airlift percolators. It is made up of a glass tube with a sieve plate at the bottom filled with ore particles. These ore particles are continuously engulfed with nutrient medium inoculated with bacterial population. The mineral-rich leach liquor which spills through the column is driven up using compressed air to the top of the column. This leach liquor is recirculated again since the process is continuous. The parameters such as pH, bacterial community and mineral concentration are evaluated by collecting required quantity of leach liquor during the process. The method is not efficient due to inadequate aeration, adverse surface ratio and slow process.

11.1.2 Agitation

Agitation method involves the movement of finer particles suspended in the liquid medium. The fine particles of size <100 μm are circulated in the liquid medium with continuous supply of aeration and agitation. High amount of air supply and controlling of process parameters help in sufficient growth of bacteria so that the yield and recovery of metals increase. This process can be carried out in a bioreactor or in an Erlenmeyer flask. Maximum extraction can be achieved in this process using airlift reactors. This process can be used for extraction of metals with high ore concentrates and waste products from industries.

11.1.3 Column

The principle and working of column is similar to percolation process. This method can be used as a replica for industrial processes such as bioheaps or dumping. The column may be of plastic, glass or steel depending on the size of the column required. The column leaching process is provided with special devices for handling samples, measuring pH, temperature, carbon dioxide and oxygen. This set-up is well suited for optimizing parameters in dump or heaping process.

11.2 Industrial Processes

11.2.1 Stirred Tank Biomining

This method is a multi-step process comprising of more number of tank bioreactors serially arranged with each other. The substrate moves starting with one reactor then onto the next, and in the final stage, it is washed with water and treated with chemicals to recuperate the metals. As the name implies, the procedure requires developing substantial circulated air through tanks that are arranged serially, so that spillover from one tank fills in as crude material for the following. By this way, the reactor can work in continuous mode, with fresh ore being added to the main tank while the overflow from the last tank is evacuated and treated. The low-grade ore to be processed is pulverized to a low molecule size, so that the solids stay suspended in the liquid medium (Rawlings et al. 2003). The stirred tank bioreactors are often specially designed with either rubber or corrosion resistant steels and insulated with cooling jackets or pipes. To ensure maximum growth and maintain population density of bacteria, nutrients are supplied in the form of ammonium sulphate or potassium dihydrogen phosphate in the medium. Major nutrients required for the microbial growth are carbon dioxide and oxygen for instance leaching of sulphide minerals using microbes (Mahmoud et al. 2017). Continuous supply of oxygen is maintained since the process is aerobic. Commonly used organism in stirred tank biomining is *Thiobacillus* sp. Recovery of copper and gold can be made efficiently through this process.

11.2.2 Heap Biomining

Bioheaps contain large quantities of low-grade ore and effluents from the process of extraction in which the minerals are present at lower concentrations (Brierley 1999). Those effluents are arranged in a large, open-space heaps treated with micro-organisms for extraction of metals. The low-grade ores are first crushed into particles of smaller size and then treated with acids. The crushed acid treated ore then agglomerates during which finer particles get attach to the coarser ones and finally treated with water or other fluids. This process is carried out for optimization of moisture content in the bacteria present in the ore to be inoculated with the liquid. The ore is then stacked in large heaps of 2–10 ft high with circulating air through tubes to give air supply to the micro-organism, therefore encouraging bio-oxidation. Chemophilic bacteria are commonly employed in bioheaping process (Pradhan et al. 2008). Nutrients may be additionally added for bacterial growth.

Dump and heap biomining involve the usage of a liquid medium (lixiviant) at the top surface of dump or heap. The metal-rich solution thus flows at the bottom phase of the dump or heap. Sulphuric acid may be added at the top surface of dump or heap for lowering the pH and encouraging the growth of acidophilic bacteria (Schippers et al. 2010). The metal-rich solution flowing at the bottom of the dump or heap is collected and transferred to the recovery process. Final metal extraction is completed by solvent extraction or electrowinning method (Brierley 2008).

11.2.3 In Situ Biomining

In this technique, the mineral is extricated specifically from the mine instead of gathering the metal and moving to a separating channel far from the site of the mine. The mining solution is added directly to the site from which the metal is to be extracted. In situ biomining is generally done to concentrate low measures of minerals present in the metals after a conventional extraction process is finished (Zammit et al. 2014). The mine is impacted to reduce the ore size and to build porousness and is then followed by water and acid treatment along with bacterial inoculums. Aeration is given with the help of pipes or shafts. Bio-oxidation occurs in situ because of developing micro-organisms and results in the extraction of mineral from the raw ore. Ore bodies must be of three types to be used for in situ biomining: deposits at the surface must be above the water table, deposits at the surface must be below the water table, and in-depth deposits must be below the water table. Uranium and copper can be extracted efficiently using this technique (Haque and Norgate 2014).

11.2.4 Vat Biomining

This method of biomining is often suitable for oxide ores where the pulverized ore materials are disintegrated in a confined tank. More process parameters can be controlled during the extraction and recovery operations. Precious metals and high

Table 6 Advantages and disadvantages of industrial biomining process

S. no.	Methods	Advantages	Disadvantages
1.	Stirred tank	The whole process is aerobic Used for substrates with high mineral concentration	Expensive Time-consuming Efficiency is high for valuable metals only Not suitable for lower substrate concentration
2.	Bioheaps	Cost-effective Easy construction, operation and maintenance Efficient process for extraction from low ore concentration	Time-consuming Requires large area for construction Low yield of mineral High risk of contamination Growth of bacteria is not uniform
3.	In situ	This process can occur at the site of ore present and there is no need for transferring ore to a separate place Minimum impact on environment	Low recovery rates Time-consuming Efficient only when the ore possesses unique characteristics
4.	Vat	Recovery can be increased by making controls	Expensive

ore concentrations can be effectively extracted using this process (Siddiqui et al. 2009).

The advantages and disadvantages of industrial biomining process are discussed in Table 6.

12 Factors Affecting Biomining Process

The efficiency of biomining process is determined by certain factors. Any alteration in the factors will influence product recovery, thus creating negative impact on biomining process. Few factors are discussed below in Table 7.

Table 7 Factors influencing biomining process (Jerez 2009)

S. no.	Factors	Parameters
1.	Physicochemical	Temperature pH Aeration
2.	Microbial	Microbial density Metal tolerance Choice of bacteria Mixed microbial diversity
3.	Mineral	Ore concentration Composition Surface area Particle size Porosity

12.1 Type of Bacteria

The choice of bacteria greatly influences the efficiency of biomining. Bacteria are selected based on their ability to survive at high temperatures, acid concentrations and growth at high concentration of metals. The bacteria possessing the above characteristics are more suitable to make the biomining process successful.

12.2 Composition of Ore

The concentration of ore to be purified plays a major role in influencing biomining process. The rate of contamination, concentration of sulphides and quantity of mineral present have direct effect on bio-oxidation process.

12.3 Temperature

Mesophilic (e.g. *Thiobacillus novellus*) or thermophilic (e.g. *Thermococcus litoralis*) bacteria are suitable for biomining process. Optimum temperature varies depending on the type of ore being selected. The optimum temperature varies between 25 and 35 °C. Optimum temperature must be maintained for the bio-oxidation process to proceed at a faster rate. The bio-oxidation rate alters depending on the optimum temperature.

12.4 Leach Solution PH

Acidophilic bacteria used in biomining process require acidic environment for their effective growth. Bio-oxidation process takes place at maximum speed at pH 2.5–3. Bio-oxidation rate decreases when the pH alters above or below the acidic range.

12.5 Aeration

Aerobic bacteria are mostly involved during biomining process. These aerobic bacteria require abundant quantity of oxygen for their growth and development. Hence, aeration plays a significant role in this process. Air supply is provided by means of aerating pipes, and mechanical agitation is done for proper blending of liquids.

12.6 Solid–Liquid Ratio

The ratio of solid (ore) to the liquid (leaching solution) must be maintained at optimum level for the bio-oxidation process to occur at standard rate. The leaching solution is a combination of water, acid solution and bacterial inoculums. After the process, the leach solution will be rich in mineral content which must be expelled periodically and replaced with fresh leach solution.

12.7 Crystal Lattice Energy

The mechanical stability and the extent to which sulphide solubilization occurs will be determined by crystal lattice energy. The solubility of sulphide ores is inversely proportional to crystal lattice energy, i.e. solubility of sulphide ores increases with lower crystal lattice energy; thus, the minerals are easily extracted into the solution by bacterial action.

12.8 Surface Area

Bio-oxidation rate by bacterial activity greatly depends upon the particle size of the ore. Surface area of the ore can be increased by crushing or grinding. Size of the ore particle and rate of bio-oxidation are inversely proportional to each other. The rate increases with lower particle size.

12.9 Surfactants

Surfactants are compounds added to reduce the interfacial tension between solid and liquid or two liquids. In biomining process, adequate quantities of surfactants are added to enhance the bio-oxidation rate of minerals from sulphide ores. Addition of surfactants lowers the surface tension of the leach solution, thus making the ore wet resulting in increased contact with the bacterial population which eventually builds the rate of bio-oxidation and increases mineral extraction.

13 Recovery of Metals by Biomining

13.1 Copper

Copper is one of the common metal resources mostly available in its ore form. Copper is found to be in piping, electronics and also in antibiotics. Copper Development Association reported that in 1999, total global consumption of copper was found to be 21 billion tons. Major copper ores are chalcopyrite ($CuFeS_2$), covellite (CuS), bornite (Cu_5FeS_4) and chalcocite (Cu_2S). Copper ores are composed of other metals other than copper. For example, one of the copper ore chalcopyrite is composed of 26% copper, 26% iron, 33% sulphur and 2.5% zinc. The chemical methods available for extraction of copper from its ore (chalcopyrite) are complex and time-consuming. Few chemical agents such as sulphur dioxide and cyanide used for this extraction process are toxic and cause environmental pollution. Strip-mining or blasting method of copper extraction requires several months to complete the process. Copper extraction performed by using microbes causes no toxic effects and simple. The bacteria commonly used for biomining of copper are *T. ferrooxidans.* Biomining of copper can be performed using either bioheap technique or in situ method. The organism *T. ferrooxidans* oxidizes the ore-insoluble chalcopyrite ($CuFeS_2$) and converts it into soluble copper sulphate ($CuSO_4$) (Eq. 10). During this process, sulphuric acid is formed as a by-product which helps to maintain acidic environment required for the growth and development of bacteria. The above chemical reaction can be represented as:

$$\begin{aligned} &Cu_2S(\text{chalcocite}), CuS(\text{covellite}), CuFeS_2(\text{chalcopyrite}), Cu_5FeS_4(\text{bornite}) \\ &\quad + Fe_2(SO_4)_3 \rightarrow CuSO_4 + FeSO_4 + S^\circ \end{aligned} \tag{10}$$

This method of copper extraction is efficient and eco-friendly. A total of 5% of global copper extraction is obtained through biomining approach. Genetic modifications of the organism can be done to improve the process time and efficiency. In situ method of biomining suits well for copper extraction since the organisms that grow well on the ore can be used for this process.

13.2 Uranium

Uranium was first produced as a by-product of gold production. Uranium biomining is commonly carrier out in countries such as India, Canada and the USA. Uranium can also be recovered from very low-grade ores of range 0.01–0.5%. Microbial mining of uranium is a cost-effective and eco-friendly technique. *Thiobacillus ferrooxidans* is employed for uranium extraction. In situ or bioheap methods are suitable for uranium biomining. The list of uranium ores used for extraction process are uraninite (UO_2), gummite (UO_3–nH_2O), becquerelite (CaU_6O_{19}–$11H_2O$), coffinite ($U(SiO_4)_{1-x}(OH)_{4x}$), brannerite (U, Ca, Ce)(Ti, Fe)O_6, etc. During the process, uraninite reacts with ferric sulphate to produce uranyl sulphate and ferrous sulphate as shown in Eq. 11.

$$UO_2 + Fe_2(SO_4)_3 \rightarrow UO_2SO_4 + 2FeSO_4 \qquad (11)$$

This reaction requires rich source of iron and high redox potential. Biomining of uranium is an indirect method of extraction since the micro-organism acts on iron and not directly on uranium. Sulphuric acid and ferrous sulphate are produced by the organism during the reaction with pyrite present within the uranium ore. The optimum conditions for effective microbial mining of uranium occur at pH 1.5–3.5, temperature 45–50 °C and CO_2 content approximately 0.2% of air entering the process. Bioheap method is preferred compared to in situ because the efficiency and recovery are high in heaping compared to in situ. The factors affecting uranium biomining are microbial population, toxicity and resistance to metals, pH, temperature, pyrite content and aeration. Recent activities on use of thermophilic bacteria for uranium biomining are being conducted. Under optimized conditions such as pulp density, proper aeration and agitation, uranium extraction was found to be high in stirred tank reactor also (Eisapour et al. 2013).

13.3 Nickel

Canada is the largest supplier of nickel. It is primarily made up of sulphidic and lateritic ores. Nickel production can be increased by biomining technique even with very low-grade ores and mine dumps. *Thiobacilli* sp. is commonly used for nickel extraction. This bacterium oxidizes sulphides to sulphuric acid, and thus, the nickel can be extracted from the mineral-rich solution. The solution pH changes due to presence of other metals in the surrounding medium, thus altering the stability and lifespan of the bacterium. New strains have been isolated resistant to nickel metals capable of degrading metal sulphides with high stability (Clercq et al. 2001). Bioheap technique shows best results for nickel extraction. The optimum conditions for effective extraction of nickel are as follows: temperature 20–80 °C, particle size of 8 mm, acid consumption of 15 kg and duration up to 14 months. The nickel and iron minerals found in nickel sulphide ores are pentlandite (nicopyrite), millerite, violarite, heazlewoodite, pyrite, etc. Nickel sulphate is formed at the end of the process, and the reaction of formation proceeds as in Eq. 12.

$$8(NiFe)_9S_8 + 141O_2 + 26H_2SO_4 \rightarrow 36NiSO_4 + 18Fe_2(SO_4)_3 + 18H_2O \qquad (12)$$

The main factors affecting nickel extraction are choice of bacteria, temperature, aeration, ore composition, agitation, acidity and solid/liquid phase relation.

13.4 Gold

Gold is one of the precious metal resources occurring in elemental form or allowed with silver in combination with other metals. There are two different natural occurrences of gold: free-milling ores and refractory ores (Natarajan 1993). Free-milling ores are those in which the gold occurs dispersed inside the quartzite. After reducing the size, it is directly subjected to cyanidation. In refractory ores, gold particles are enclosed in a sulphide matrix or carbonaceous compound. Beneficial organisms suitable for effective extraction of gold are *T. ferrooxidans*, *T. thiooxidans* and *L. ferrooxidans*. Gold bimining can also be effectively processed using fungi (Bindschedler et al. 2017). The optimum parameters for gold extraction occur at pH 1.7–2, temperature 30–35 °C, proper aeration and agitation, removal of ferric and other products during the reaction. Bioheap technique suits more for gold extraction from low-grade ores as well as from mine tailings. Gold particles are embedded in sulphide minerals such as pyrite, chalcopyrite, arsenopyrite and pyrrhotite. Pyrite and arsenopyrite gets oxidized by the bacteria to form ferric sulphate as shown in Eqs. 13 and 14.

$$4FeAsS + 13O_2 + 6H_2O \xrightarrow{T.ferrooxidans} 4H_3AsO_4 + FeSO_4 \quad (13)$$

$$2FeS_2 + 7O_2 + 2H_2O \xrightarrow{T.ferrooxidans} 2FeSO_4 + 2H_2SO_4 \quad (14)$$

The bacteria break down the sulphide minerals thus releasing gold followed by continuous cyanidation. The gold particles are released as depicted in Eq. 15.

$$2Au + 4NaCN + O_2 + 2H_2O \rightarrow 2NaAu\ (CN)_2 + 2NaOH + 2H_2O_2 \quad (15)$$

13.5 Manganese

Manganese is hard, silvery metal and brittle, used in alloys mainly in steel. Manganese is used to increase the strength, resistance to wear and also for increasing the workability. Many research works are being conducted for biomining of manganese using microbes (Das et al. 2011). Anaerobic bacteria are used for their ability to reduce manganese oxides. Reduction process is carried out either by production of acids or by using reducing substrates like sulphides. Microbes like bacteria, fungi, algae and yeast are also capable of reducing manganese oxides. The enzyme cellobiose dehydrogenase (CDH) produced by anaerobic bacteria effectively catalyses Mn redox reactions. Manganese biomining was performed in batch process using *T. thiooxidans* in presence of excess amount of sulphur. During the process, the leaching efficiency of *T. thiooxidans* decreases with increase in manganese concentration. Manganese biomining can also be done using fungi *Penicillium citrinum* (Acharya et al. 2004). Fungal biomining is found to be effective only in presence of certain energy substrates like sucrose. Manganese biomining occurs at a faster rate with high sucrose concentration. In situ method of mining was found suitable, and the factors affecting the process are pulp density, inoculums and particle size. International Manganese Institute reports that the total manganese produced in the world contribution of manganese produced from India is about 6%.

14 Traditional Mining (Vs.) Biomining

Conventional mining systems affect, reveal and then attract the ores to the surface, where it is pulverized and ground into more minute particles. The vitality produced all inclusive from fossil and atomic powers and inexhaustible resources is believed to be utilized for this reason. However, metals make up just around 1% of the unearthed undigged material, implying that the vitality spent on crushing the remaining 99% of material that has no esteem is squandered, and the broad area are spent in storage of possibly toxic wastes. In biomining, liquids produced by and containing micro-organisms and archaea that flourish in numerous acidic conditions are utilized to separate rocks, filter metals from minerals or generally make them available for substance extraction usually chemicals. Currently, biomining is a specialty innovation utilized for the most part where traditional mineral handling options are not reasonable or are too expensive, yet adjusting this procedure for the profound sub-surface could minimize vitality costs and the negative impact related with conventional mining techniques.

15 Biomining in India

The mining and mineral industry gives a vital help to the economy and is of key significance to each nation. The mining division in India has achieved an abnormal state of improvement however despite everything it has a great deal of hidden potential. Due to the utilization of innovation, Indian organizations are now and again unfit to investigate and abuse many resources. The present mining innovation and works on being utilized enable mining just to the extent of few metres, which implies that major part of the resources of India still stay unexploited. There are open doors for revelations of sub-surface reserves with the use of current methods and procedures. Minerals and mining area contributes fundamentally to the economic development and improvement of India through export of crude material to different ventures in the nation and lastly resulting in tremendous employment openings. The Indian mining industry at present utilizes more number of individuals and offers an extensive variety of chances to mining engineers in the public and private sectors. The administration offers an extensive variety of modifications to financial specialists occupied with the mining action.

16 Advantages and Limitations

The major advantages of biomining are as follows:

- It does not involve the production of toxic gases and pollution such as noise and air.
- Self-generating process can be carried out in situ.
- Efficient removal of impurities works more on low-grade ores.
- More number of metals can be mined easily.

Limitations of biomining include:

- Biomining is a slow process.
- Sometimes microbial growth may be slow and needs optimum conditions.
- Time-consuming process.

17 Future of Biomining Process

In spite of the fact that mining is one of mankind's most established practices, the strategies used to concentrate minerals have not changed significantly for a considerable length of time. Metals are burrowed from the earth, smashed, and afterwards, minerals, for example copper and gold, are extracted by extraordinary heat or lethal chemicals. The ecological and well-being impacts of conventional mining innovations have been pernicious. In the previous couple of years, the mining business has been swinging to a more productive and earth beneficial strategy for removing minerals from metals: micro-organisms that mine them out. Utilizing a bacterium, for example *T. ferrooxidans,* to extract copper from mine tailings has enhanced recovery rates and lessened working expenses. Besides, it grants extraction from poor low-grade ores thereby avoiding high utilization of high-grade ores. Low-quality copper metal, which is bound up in a sulphide framework, is dumped outside a mine and treated with sulphuric acid to enhance the development of *T. ferrooxidans*. As the microscopic organisms feed on the metal, copper is discharged and collected. The sulphuric corrosive is reused. Biomining is likewise being utilized to remove gold from poor-quality sulphide gold metals, once thought to be useless. To expand the proficiency of biomining, the research is on for bacterial strains that are more qualified to substantial scale operations. Researchers are swinging to thermophilic microscopic organisms found in hot areas to tackle the issue of heat tolerance. These microbes flourish in temperatures up to 100 degrees Celsius or higher and could work in a high-temperature oxidative condition. At present, 25% of all copper around the world, worth more than 1 billion dollar every year, is created through biomining. This makes biomining an innovative green approach in biotechnology. Work is also progressing in the field of isolating bacteria and engineering them so that they exhibit tolerance to heavy metals such as mercury, chromium, arsenic.

18 Conclusion

Biomining is the manageable, biotechnological approach using micro-organisms to expel metals from sulphide mineral metals. The improvement of biomining has advanced from inadequately composed dumps to exceedingly built heaps and stirred tank reactors in an industrially significant biotechnological method. The extract of metals from sulphide minerals is catalysed by iron-oxidizing acidophilic (ideal pH for development <3) micro-organisms that function in consortia with heterotrophic and sulphur-oxidizing acidophiles in a mixed culture. The micro-organisms catalyse metal discharge by recovering ferric iron that oxidizes the mineral sulphide bound to create metals. Alternate micro-organisms in the mixed culture may oxidize the decreased inorganic sulphur mixes to sulphuric compounds. Presently, substantial enhancements in biomining are preceded with research in distinguishing bacterial strains more efficient for individual applications and also in the hereditary modification of bacterial strains that can tolerate high temperature and heavy metals, for example arsenic, mercury or cadmium. Biomining has turned out to be one of the leading mining innovations, and the future mining process greatly depends on this methodology. Research activities bring about innovation usage for more prominent effectiveness and efficiency or novel solutions for various mining issues. In addition, research work is also conducted to incorporate the biomining of metals from sulphide-containing compounds, phosphate metal bioprocessing and the bioconcentration of metals from fluids. From in situ-mining to mineral extraction and purifying innovation, biotechnological invention provides more advantages in making the extraction process cost-effective for mining industries and also to the environment in lowering the pollution and contamination caused by usage of chemical reagents for the extraction process. The potential uses of biotechnology to mining and mechanism involved are numerous. Thus, biomining serves as the safest and energy-saving process compared to other conventional techniques.

References

Mehta KD, Kumar V et al (2011) Bioleaching—an alternate uranium ore processing technology for India. Energy Proc 7:158–162

Acharya C, Kar RN, Sukla LB et al (2004) Fungal leaching of manganese ore. Trans Indian Inst Metals 57(5):501–508

Bindschedler S, Bouquet TQTV, Job D et al (2017) Fungal biorecovery of gold from e-waste. Adv Appl Microbiol 99:53–81

Bridge G (2009) Natural resources. Int Encycl Human Geol, pp 261–268

Brierley CL (1999) Bacterial succession in bioheap leaching. Process Metall 9:91–97

Brierley CL (2008) How will biomining be applied in future? Trans Nonferrous Metals Soc China 18(6):1302–1310

Calmoi R, Cecal A (2007) Bioleaching of uranyl ions from uranium ores by some algae. Environ Eng Manag J 6(1):27–30

Campodonico MA, Vaisman D, Castro JF et al (2016) *Acidithiobacillus ferrooxidans's* comprehensive model driven analysis of the electron transfer metabolism and synthetic strain design for biomining applications. Metaboll Eng Commun 3:84–96
Clercq M, Adschiri T, Arai K (2001) Hydrothermal processing of nickel containing biomining or bioremediation biomass. Biomass Bioenerg 21(1):73–80
Das AP, Sukla LB, Pradhan N et al (2011) Manganese biomining: a review. Biores Technol 102:7381–7387
Das BK, Roy A, Koschorreck et al (2009) Occurrence and role of algae and fungi in acid mine drainage environment with special reference to metals and sulfate immobilization. Water Resour 43:883–894
Davis-Belmar CS, Cautivo D, Demergasso C et al (2014) Bioleaching of copper secondary sulfide ore in the presence of chloride by means of inoculation with chloride-tolerant microbial culture. Hydrometallurgy 150:308–312
Orell A, Navarro CA, Arancibia R et al (2010) Life in blue: copper resistance mechanisms of bacteria and Archaea used in industrial biomining of minerals. Biotechnol Adv 28(6):839–848
Pakostova E, Grail BM, Johnson DB (2017) Indirect oxidative bioleaching of a polymetallic black schist sulfide ore. Miner Eng 106:102–107
Pradhan N, Nathsarma KC, Rao KS et al (2008) Heap bioleaching of chalcopyrite: a review. Miner Eng 21(5):355–365
Rawlings DE (1997) Biomining: theory, microbes and industrial process. Springer, Berlin
Rawlings DE, Dew D, du Plessis C (2003) Biomineralization of metal-containing ores and concentrates. Trends Biotechnol 21(1):38–44
Schippers A, Breuker A, Blazejak A et al (2010) The biogeochemistry and microbiology of sulfidic mine waste and bioleaching dumps and heaps, and novel Fe(II)-oxidizing bacteria. Hydrometallurgy 104(3–4):342–350
Siddiqui MH, Kumar A, Kesari KK et al (2009) Biomining—a useful approach toward metal extraction. Am Euras J Agron 2(2):84–88
Thosar A, Satpathy P, Nathiya T et al (2014) Bio-mining: a revolutionizing technology for a safer and greener environment. Int J Recent Scientific Res 5(9):1624–1632
Zammit CM, Brugger J, Southam G et al (2014) In situ recovery of uranium—the microbial influence. Hydrometallurgy. https://doi.org/10.1016/j.hydromet.2014.06.003

Author Biographies

Dr. P. Senthil Kumar is currently working as Associate Professor in the Department of Chemical Engineering, Chennai, Tamilnadu, India. He has 11 years of teaching experience, 10 year of research experience and 1 year of industrial experience. He is a recognized Anna University Research Supervisor to guide Ph.D./M.S. candidates. Under his guidance, 03 Ph.D Scholars have completed and 10 Ph.D. scholars are currently doing their active research in the area of Waste Water Treatment and Alternative Fuels. He has published more than 120+ research papers in highly reputed International Journals and 70 research papers in National/International Conferences. He has also published more than 25+ book chapters in the reputed publishers by Elsevier, Springer, etc. He recently received “IEI Young Engineers Award 2017-2018″ in Environmental Engineering Discipline at 33rd National Convention of Environmental Engineers held at Dhanbad during August 24-25, 2017 and received “Hiyoshi Think of Ecology Award 2017” with a Cash award of Rs. 25000/-, Plaque and Certificate from Hiyoshi Corporation, Japan and Hiyoshi India Ecological Services Pvt. Ltd., India.

P. R. Yaashikaa is a Research Scholar in Department of Chemical Engineering, SSN College of Engineering, Chennai, India. Her area of research emphasizes on application of microbes in the treatment of industrial wastewater.

Prof. G. Baskar is currently working as Professor of Biotechnology at St. Joseph's College of Engineering, Chennai, India. He has published more than 100 papers in national and international journals and 5 book chapters. His research expertise is in biofuels, therapeutic proteins, microbial enzymes, nanomedicine and nanocatalysis. He has received the International Bioprocessing Association's Young Scientist Award (2015) and the Indian Society for Technical Education (ISTE)-Syed Sajed Ali National Award (2016).

Chapter 18
Anaerobic Digestion: Factors Affecting Anaerobic Digestion Process

Sunil P. Lohani and Jouni Havukainen

Abstract Anaerobic digestion (AD) is a biological decomposition process that occurs in the absence of oxygen. The decomposition of organic matter is a multi-step process of series and parallel reactions namely hydrolysis, acidogenesis, acetogenesis and methanogene. Most of the control in anaerobic digestion is undertaken directly by the microorganisms themselves; however, the operational conditions such as temperature, pH, essential trace nutrients and toxicants can play a major role in modifying reaction rates of individual sub-processes. The energy performance of the anaerobic digestion is depending mainly on the biogas production technology, raw materials and geographic location (ambient temperature). Since the feedstocks coming to anaerobic digestion have usually lower heating value as received, the usual energy efficiency calculation used for incineration plant is not useful. Most commonly used method is the input/output method, and the estimation is dependent upon the chosen system boundary.

Keywords Anaerobic digestion · Operational conditions · Reaction rate
Energy performance · System boundary

1 Introduction

Anaerobic digestion (AD) is a biological process that occurs in the absence of oxygen when organic materials are available. The process is accomplished with a consortium of microorganisms such as fermentative bacteria, hydrogen-producing acetogenic bacteria, hydrogen-consuming acetogenic bacteria, carbon

S. P. Lohani (✉)
School of Engineering, Kathmandu University, P.O. Box 6250,
Dhulikhel, Nepal
e-mail: splohani@ku.edu.np

J. Havukainen
Sustainability Science, Lappeenranta University of Technology,
P.O. Box 20, Lappeenranta 53851, Finland

S. J. Varjani et al. (eds.), *Waste Bioremediation*, Energy, Environment,
and Sustainability, https://doi.org/10.1007/978-981-10-7413-4_18

dioxide-reducing methanogens and aceticlastic methanogens (Appels et al. 2008). AD process makes use of these anaerobes to breakdown organic substances to biogas mainly composed of methane (CH_4) and carbon dioxide (CO_2). The amount of excess sludge production is very small (Mittal 2011).

Biogas from organic waste usually contains 60–70% methane, 30–40% carbon dioxide and <1% nitrogen (Jonsson et al. 2003) in an ideal condition, whereas some amount of hydrogen sulphide and ammonia is also produced otherwise (Jansen and Jensen 2000).

The energy performance of AD is mainly dependent on the biogas production technology (wet or dry technology, mesophilic or thermophilic) and geographic location (ambient temperature). The used feeding materials have also an effect because the biogas yield is varying from different feeding materials. Moreover, the process of obtaining the feeding materials and types of digestion technology, for instance, wet digestion can consume significant amount of energy.

2 AD of Organic Material into Methane

The decomposition of organic matter is a multi-step process of series and parallel reactions. This successive degradation process occurs in four stages, namely (i) hydrolysis, (ii) acidogenesis, (iii) acetogenesis and (iv) methanogenesis as shown in Fig. 1. A brief discussion of each stage is presented below.

2.1 *Hydrolysis*

Hydrolysis of the complex organic matter is an important step of the anaerobic biodegradation process. During hydrolysis, the first stage of anaerobic digestion, bacteria transform the insoluble complex organic substrate (carbohydrates, proteins, lipids, etc.) into soluble monomers and polymers. This process is catalysed by enzymes like cellulase, protease and lipase excreted by the microorganisms responsible for fermentation for the conversion of proteins to amino acids; lipids to long-chain fatty acids (LCFA), polysaccharides, to simple sugars (Parawira 2012; Ostrem et al. 2004). This group of microorganisms is considered to be composed of a large group of facultative bacteria that can thrive with or without oxygen (Botheju et al. 2010; Schluter et al. 2008). Hydrolysis is the rate-limiting process for the overall digestion of substrates with high suspended solids (SS)/chemical oxygen demand (COD) ratio. It is usually not due to a lack of enzyme activity but to the availability of free accessible surface area of the particles and the overall structure of the solid substrate (Zeeman and Sanders 2001; Van Lier et al. 2008). Moreover, at low temperature, hydrolysis may limit the overall process (Lew et al. 2011) and thereby determining the required reactor design. The products of hydrolysis are the substrates for acidogenic bacteria. Equation (1) shows an example of hydrolysis

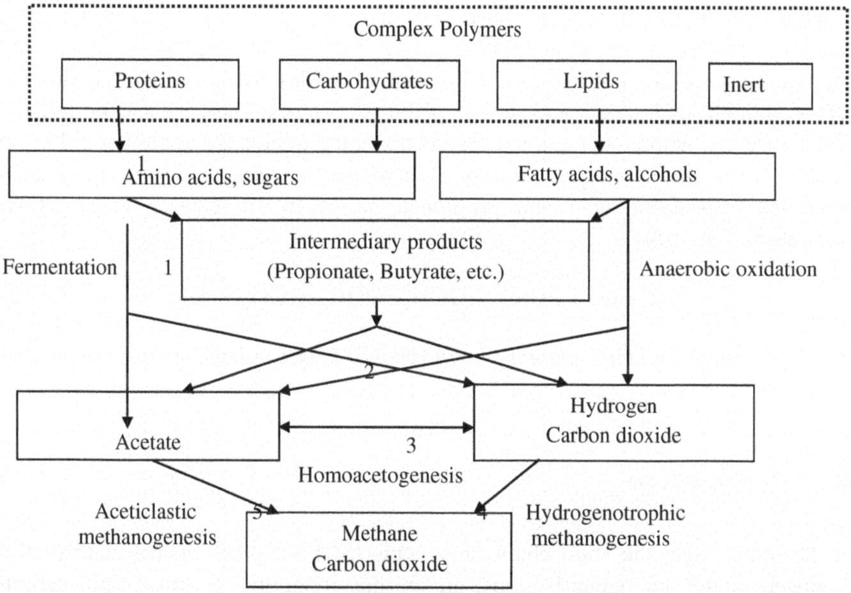

Fig. 1 Reaction of the anaerobic digestion of polymeric materials (numbers indicate the bacterial groups involved): (1) Hydrolytic and fermentative bacteria (2) Acetogenic bacteria (3) Homo-acetogenic bacteria (4) Hydrogenotrophic methanogens (5) Aceticlastic methanogens. Adapted from Gujer and Zehnder (1983)

reaction where organic waste is broken down into a simple sugar, in this case, glucose (Ostrem et al. 2004).

$$C_6H_{10}O_4 + 2H_2O \rightarrow C_6H_{12}O_6 + 2H_2 \tag{1}$$

2.2 Acidogenesis

In the second stage, the hydrolysis products (amino acids, LCFA and simple sugars) which are relatively soluble compounds are converted into variety of small organic compounds mainly volatile fatty acids (VFAs), that is, acetate (CH_3COOH) and organic acids such as propionate (CH_3CH_2COOH), butyrate ($CH_3CH_2CH_2COOH$), valeric ($CH_3CH_2CH_2CH_2COOH$), formic ($HCOOH$), lactic ($C_3H_6O_3$) as well as H_2, CO_2 and ammonia (Zeeman et al. 1996; Ostrem et al. 2004; WtERT 2014). This process is mainly performed by a fermentative microorganism, and the nature of end products depends on the conditions of reactor medium. For instance, acetate will be the main end product if H_2 is effectively removed by H_2-scavenging organisms such as methanogens (Van Lier et al. 2008). However, if methanogenesis is retarded and H_2 accumulates, more reduced products such as propionate and butyrate are likely to appear. Therefore, effluents of overloaded or disturbed anaerobic reactors often contain these more reduced intermediate products and become acidic (Van Lier et al. 2008). Out of acidogenesis product, hydrogen, carbon dioxide and acetic acid will skip the acetogenesis process and be utilized directly by the methanogenic microorganisms in the final stage as shown in Fig. 1. Equations (2) and (3) represent typical acidogenic reactions where glucose is converted into acetic acid and propionate, respectively (Ostrem et al. 2004; Bilitewski et al. 1997).

$$C_6H_{12}O_6 + 2H_2O \rightarrow 2CH_3COOH + 2CO_2 + 4H_2 \tag{2}$$

$$C_6H_{12}O_6 + 2H_2 \rightarrow 2CH_3CH_2COOH + 2H_2O \tag{3}$$

2.3 *Acetogenesis*

In the third stage, the short-chain fatty acids (SCFA), other than acetate that is produced in the acidogenesis steps, are further converted to acetic acid, carbon dioxide and hydrogen by the acetogenic bacteria as shown in Fig. 1. There are two types of acetogenic bacteria namely hydrogen-producing acetogens and homoacetogens (Parawira 2012; Cavinato 2011). Equations (4) and (5) show the production of acetic acid from butyrate and propionate and by utilizing hydrogen-producing bacteria (Ostrem et al. 2004).

$$CH_3CH_2CH_2COOH + 4H_2O \rightarrow CH_3COOH + 2CO_2 + 6H_2 \tag{4}$$

$$CH_3CH_2COOH + 2H_2O \rightarrow CH_3COOH + CO_2 + 3H_2 \tag{5}$$

$$2CO_2 + 4H_2 \rightarrow CH_3COOH + 2H_2O \tag{6}$$

Homoacetogenesis is the generation of acetic acid from dissolved H_2 and CO_2 by homoacetogens as shown in Eq. (6).

2.4 *Methanogenesis*

The final stage of overall anaerobic conversion is called methanogenesis in which stage the degradable organic material is finally converted to a gaseous form that automatically leaves the reactor system. Acetoclastic and hydrogenotrophic methanogens are responsible for this conversion (Parawira 2012; Cavinato 2011). Acetoclastic methanogenesis is the final stage of anaerobic digestion in which acetic acid is converted into CH_4 and CO_2 by a group of archaea known as acetoclastic methanogens. This is responsible for the production of about two-third of methane as shown in Eq. (7) (Cavinato 2011; Ostrem et al. 2004).

$$CH_3COOH \rightarrow CH_4 + CO_2 \tag{7}$$

$$CO_2 + 4H_2 \rightarrow CH_4 + 2H_2O \tag{8}$$

Hydrogenotrophic methanogenesis is the production of CH_4 from dissolved H_2 and CO_2 by a group of slow-growing hydrogenotrophic methanogens. These methanogens produce the remaining one-third of methane by the reaction shown in Eq. (8) (Cavinato 2011; Ostrem et al. 2004). Methanogenic microorganisms may compete with sulphate-reducing microorganisms if sulphate is present at sufficiently high concentrations (Speece 1996).

3 Factors Affecting the Rate of Anaerobic Digestion

Each of the four sub-processes has different rates depending on operating conditions and substrate concentration. The overall rate of stabilization therefore will be limited by the slowest or rate-limiting step. The rate-limiting step may change from one sub-process to another with time within a system dependent upon the substrate characteristics (Ma et al. 2013; McCarty and Mosey 1991). In the case of high solid content, an initial hydrolysis step to convert particulate matter into soluble substrate is required to obtain efficient AD. The hydrolysis step is appreciably affected by temperature and is usually the rate-limiting step for low-temperature conditions (Xia et al. 2016; Lew et al. 2011; Zeeman 1991).

For predominantly dissolved organic waste, the rate-limiting steps are the acetogenesis and the methanogenesis as these bacteria groups have the slowest growing rates (Xia et al. 2016; Gujer and Zehnder 1983). Most of the control in anaerobic digestion is undertaken directly by the microorganisms themselves. The environmental conditions such as temperature, pH, essential trace nutrients and toxicants can play a major role in modifying reaction rates of individual sub-processes (Xia et al. 2016; Mckeown et al. 2012; Cavinato 2011). The international water association (IWA) task group for mathematical modelling of anaerobic digestion processes defined two categories of inhibition for microorganism: biocidal and biostatic inhibition. Biocidal inhibition describes the toxicity experienced by the microorganism due to normally irreversible conditions, whereas, in biostatic inhibition, the growth of the microbes ceases during exposure to inhibitory conditions, but resumes growth after re-establishment (Batstone et al. 2002). Some of the inhibition factors are discussed below.

3.1 pH

Acetoclastic methanogenesis is particularly vulnerable to low-pH conditions and quickly inhibited the process if the pH drops below 6.5 (Van Lier et al. 2008), which halts the removal of acids from the system. This happens when there is an increase in acid-producing rate (due to high organic loading rate) and decrease in acid-removing rate (decrease in buffer) causing souring (Yuan and Zhu 2016; Speece 1996).

There are three principal bacteria types involved in biogas production: bacteria responsible for hydrolysis, fermentative bacteria and methane-producing archaea. The fermentative bacteria can function in pH range from 8.5 down to pH 4 with their optimal pH range of 5.0–6.0 (Hwang et al. 2004); on the other hand, methanogenic archaea can function in pH interval from 5.5 to 8.5 with an optimal range of 6.5–8.0 (Boe 2006). pH inhibition occurs as a result of disruption of homeostasis and increase levels of non-dissociated VFA (Batstone et al. 2002). The bicarbonate produced by the methane-producing bacteria normally neutralizes the pH reduction caused by acid-producing bacteria (Liu and Tay 2004).

The greatest risk for digester failure is a result of acid accumulation which would occur if the amount of volatile solids loaded into the digester increased sharply. The acidogenic bacteria would then flourish, producing high volumes of organic acids and further lowering the pH to below 5.0 which is lethal to methanogens. pH values above 8 are toxic to most anaerobic organisms which results in the inhibition of biological functions. High pH could be due to prolific methanogenesis, resulting in a higher concentration of ammonia that would impede acidogenesis (Lusk 1999). This can now be opposed by adding a greater amount of fresh feedstock (Ostrem et al. 2004).

Systems with low potential for generating alkalinity through metabolism may necessarily add alkalinity in the form of lime (CaO), carbonate, hydroxide or bicarbonate for buffering digestion (Speece 1996).

3.2 Temperature

Anaerobic process can occur in a wide range of temperature that is psychrophilic (<20 °C), mesophilic (25–40 °C) and thermophilic (45–60 °C) (Khalid et al. 2011; Mathew et al. 2014). The anaerobic process temperature of the reactor has influence to the physical and chemical properties of the substrate which in turn affects the thermodynamic and kinetic reaction of the biological processes. There are several advantages with increasing temperatures (Abdelgadir et al. 2014; Van Lier et al. 1996), for instance, increase hydrolysed soluble product which in turn makes them more accessible for microorganism, increase reaction kinetics in both chemical and biological process that shorten the reaction time and hence the hydraulic retention time (HRT) of the reactor, and it also makes physical–chemical properties of the soluble substrate favourable and improves diffusivity, increase liquid-to-gas transfer rate because of lower gas solubility and improve liquid–solid biomass separation (Van Lier et al. 1996). Moreover, increased temperature also increases death rate of pathogenic bacteria, reducing time required for pathogen destruction in AD process (Bendixen 1994; Smith et al. 2005). However, high temperature (thermophilic) can have negative effects as well. Increasing temperature increases the fraction of free ammonia (NH_3) that is inhibitory to microorganisms. Ammonia inhibition could result process disturbance in thermophilic process. The stability of the mesophilic process makes it more acceptable in current AD facilities, but achieved at longer retention times (Ostrem et al. 2004).

3.3 C:N Ratio

All microorganisms present in anaerobic digestion process need essential elements for their growth. One of the main nutrient elements is nitrogen that is required for the synthesis of amino acids, proteins, etc., which in turn could be converted into ammonia, a buffer compound for the neutralization of the acidification process. Thus, all feedstock should contain nutrients and essential trace elements for the efficient anaerobic digestion process. It is reported that C:N:P ratio of 100:3:1 is suitable for high methane yield (Rajeshwari et al. 2000). Significantly deviation of C:N:P ratio could lead to deficiency of buffering capacity or insufficient nutrients for microorganism growth.

3.4 Effect of Moisture Content in Feedstock

The amount of moisture content in feedstock can have influence in anaerobic digestion process and methane yield. A study on the effect of moisture content ranged from 97 to 89% on anaerobic digestion of sludge showed that the amount of methane yield was decreased from 330 to 280 mL/g-VSS (Fujishima et al. 2000). The reason behind this reduction of methane yield was the decreased in removal efficiency of carbohydrate from 71 to 28% due to decrease in moisture content in the feedstock during anaerobic digestion process (Fujishima et al. 2000). This suggests that optimum level of moisture content is necessary for AD process.

3.5 Effect of Particle Size

The particle size of feedstock has strong influence in process kinetics of anaerobic digestion process. A study on the effect of particle size increased from 1.02 to 2.14 mm on anaerobic thermophilic food waste digestion carried out by Kim et al.

in 2000 showed that a strong effect on the substrate utilization rate coefficient which is decreased from 0.0033 to 0.0015 h^{-1}. This clearly indicated that the smaller the particle size the better the kinetic process and hence methane yield.

3.6 Organic Loading Rate (OLR)

The degree of starvation of microorganisms in biological systems is dependent on the OLR. At a high OLR, a fast microbial growth (but intoxication may occur with high quantities of organic matter) takes place whereas at a low OLR microorganism starvation takes place. However, if the applied OLR is too high, microorganism could not use up all produced organic acids and causes acidic state of the digester (Liu and Tay 2004). OLR is mainly determined based on feeding materials and reactor temperature.

3.7 Solid Retention Time (SRT)

SRT is a key parameter that affects biochemical properties of organic materials. The SRT plays an important role in anaerobic digestion especially for methanogens at low operational temperatures (Halalsheh et al. 2005). The SRT should be long enough to provide sufficient methanogenic activity. Methanogenesis starts at SRT between 5 and 15 days at 25 °C and between 30 and 50 days at 15 °C (Halalsheh et al. 2005); however, it again depends on characteristics of feeding materials.

3.8 Sulphate Reduction

The effect of sulphate reduction on anaerobic systems is complicated by the fact that the reduced product sulphide has an inhibitory effect on almost all the microbial groups (Batstone et al. 2002). The methanogenic microorganism competing with sulphate-reducing microorganism for the common intermediate acetic acid, due to the presence of sufficiently high concentrations of sulphur (Speece 1996).

However, reduction of sulphate leads to an increase of pH and the buffer capacity and leaves the system with H_2S gas as shown in Eqs. (9) and (10), respectively (Arceivala and Asolekar 2007).

$$SO_4{}^{--} + 4H_2 \rightarrow H_2S + 2H_2O + 2OH^- \quad (9)$$

$$SO_4{}^{--} {}_{+} CH_3COOH \rightarrow H_2S + 2HCO^{3-} \quad (10)$$

3.9 Denitrification

Denitrifying microorganisms have a higher cell growth yield per unit substrate consumed than methanogenic microorganism and compete for the same carbon source and electron source (e.g. acetate or H_2). Thus in anaerobic digestion, the presence of nitrate has significant impact in the form of microbial competition which leads to inhibition of CH_4 production. The reaction (Eq. 11) shows the overall reduction of nitrate by acetic acid to produce N_2 (Batstone et al. 2002; Foxon et al. 2006).

$$5CH_3COOH + 8NO_3^{-} + 8H_2 \rightarrow 4N_2 + 10CO_2 + 19H_2O \quad (11)$$

3.10 Ammonia

Nitrogen in the form of NH4-N is required by bacteria for their cell mass synthesis. The major nitrogen compound is obtained from nitrogenous materials available in organic matter usually proteins and urea. Ammonia is produced during hydrolysis of proteins and urea. Urea is readily hydrolysed to ammonia and carbon dioxide by the enzyme urease present in organic matter (Arceivala and Asolekar 2007). Urea is decomposed by bacteria via the following enzymatic catalysed reaction as shown in Eq. (12) (Fidaleo and Laveccio 2003).

$$CO(NH_2)_2 + 2H_2O \rightarrow 2NH_4^{+} + CO_3^{2-} \quad (12)$$

Also, hydrolytic bacteria further hydrolyse amino acids to form ammonia, H_2, CO_2 and VFAs (Tchobanoglous et al. 2003). The release of ammonia through the decomposition of urea, hydrolysis of amino acids is the primary parameter that causes a rise in the bicarbonate-ammonia buffer (alkalinity) and controlling the pH and the process stability of the digester (Shanmugam and Horan 2008). Consequently, a dramatic pH falls below 6 as a critical value hardly occurs (Wendland 2008). Even if the formation of VFA (HAc-acetic acid) decreases the buffer capacity but the formation of NH_4^+ increases the bicarbonate concentrations and the process stabilities.

Ammonia inhibits predominantly the methanogenesis (Wendland 2008). Acetate utilizing methanogenic bacteria was found to be more sensitive to ammonia than hydrogen-consuming ones (Fotidis et al. 2013). Two different mechanisms were attributed to ammonia inhibition: firstly methanogens are directly inhibited by free ammonia, and secondly in the bacterial cell wall, free ammonia is rapidly converted to ammonium ion as shown in Eq. (13) (Kadam and Boone 1996).

$$NH_3 + H_2O \rightarrow NH_4^{+} + H_2O \quad (13)$$

4 Energy Performance of AD

4.1 Methods Used

The energy performance of the anaerobic digestion is depending mainly on the biogas production technology (wet or dry technology, mesophilic or thermophilic) and geographic location (ambient temperature). The feedstock of course has also an effect because the biogas yield is varying. In addition when utilizing cultivated feedstock, the processes of obtaining the feedstock can consume significant amount of energy. Especially when utilizing wet digestion technology, the main part of the parasitic energy demand is the heat required for heating the feedstock to the desired temperature (Havukainen et al. 2014).

The energy performance of AD is difficult to calculate similarly to incineration since the lower heating value as received (LHV_{as}) can be even negative especially with feedstock coming for wet digestion. This means that the energy efficiency defined as produced energy (electricity and/or heat) divided by the fuel energy of the incoming waste would be negative. Therefore, other methods have needed to be developed to ascertain the energy performance of AD. Table 1 describes some methods which have been used in obtaining information about the different methods which have been used for energy performance calculation. These studies on energy performance include waste as well as energy crops as feedstocks and have used varying system boundaries. Energy performance has been calculated as energy output divided by energy input (Prade et al. 2012; Tanaka 2008) as well as energy input divided by energy output. However, it seems that the output/input is most commonly used method among these studies.

Energy output divided by energy input has been used for example by Berglund and Börjesson (2006). Berglund and Börjesson (2006) studied the energy performance of wet anaerobic digestion operating at mesophilic temperature of energy crops, harvest residues, manure, industrial organic waste and municipal organic waste. The input/output range was calculated by using the primary energy used for the unit processes as input and biogas energy content as output. The input/output ratios ranged from 20 to 50% being lowest for grease trap sludge and highest for ley crops. The differences were mainly due to varying properties of raw materials, system design and allocation method. The heat and electricity consumption of biogas production was responsible for approximately 40–80% of the net energy consumption. Similarly, Pöschl et al. (2010) used primary energy to calculate primary energy input output (PEIO) ratio wet digestion at mesophilic digestion in two-stage digester. The feedstocks include agricultural waste, energy crops, municipal solid waste and food industry residues. PEIO was 11–64% for single feedstock digestion and 34–55% for co-digestion.

Salter and Banks (2009) used output/input ratios for estimating energy performance of anaerobic digestion of energy crops (maize, fodder beet, lupin and

Table 1 Methods for calculating energy balance of anaerobic digestion

Method	Inputs and outputs included	Result	References
Input/output 1	Input: Primary energy for obtaining raw material, transport, operation of biogas plant Output: Biogas energy content	20–40%	Berglund and Börjesson (2006)
Input/output 2	Input: Crop cultivation, collection, transport, biogas plant operation, digestate processing Output: Energy produced from biogas	10.5–64%	
Input/output 3	Input: Production of inputs, cultivation, digestion, biogas processing and transport fuel delivery Output: Biomethane energy	22–37%	Tuomisto and Helenius (2008)
Output/input 1	Output: Methane Input: Energy for cultivation, transport, fertilizer and pesticides	7–25	Gerin et al. (2008)
Output/input 2	Output: Heat, power and biomethane Input: Crop production, transport, biogas production and upgrading	3.5–8.2	Seppälä et al. (2008)
Output/input 3	Output: Heat, power and biomethane Input: Crop production and digestion, biogas and digestate use (direct and indirect energy)	1.8–3.3	Salter and Banks (2009)
Output/input 4	Output: Heat, power and biomethane Input: Crop production and processing, reactor	4.04–6.5	Salter et al. (2005)
Output/input 5	Output: Electricity and heat Input: Cultivation, harvesting, digestion, digestate	5.5–6.8	Navickas et al. (2012)
Biomethane yield (BMY)	BMY_1 = (methane potential of input biomass − methane potential of the digestate)/methane potential of the input biomass BMY_2 = effective specific methane produced/biomethane potential of input	BMY_1 and BMY_2 84–93%	Schievano et al. (2011)
Energy efficiency	Mechanical energy of the tractor/(biogas energy + energy produced outside system, e.g. electricity, diesel)	5.8–13%	Lacour et al. (2012)
Relative biogas yield	Measured biogas yield/theoretical biogas yield	90–161%	Djatkov et al. (2012)
Total annual efficiency	(produced electricity + used heat)/biogas energy	30.5–73%	Laaber et al. (2007)
Electricity use	Parasitic electricity use/produced electricity	30.4%	Banks et al. (2011)

Modified from Havukainen et al. (2014)

perennial ryegrass) and found that ratio was 1.8–3.3 for energy crops being lowest for lupin and highest for maize. Navickas et al. (2012) also studied energy crops (fresh grass, hay and reed canary grass), and output ratio was highest for hay (6.8) and lowest for fresh grass (5.5).

4.2 System Boundary

The comparison of energy balance values in Table 1 is difficult since the system boundaries around the anaerobic digestion systems are varying a lot. The system boundary can stop to the produced biogas (Berglund and Börjesson 2006), or it can also include the energy produced from biogas (Salter et al. 2005). There are also significant differences in which energy consumptions are included. Gerin et al. (2008) excluded electricity and heat consumption of anaerobic digestion when studying energy crop biogas system, even though according to Berglund and Börjesson (2006) anaerobic digestion is the most energy-consuming process also in energy crop biogas system. In most cases, the indirect energy consumption in construction is excluded. However, at least Salter and Banks (2009) included the indirect energy use of construction and maintenance of digester and auxiliary equipment.

Comparing energy balances of different anaerobic digestion systems would require that some general system boundaries could be set. The energy performance of anaerobic digestion would be better estimated with utilizing few different system boundaries. Figure 2 presents four system boundaries which can be used in calculating the different energy performance values which can be used in following the energy performance of given anaerobic digestion system and to compare to other systems.

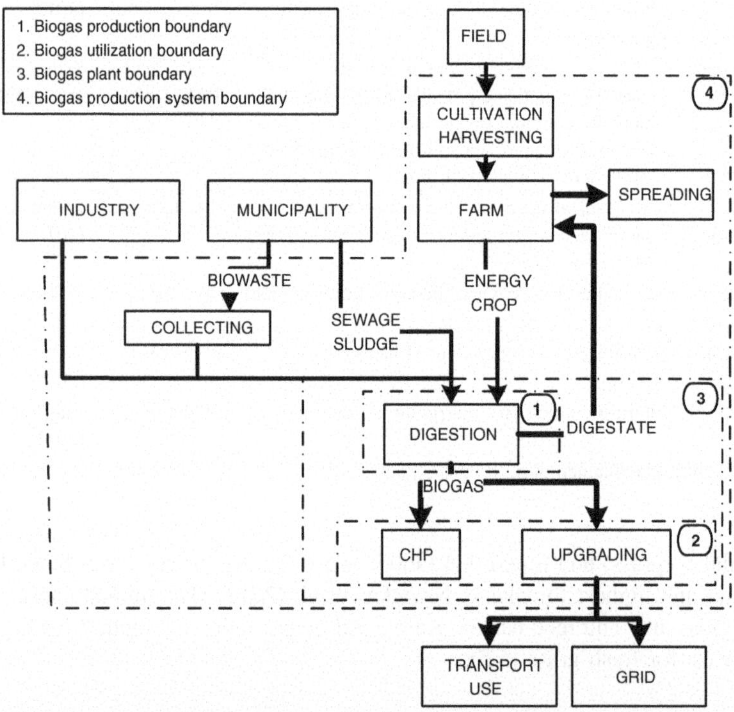

Fig. 2 Different system boundaries for estimating energy performance of anaerobic digestion system. Modified from Havukainen et al. (2014)

4.3 Output/Input Ratio Calculation

The energy performance calculation can be done on biogas production alone, utilization of produced biogas from anaerobic digestion, for the anaerobic digestion plant or for the anaerobic digestion system. Energy performance of biogas production can be calculated by Eq. (14) (Havukainen et al. 2014) utilizing the system boundary 1 in Fig. 2

$$R_{\mathrm{pr2}} = \frac{E_{\mathrm{bg}}}{E_{\mathrm{el,par}} + E_{\mathrm{h,par}}} \tag{14}$$

where E_{bg} is the energy content of the produced biogas (MWh), $E_{\mathrm{el,par}}$ is the parasitic electricity used for biogas production (MWh) and $E_{\mathrm{h,par}}$ is the parasitic heat needed in biogas production processes (MWh). The energy performance of biogas utilization can be calculated by Eq. (15) (Havukainen et al. 2014) utilizing system boundary 2 in Fig. 2.

$$R_{\mathrm{ut}} = \frac{E_{\mathrm{el,prod}} + E_{\mathrm{h,prod}} + E_{\mathrm{bm}} - E_{\mathrm{el,par,CHP}} - E_{\mathrm{el,par,up}} - E_{\mathrm{h,par,up}}}{E_{\mathrm{bg}}} \tag{15}$$

where $E_{\mathrm{el,prod}}$ is the produced electricity (MWh), $E_{\mathrm{h,prod}}$ is the produced heat (MWh), E_{bm} is the energy content of produced biomethane (MWh), $E_{\mathrm{el,par,CHP}}$ is the parasitic electricity need of the (CHP) equipment (MWh), $E_{\mathrm{el,par,up}}$ is the parasitic electricity need of the upgrading process (MWh) and $E_{\mathrm{h,par,up}}$ is the parasitic heat need of the upgrading process (MWh).

Equation (16) (Havukainen et al. 2014) can be used for calculating energy performance for the anaerobic digestion plant (R_{pl}) utilizing system boundary 3 in Fig. 2.

$$R_{\mathrm{pl}} = \frac{E_{\mathrm{h,s}} + E_{\mathrm{bm}} + E_{\mathrm{el,s}}}{E_{\mathrm{el,par,pl}} + E_{\mathrm{f}} + E_{\mathrm{h,par,pl}}} \tag{16}$$

where E_{f} is the energy content of other fuels used in the production of energy in the biogas plant, $E_{\mathrm{h,s}}$ is the heat energy supplied to processes outside the biogas plant boundary, $E_{\mathrm{el,s}}$ is the electricity supplied to the grid, $E_{\mathrm{el,par,pl}}$ is the electricity need from the electricity grid and $E_{\mathrm{h,par,pl}}$ is the heat need from outside the biogas plant.

The energy performance for the whole anaerobic digestion system (R_{sy}) can be calculated with Eq. (17) (Havukainen et al. 2014).

$$R_{\mathrm{sy}} = \frac{E_{\mathrm{h,s}} + E_{\mathrm{bm}} + E_{\mathrm{el,s}}}{E_{\mathrm{t,d}} + E_{\mathrm{t,}fs} + E_{\mathrm{ch}} + E_{\mathrm{c}} + E_{\mathrm{sd}} + E_{\mathrm{el,o}} + E_{\mathrm{f}} + E_{\mathrm{h,o}}} \tag{17}$$

where the fuel need is $E_{\mathrm{t,d}}$ for transporting the digestate (MWh), E_{sd} for spreading the digestate (MWh), $E_{\mathrm{t,fs}}$ for transporting the feedstock (MWh), E_{c} for the collection of biowaste (MWh) and E_{ch} for the cultivation and harvesting of the energy crop (MWh).

4.4 Conclusion

Anaerobic digestion process is the biochemical and physic chemical process of organic materials. In the process, there are several factors that affect the biochemical as well as physiochemical process. The most important factors are temperature, pH, OLR, C:N ratio, etc., that has to be controlled to be the adequate level for the efficient anaerobic digestion process and the methane yield. As with other energy production methods, it is also necessary to determinate the energy performance of biogas plant. The output/input ratio method which can be compared to other biogas production systems or biogas plants has been chosen for this case. Since the output/input ratio results are directly dependent on the used system boundary, the estimated system boundary has to be clearly described. After this method, to calculate energy performance of biogas plants and systems, an improvement in the production systems might be possible.

References

Abdelgadir A, Chen X, Liu J, Xie X, Zhang J, Zhang K, Wang H, Liu N (2014) Characteristics, process parameters, and inner components of anaerobic bioreactors. BioMed Res Int 2014:1–10, http://dx.doi.org/10.1155/2014/841573

Appels L, Baeyens J, Degreve J, Dewil R (2008) Principles and potential of the anaerobic digestion of waste-activated sludge. Prog Energy Combust Sci 34:755–781

Arceivala SJ, Asolekar SR (2007) Wastewater treatment for pollution control and reuse. Tata McGraw-Hill, New Delhi

Banks C, Chesshire M, Heaven S, Arnold R (2011) Anaerobic digestion of source-segregated domestic food waste: performance assessment by mass and energy balance. Bioresour Technol 102:612–620

Batstone DJ, Keller J, Angelidaki RI, Kalyuzhnyi SV, Pavlostathis SG, Rozzi A, Sanders WTM, Siegrist H, Vavilin VA (2002) Anaerobic digestion model no. 1 (ADM1). IWA Publishing, London

Bendixen HJ (1994) Safeguards against pathogens in Danish biogas plants. Water Sci Technol 30:171–180

Berglund M, Börjesson P (2006) Assessment of energy performance in the life-cycle of biogas production. Biomass Bioenergy 30:254–266

Bilitewski B, Härdtle G, Marek K (1997) Waste management. Springer, Berlin. ISBN 3-540-59210-5

Boe K (2006) Online monitoring and control of the biogas process. Ph.D. thesis. Institute of Environment and Resources. Technical University of Denmark (DTU)

Botheju D, Lie B, Bakke R (2010) Oxygen effects in Anaerobic Digestion—II. Modell Ident Control 31:55–65

Cavinato C (2011) Anaerobic digestion fundamentals 1. Summer School on biogas technology for sustainable Second Generation Biofuel Production. http://www.valorgas.soton.ac.uk/Pub_docs/JyU%20SS%202011/CC%201.pdf. Accessed 15 Apr 2015

Djatkov D et al (2012) New method for assessing the performance of agricultural biogas plants. Renew Energy 40:104–112

Fidaleo M, Laveccio R (2003) Kinetic study of enzymatic urea hydrolysis in the pH range of 4–9. Chem Biochem Eng 17(4):311–318

Fotidis IA, Karakashev D, Kotsopoulos TA, Martzopoulos GG, Angelidaki I (2013) Effect of ammonium and acetate on methanogenic pathwayand methanogenic community composition. FEMS Microbiol Ecol 83:38–48

Foxon KM, Brouckaert CJ, Remigi E, Buckley CA (2006) The anaerobic baffled reactor: an appropriate technology for onsite sanitation. Water SA 30:44–50

Fujishima S, Miyahara T, Noike T (2000) Effect of moisture content on anaerobic digestion of dewatered sludge: ammonia inhibition to carbohydrate removal and methane production. Water Sci Technol 41(3):119–127

Gerin PA, Vliegen F, Jossart J-M (2008) Energy and CO_2 balance of maize and grass as energy crops for anaerobic digestion. Bioresour Technol 99:2620–2627

Gujer W, Zehnder AJB (1983) Conversion processes in anaerobic digestion. Water Sci Technol 15 (8/9):127–167

Halalsheh M, Koppes J, Den Elzen J, Zeeman G, Fayyad M, Lettinga G (2005) Effect of SRT and temperature on biological conversions and the related scum-forming potential. Water Res 39 (12):2475–2482

Havukainen J, Uusitalo V, Niskanen A, Kapustina V, Horttanainen M (2014) Evaluation of methods for estimating energy performance of biogas production. Renew Energy 66. https://doi.org/10.1016/j.renene.2013.12.011

Hwang MH, Jang NJ, Hyum SH, Kim IS (2004) "Anaerobic biohydrogen production from ethanol fermentation: the role of pH. J Biotechnol 111:297–309

Jansen JK, Jensen AB (2000) Biogas and natural gas fuel mixture for the future. In: 1st world conference and exhibition on biomass for energy and industry, Sevilla

Jonsson O, Polman E, Jensen JK, Eklund R, Schyl H, Ivarsson S (2003) Sustainable gas enters the European gas distribution system. Danish gas technology center. http://www.dgc.eu/sites/default/files/filarkiv/documents/C0301_sustainable_gas.pdf. Accessed 20 Feb 2016

Kadam PC, Boone DR (1996) Influence of pH on ammonia accumulation and toxicity in halophilic, Methylotrophic Methanogens. Appl Environ Microbiol 62(12):4486–4492

Khalid A et al (2011) The anaerobic digestion of solid organic waste. Waste Manag 31:1737–1744

Kim SI, Kim HD, Hyun HS (2000) Effect of particle size and sodium ion concentration on anaerobic thermophilic food waste digestion. Water Sci Technol 41(3):67–73

Laaber M, Madlener R, Kirchmayr R, Braun R (2007) Aufbau eines Bewertungssystems für Biogasanlagen—"Gütesiegel Biogas"

Lacour S et al (2012) Energy and environmental balance of biogas for dual-fuel mobile applications. Renew Sustain Energy Rev 16:1745–1753

Lew B, Lustig I, Beliavski M, Tarre S, Green M (2011) An integrated UASB-sludge digester for raw domestic wastewater treatment in temperate climates. Bioresour Technol 102:4921–4924

Liu Y, Tay J (2004) State of the art of biogranulation technology for wastewater treatment. Biotechnol. Adv. 22:533–563

Lusk P (1999) Latest progress in anaerobic digestion. Biocycle 40(7):52–54

Ma J, Frear C, Wang Z, Yu L, Zhao Q, Li X, Chen S (2013) A simple methodology for rate-limiting step determination for anaerobic digestion of complex substrates and effect of microbial community ratio. Bioresour Technol 134:391–395

Mathew AK, Bhui I, Banerjee SN, Goswami R, Shome A, Chakraborty AK, Balachandran S, Chaudhury S (2014) Biogas production from locally available aquatic weeds of Santiniketan through anaerobic digestion. Clean Technol Environ Policy. https://doi.org/10.1007/s10098-014-0877-6

McCarty PL, Mosey FE (1991) Modelling of anaerobic digestion process (A discussion of concepts). Water Sci Technol 24:17–33

McKeown MR, Hughes D, Collins G, Mahony T, O'Flaherty V (2012) Low-temperature anaerobic digestion for wastewater treatment. Curr Opin Biotechnol 23:444–451

Mittal A (2011) Biological wastewater treatment. Water Today 33–44. http://www.watertoday.org/Article%20Archieve/Aquatech%2012.pdf. Accessed 15 Apr 2017

Navickas K et al (2012) Influence of different biomass treatment technologies on efficiency of biogas production. In: 2012. 11th international scientific conference engineering for rural development, Jelgava, Latvia, 24–25 May 2012

Ostrem MK, Millrath K, Themelis NJ (2004) Combining anaerobic digestion and waste to energy. In: 12th North America waste to energy conference. Columbia University, New York

Parawira W (2012) Enzyme research and applications in biotechnological intensification of biogas production. Crit Rev Biotechnol 32(2):172–186

Prade T, Scensson S-E, Mattsson JE (2012) Energy balances for biogas and solid biofuel production from industrial hemp. Biomass Bioenergy 40:36–52

Rajeshwari KV, Balakrishnan M, Kansal A, Lata K, Kishore VVN (2000) State-of-the-art of anaerobic digestion technology for industrial wastewater treatment. Renew Sust Energ Rev 4:135–156

Salter A, Banks CJ (2009) Establishing an energy balance for crop-based digestion. Water Sci Technol 59:1053–1060

Salter A et al (2005) Establishing an energy balance for crop-based digestion energy. In: 2005. 14th European biomass conference and exhibition: biomass for energy. Industry and Climate Protection, Paris, France, 17–21 Oct 2005

Schievano A, D'Imporzano G, Orzi V, Adani F (2011) On-field study of anaerobic digestion full-scale plants (part II): new approaches on-field study of anaerobic digestion full-scale plants (part II): new approaches. Bioresour Technol 102:8814–8819

Schluter A, Bekel T, Diaz NN, Dondrup M, Eichenlaub R, Gartemann KH, Krahn I, Krause L, Kromeke H, Kruse O, Mussgnug JH, Neuweger H, Niehaus K, Puhler A, Runte KJ, Szczepanowski R, Tauch A, Tilker A, Viehover P, Goesmann A (2008) The metagenome of a biogas-producing microbial community of a production-scale biogas plant fermenter analysed by the 454-pyrosequencing technology. J Biotechnol 136:77–90

Seppälä M et al (2008) Biogas from energy crops—optimal pre-treatments and storage, co-digestion and energy balance in boreal conditions. Water Sci Technol 58:1857–1863

Shanmugam P, Horan NJ (2008) Simple and rapid methods to evaluate methane potential and biomass yield for a range of mixed solid wastes. Bioresour Technol 100(1):471–474

Smith SR, Lang NL, Cheung KHM, Spanoudaki K (2005) Factors controlling pathogen destruction during anaerobic digestion of biowastes. Waste Manag 25(4):417–425

Speece RE (1996) Anaerobic biotechnology for industrial wastewaters. Archae Press, Nashville, TN

Tanaka K (2008) Assessment of energy efficiency performance measures in industry and their application for policy. Energy Policy 36:2887–2902

Tchobanoglous G, Burton FL, Stensel DH (2003) Wastewater engineering treatment and reuse. Tata McGraw Hill Education Private Limited

Tuomisto HL, Helenius J (2008) Comparison of energy and greenhouse gas balances of biogas with other transport biofuel options based on domestic agricultural biomass in Finland. Agric Food Sci 17:240–251

Van Lier JB, Sanz Martin JL, Lettinga G (1996) Effect of temperature on the anaerobic thermophilic conversion of volatile fatty acids by dispersed and granular sludge. Water Res 30:199–207

Van Lier JB, Mohmoud N, Zeeman G (2008) Anaerobic wastewater treatment. In: Biological wastewater treatment, principles, modelling and design, IWA Publishing

Wendland C (2008) Anerobic digestion of blackwater and kitchen refuse. Ph.D. thesis. Technical University Hamburg-Hamburg, Germany

WtERT (2014) Waste to energy research and technology council. http://www.wtert.eu/default.asp?Menue=13&ShowDok=12. Accessed 30 May 2015

Xia A, Cheng J, Murphy DJ (2016) Innovation in biological production and upgrading of methane and hydrogen for use as gaseous transport biofuel. Biotechnol Adv (in press). http://dx.doi.org/10.1016/j.biotechadv.2015.12.009

Yuan H, Zhu N (2016) Progress in inhibition mechanisms and process control of intermediates and by-products in sewage sludge anaerobic digestion. Renew Sustain Energ Rev 58:429–438

Zeeman G (1991) Mesophilic and psychrophilic digestion of liquid manure. Ph.D. thesis. Department of Environmental Technology, Agricultural University, Wageningen

Zeeman G, Sanders W (2001) Potential of anaerobic digestion of complex waste (water). Water Sci Technol 44:115–122

Zeeman G, Sanders WTM, Wang KY, Lettinga G (1996) Anaerobic treatment of complex wastewater and waste activated sludge—application of up-flow anaerobic removal reactor for the removal and pre-hydrolysis of suspended COD. In: Proceeding of the IAWQ-NVA conference on advanced wastewater treatment, Amsterdam, pp 225–232, 23–25 Sept 1996

Author Biographies

Dr. Sunil Prasad Lohani is an Assistant Professor at Department of Mechanical Engineering, Kathmandu University, Nepal. His area of research activities is mainly focused on biogas production from waste (water) and low carbon technologies.

Dr. Jouni Havukainen is a Postdoctoral Researcher at Sustainability Science Department at Lappeenranta University of Technology. His main research areas include biogas production, waste management utilizing system analysis and life cycle assessment as tools.

Chapter 19
Biodigester Technology for Effective and Ecofriendly Decomposition of Nightsoil

Mukesh K. Meghvansi, Prince Kumar, V. Vasudevan, Arvind Tomar, D. V. Kamboj and Lokendra Singh

Abstract Adequate sanitation and hygiene practices are essential for maintaining the good health. In developing countries, rapid increase in population and urbanization has put tremendous pressure on the existing infrastructure related to the sanitation, thereby necessitating development of more effective technologies. Among the sanitation practices, proper disposal of human excreta assumes much significance as it is one of major sources of human pathogens. Various technological solutions are available for treatment of human fecal matter. Employing the principles of anaerobic biodegradation, biodigester technology has been developed by Defence R&D Organisation, India, for effective decomposition of human fecal matter under varied geoclimatic conditions. The present article deals with the biodigester technology, its uses, and possible impacts on overall sanitation scenario in India. Chemistry and biology of anaerobic degradation of nightsoil using a unique anaerobic microbial inoculum and a bioreactor are also discussed. In addition, we highlight the factors governing the biodegradation efficiency, effluent quality as well as possible recycling/reuse of water and harvesting, and use of

---

M. K. Meghvansi (✉) · P. Kumar · A. Tomar · D. V. Kamboj · L. Singh
Biotechnology Division, Defence Research & Development Establishment,
Gwalior 474002, Madhya Pradesh, India
e-mail: mk_meghvansi@yahoo.co.in

P. Kumar
e-mail: princemsc05@gmail.com

A. Tomar
e-mail: arvindtomar@drde.drdo.in

D. V. Kamboj
e-mail: dvkamboj@gmail.com

L. Singh
e-mail: lst2397@rediffmail.com

Present Address:
V. Vasudevan
Defence Food Research Laboratory, Mysuru 570011, Karnataka, India
e-mail: fly2vass@gmail.com

S. J. Varjani et al. (eds.), *Waste Bioremediation*, Energy, Environment, and Sustainability, https://doi.org/10.1007/978-981-10-7413-4_19

methane generated from it. Also, based on the critical and in-depth analysis of existing scientific information, we further deliberate on future prospects of this technology for sustainable and ecofriendly treatment of nightsoil in developing countries.

Keywords Biodigester · Anaerobic degradation · Fecal matter
Nightsoil · Anaerobic microbial inoculum

1 Introduction

The lack of clean water and poor sanitation are the etiological agents of many diseases and the spread of diseases affecting poorer members of the society disproportionately. As the human excreta is one of the major sources of pathogenic organisms, its improper disposal leads to contamination of water bodies, soil, and food crops, thereby posing a serious health hazard to the environment. A study on Global Burden of Disease (GBD 2016) estimated that diarrhea caused more than 1.3 million deaths globally and was the fourth leading cause of death among children younger than 5 years. Further, the study emphasized that the improved access to safe water, sanitation, and childhood nutrition will be important in reducing the global burden of diarrhea (GBD 2016). The agenda of United Nations for 2030 has highlighted significantly water and sanitation, with a dedicated Sustainable Development Goal (SDG)-6 on water and sanitation. It also has clear linkages to goals pertaining to health, food security, climate change, resiliency to disasters, and ecosystems. The biodigester technology could play a significant role in achieving sanitation target under Sustainable Development Goals (SDGs) of United Nations.

The anaerobic degradation of organic matter is one of the most commonly used strategies of waste treatment and renewable energy production. During this process, bacteria, in the absence of oxygen, degrade macromolecules such as carbohydrates, lipids, and proteins into short-chain fatty acids, hydrogen, and carbon dioxide. Subsequently, the archaea use these degraded macromolecules to produce methane, carbon dioxide, and small amounts of other gases. Using this principle, a technology has been developed by Defence R&D Establishment, Defence R&D Organisation (DRDO), Gwalior, India, for degradation of the nightsoil. After optimizing the process parameters, the reactor design, and the anaerobic microbial consortium, the biodigesters have been built and installed on varied scales in different geoclimatic conditions in India. Primarily, this technology was developed by the DRDO for the armed forces deployed in high-altitude Himalayan region to cater the problem of nightsoil disposal at low temperature. Subsequently, a modular version of biodigester (i.e., biotoilet) has been developed for the Indian Railways and is currently being used very successfully in passenger coaches for more than 900 trains for the onboard treatment of human fecal matter. Indian Railways plan to install biotoilets in all its trains by 2019. When completed, Indian Railways would be the first transport system in the world having onboard human waste management system based on DRDO-biodigester technology. Similarly, biodigester-based biotoilets are being installed in urban and rural areas of India as a part of sanitation improvement program.

The present article attempts to provide an overview of DRDO-biodigester technology, its uses, and possible impacts on overall sanitation scenario in India. After providing a brief account on historical perspectives of nightsoil disposal worldwide, we discuss it in Indian context. Also, we provide a glimpse of commonly used sewage treatment technologies employed in sewage treatment plants (STPs) in India. Further, the various aspects pertaining to chemistry and biology of anaerobic degradation of nightsoil using a unique anaerobic microbial inoculum and a bioreactor are discussed. In addition, we highlight the factors governing the degradation efficiency, effluent quality as well as possible recycling/reuse of water and harvesting, and use of methane generated from it. Also, based on the critical and in-depth analysis of existing scientific information, we further deliberate on future prospects of this technology for sustainable and ecofriendly treatment of nightsoil.

2 Brief History of Nightsoil Disposal

Jackson et al. (2015), in a publication of World Health Organization (WHO), have defined 'Nightsoil' as untreated excreta transported without water (e.g., via containers or buckets). The term was used in a sense of Victorian euphemism, because in nineteenth century, in pre-industrial cities of western countries, it was a practice of removing the human waste from city's cesspools and septic tanks during the night time by the 'Nightsoil Men' in order to avoid sight of it by the public. This term is not frequently used in current scientific literature and has been replaced by 'fecal sludge.' Some authors have used the term 'Nightsoil' sensu lato to represent a mixture of human feces. For instance, Choi et al. (1997) have used the term for a mixture of human feces and urine. Similarly, Caincross and Feachem (1993) mentioned that nightsoil comprises only feces and urine plus small volumes of water if it is used for anal cleansing and pour flushing.

The Nightsoil Men, having lower social status, used to collect the human excreta in buckets or containers and carted it away from the residential area. The evacuation and cleaning of a typical household privy were done 2–3 times year. This method of human excreta disposal in the western world was very unhygienic and used to result in epidemics of cholera, typhoid, diarrhea, and many other diseases. On the contrary, the ancient Indus valley civilization of South Asia had much advanced sanitation system wherein the sewage was disposed through underground drains built with precisely laid bricks. Also, the drains from houses were connected to wider public drains (Rodda and Ubertini 2004). Moreover, between sixteenth and eighteenth century, many Asian cities belonging to China and Japan had put considerable efforts and resources on the collection and processing of human excreta and urine. Also these countries explored possibilities of using it as manure for agricultural purposes. In these countries, regional markets in nightsoil flourished, with varying degree of nightsoil collection depending upon the city and region (Ferguson 2014). In China, the nightsoil trade flourished from at least the Ming era. In Chinese cities, during the late Ming era, an intensive nightsoil trade flourished in the course of rapid urbanization (Xue 2007). Similarly, a Japanese study estimated that perhaps 36% of the manure nitrogen that was consumed yearly in Japan in the early twentieth century came from nightsoil (Sakuta 1914). Agricultural treatises of late seventeenth and early eighteenth century provide a detailed account on best practices to be followed for constructing toilets to best collect feces, the best times to fertilize the fields, and proper methods for applying nightsoil. This literature also identified specific quarters where the best nightsoil could be obtained and provided detailed information on the kinds of tools and buckets most appropriate for collecting nightsoil (Tajima 2005).

3 Status of Nightsoil Disposal and Associated Challenges in India

Since ancient times, in India, human excreta have been collected and disposed manually by a specific section of people in the society as there were no modernized toilet systems. In ancient literature and scripture, this section of people is mentioned as *Shudras*. Those engaged in manual scavenging were considered as untouchables and were forced to remain secluded from the main society. As the civilization progressed, enforcement of equality without any sorts of discrimination was envisaged. Since the independence, many legislations have been passed by the Indian government to prohibit manual scavenging. In 1993, the practice of manual scavenging was banned in the country through passing a legislation named 'The Employment of Manual Scavengers and Construction of Dry Latrines (Prohibition) Act, 1993.' The act has further been strengthened in 2013. Currently in Indian cities, sewage is collected and transported by a network of pipes and pump stations to an STP where it undergoes a series of treatment processes that ultimately produce the effluent which is minimally harmful to the humans and environment.

India is second-most populous country in the world with a population of 1.2 billion, which is around 17.5% of the total world population living in less than 2% of the total landscape in the world. According to a recent Swachhta (Sanitation) Status Report on ambitious Clean India Mission of Government of India as submitted by National Sample Survey (NSS) Office, more than half the rural population in India still opts for open defecation. In rural areas, only 45.3% households reported to have sanitary toilets, whereas in urban areas, 88.8% households reported to have sanitary toilets. In rural areas, the percentage of persons going for open defecation was estimated to be 52.1% while this figure for urban India was estimated to be 7.5%. In rural areas, 55.4% households contributed to open defecation (http://mospi.nic.in/Mospi_New/upload/Swachhta_%20Status_Report2016.pdf; accessed July 7, 2017). According to Central Pollution Control Board (CPCB) of India, during 2015, the estimated sewage generation in the country was 61,754 million liters per day (MLD) as against the developed sewage treatment capacity of 22,963 MLD. About 38,791 MLD of untreated sewage (62% of the total sewage), is discharged directly into nearby water bodies. Further, the five states, viz., Maharashtra, Tamil Nadu, Uttar Pradesh, Delhi, and Gujarat have lion's share contributing approximately 50% of the total sewage generated in the country, among which Maharashtra alone accounts for 13% of the total sewage generation in the country. As per CPCB, only Himachal Pradesh and Sikkim have sewage treatment plants of adequate capacity to treat the total quantity of sewage generated in these states (CPCB 2016).

A typical STP provides treatment at three stages, called as primary, secondary, and tertiary treatment. In primary treatment, heavier solid matter is allowed to settle at bottom while lighter solids float on the surface. The settled and floating materials are subsequently removed using mechanical devices, and the remaining waste is subjected to secondary treatment. In secondary treatment, dissolved and suspended biological matter is degraded typically using microorganisms or by chemical means. The tertiary treatment includes chemical or physical disinfection of waste which further improves the effluent quality, thereby making it safe for discharge into more sensitive ecosystem. Taking into account the type of sewage and effluent quality standards to be achieved, various technologies have been developed. Some of the commonly used sewage treatment technologies employed in STPs in India are listed in Table 1. Moreover, some of the technologies are used in combination for improving the efficiency of the process. At individual household level, usually the nightsoil is collected in septic tank which is either connected to sewer system of the city or the tanks are evacuated at regular interval and the waste is disposed untreated in distant places.

Discharge of untreated sewage into water bodies being cause of environmental pollution has severe implications for humans, animals, and environment. The STPs installed in India are helpful in reducing the pollution level. Nevertheless, as mentioned above, the current installed capacity of STPs in India is inadequate. Moreover, according to CPCB, 39% plants are not conforming to the general regulatory standards prescribed under the environment (protection) rules for discharge into streams. Besides, there are several demerits associated with use of STPs which include high capital and operational cost, large footprint, need for disposal of excess sludge production, and requirement of highly skilled manpower for operation and maintenance. This necessitates development of cost-effective and decentralized fecal waste management technologies which are practically more suitable

Table 1 Some of the commonly used sewage treatment technologies employed in STPs in India

Technology	Salient features
Activated sludge process (ASP)	• Most commonly used in class I cities (i.e., with population 100,000 and above) • Requires comparatively less space than UASB and WSP • In conventional ASP, both primary and secondary sludges are commonly treated in anaerobic sludge reactors • Efficiency of the process can be further improved using extended aeration
Fluidized aerobic bioreactor (FAB)	• Suitable as decentralized STP for residential complexes, small townships, industries, etc. • Requires less space due to high surface area and loading rate of FAB media • Reduced power and operating costs • No sludge recycle required
Trickling filters or biofilters	• It comprises of a tower filled with support media such as stones, plastic shapes, or wooden slats • A layer of microbial biofilm grows on the bed, covering the bed of media • Aerobic conditions are maintained through various mechanical means
Upflow anaerobic sludge blanket (UASB)	• Mostly used as the main treatment unit of a scheme including other primary and tertiary treatment units • It comprises of grit chamber as preliminary treatment unit and one-day retention time pond as the terminal polishing unit • Operation cost is comparatively less
Waste stabilization ponds (WSP)	• Most commonly used in class II towns (i.e., with population range 50,000–99,999) • Operation is comparatively less sensitive with greater improvement in bacteriological quality • Most common configuration consists of two parallel streams of at least three stages of ponds. The first stage is anaerobic ponds; the second stage is facultative pond and; and the third stage is maturation pond
Karnal technology (for plantation)	• In this technology, forest trees having utility as fuel wood, timber or pulp are grown on ridges with fixed dimension, and untreated sewage is disposed in furrows • This technique utilizes the entire biomass as living filter for supplying nutrients to soil and plant
Biodigester technology[a]	• Uses specially adapted anaerobic microbial inoculum and fermentation tank • Customize variants available for use in mobile as well as stationary applications • Suitable for use as centralized waste treatment system for residential complexes as well as decentralized waste treatment system for individual households

Source Status of Sewage Treatment in India, Central Pollution Control Board, Government of India (2005)

[a]A DRDO-developed technology presently being used in passenger coaches of Indian Railways and in high-altitude low-temperature areas, coastal areas, plain areas, and islands in India as decentralized human fecal waste treatment system

for Indian conditions. In this perspective, the biodigester technology developed by DRDO may be more appropriate, particularly for the developing countries.

4 Composition of Nightsoil

Nightsoil is a complex material containing a great variety of biological and chemical entities. It is well known that the gut contains the highest density of microbes in the human body. Recently, Sender et al. (2016) have estimated the bacteria-to-human cell ratio to be 1.3:1. In feces, bacteria constitute 25–54% of solid matter (Stephen and Cummings 1980) with a clear distinction between the viable (49%), injured (19%), and dead (32%) bacterial cells collected from fresh fecal samples under anaerobic conditions (Ben-Amor et al. 2005) implying that only 3.0–6.6% of total fecal matter may be composed of viable bacteria (Bojanova and Bordenstein 2016).

For developing appropriate fecal biodegradation strategies, it is very important to ascertain the composition of fecal matter. However, in scientific literature, intensive studies on fecal matter composition are scanty. Moreover, there have been observed considerable variations in composition of fecal matter with respect to geographical location, diet, age, and health status of the people (Rose et al. 2015). Generally, the fecal matter comprises water, protein, undigested fats, polysaccharides, bacterial biomass, ash, and undigested food residues. Further, the key elements in feces on percentage of wet weight basis are oxygen 74%, hydrogen 10%, carbon 5%, and nitrogen 0.7%, including the hydrogen and oxygen present in the water fraction of the feces (Snyder et al. 1975). On yearly basis, it has been estimated that one person produces approximately 5.7 kg of nitrogen, 0.6 kg of phosphorus, and 1.2 kg of potassium through excreta (Wolgast 1993). Interestingly, the fecal pH, although neutral usually, tends not only to vary between different populations but has also been found to differ between individuals consuming the same diet and with time (Silvester et al. 1997). A study conducted in southern Thailand estimated that the generation rate of the elements in the excreta was 7.6–7.9 g N/capita/day, 1.6–1.7 g P/capita/day, 1.8–2.7 g K/capita/day, 1.0–1.1 g S/capita/day, 0.75–1.5 g Ca/capita/day, 0.25–0.4 g Mg/capita/day, 9–16 mg Zn/capita/day, 1.4–1.5 mg Cu/capita/day, 0.3 mg Ni/capita/day, 0.02–0.03 mg Cd/capita/day, 0.07–0.14 mg Pb/capita/day, 0.01 mg Hg/capita/day, and 0.8–1.1 mg B/capita/day (Schouw et al. 2002). Rose et al. (2015) have conducted a similar study in UK wherein data regarding the generation rate and the chemical and physical composition of fresh feces and urine were collected from the medical literature as well as the treatability sector and then subjected to statistical analysis. Based on the detailed analysis, these researchers have listed the key criteria and values that may be very useful in the operation of existing on-site sanitation systems (Table 2).

Table 2 Feces and urine characteristics providing on-site sanitation design criteria

Key design criteria	Median value
Feces	
Fecal wet weight (g/capita/day)	128
Fecal dry weight (g/capita/day)	29
Stool frequency (motions/24 h)	1.1
Total solids (%)	25
VS (% of TS)	89
COD (g/capita/day)	71
Nitrogen (g/capita/day)	1.8
Protein (g/capita/day)	6.3
Lipids (g/capita/day)	4.1
Carbohydrate (g/capita/day)	9
Fiber (g/capita/day)	6
Calorific value (kcal/capita/day)	132
pH	6.6
Urine	
Urine wet weight (L/capita/day)	1.4
Urine dry weight (g/capita/day)	59
Urination frequency (urinations/24 h)	6
Nitrogen (g/capita/day)	11
Calorific value (kcal/capita/day)	1701
pH	6.2

Source Rose et al. (2015), reproduced with permission

5 Anaerobic Biodegradation of Nightsoil

The human intestinal tract contains a complex microbial community with the dominant bacterial phyla in healthy subjects belonging to Bacteroidetes, Firmicutes, and Actinobacteria, together with Verrucomicrobia and Proteobacteria groups (Flint et al. 2012). Nightsoil can be degraded efficiently in the absence of oxygen by anaerobic microorganisms. It is a series of biological process mediated by diverse groups of bacteria through which the nightsoil is ultimately converted into methane and carbon dioxide. This process starts with bacterial hydrolysis of the substrate during which carbohydrates and other insoluble organic compounds are broken down to soluble compounds that subsequently are utilized by other bacteria as substrate. Hence, the hydrolysis of these high-molecular-weight polymeric components is the necessary first step in anaerobic digestion (Sleat and Mah 2006). Various studies have reported isolation of cellulolytic strains belonging to *Ruminococcus* sp, *Clostridium* sp, *Eubacterium* sp, and *Bacteroides* sp from human feces (Tap et al. 2009; Kabel et al. 2011; Qi et al. 2011). *Bacteroides ovatus*, *B. thetaiotaomicron,* and *B. uniformis* are known to ferment a wide range of polysaccharides. *B. thetaiotaomicron* can utilize various forms of starch, including amylose, amylopectin, and pullulan as well as the corresponding malto-oligosaccharides (Salyers et al. 1977). *Bifidobacterium* spp. have been reported to be particularly effective degraders of high-amylose starches (Macfarlane and Englyst 1986). The acidogenic bacteria then convert the amino acids and sugars into carbon dioxide, hydrogen, ammonia, and volatile fatty acids. During the third step, resultant organic acids are converted by acetogenic bacteria into acetic acid along with additional ammonia, hydrogen, and carbon dioxide. Known acetogens comprise a phylogenetically diverse group of bacteria, distributed among 22 different genera belonging to three phyla most of which are members of the Firmicutes (Mahowald et al. 2009). Two human gut isolates belonging to Firmicutes, *Blautia hydrogenotrophica* (Ding et al. 2001), and *Marvinbryantia formatexigens* (Jindou et al. 2006) are reported. In final step, the methanogenic bacteria, particularly the archaeal community, convert these products into methane and carbon dioxide (Fig. 1).

According to Van Haandel and Lettinga (1994), anaerobic digestion is the ultimate fermentative process with the end product being methane, which is the most reduced organic compound. It is also estimated that about 90% of the available energy in the organic matter is conserved in the methane produced, thereby leaving very little energy that can be utilized by the new microbial cells (NMCs) produced during the process (McInerney et al. 1980). Therefore, the relative amount of NMCs formed as excess sludge is meager (Mosey 1981). Furthermore, the anaerobic digestion process is considered ideal for the stabilization of organic sludges, treatment of concentrated organic industrial wastes, and the production of methane gas from agricultural and domestic wastes owing to a number of reasons.

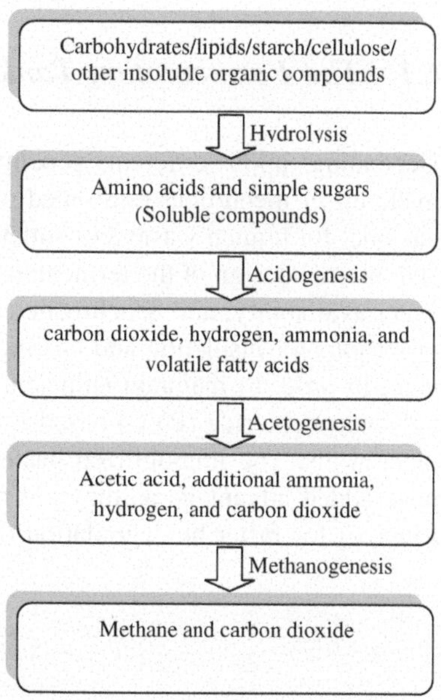

Fig. 1 Schematic representation of process of the anaerobic biodegradation of nightsoil

Here, the dissolved oxygen is not needed for the process; the methane as a combustible gas has great commercial value; and the biomass production is relatively small (Pretorius 1983). Daisy and Kamaraj (2011) have demonstrated that the potential utility of anaerobic digester to anaerobically treat nightsoil when diluted appropriately which can be incorporated in fecal wastes treatment in developing countries where majority of sanitation facilities are on-site systems.

6 The Biodigester Technology

Biodigester technology is a simple, ecofriendly, cost-effective, decentralized, odor-free on-site human waste treatment technology that was developed initially for the armed forces deployed in high-altitude low-temperature areas of Indian Himalayan region, where the process of natural biodegradation of nightsoil is very slow (Fig. 3). Subsequently, it was customized to work under varied conditions like on mobile platforms such as railways, plain areas, coastal areas, and islands. This technology has two main components: a specially designed fermentation tank and a cold-active anaerobic microbial consortium. The closed-system fermentation tank containing microbial consortium is attached to a toilet for collection of the human excreta. It provides anaerobic conditions, suitable temperature, and adequate reaction time for the bacterial activity. The anaerobic biodegradation process is mediated through four sequential steps namely hydrolysis, acidogenesis, acetogenesis, and methanogenesis, as described in preceding section. Microbial consortium thus degrades the human waste into biogas.

6.1 The Fermentation Tank

Depending upon different geoclimatic conditions, fermentation tank can be made-up of metal/fiber reinforced or plastic/reinforced cement concrete. The tank has inlet for human waste and outlet for treated effluent and biogas. The dimension and internal design of the fermentation tank may vary according to number of users, water availability, and geoclimatic conditions. For low-temperature areas, adequate provisions for insulation and heating have been made in the design of fermentation tank. In order to maintain sufficient bacterial population in the fermentation tank, poly vinyl chloride (PVC)-based immobilization matrix has been provided on the inner walls. The immobilized bacteria of diverse group residing in close vicinity have added advantage as they are retained in the tank as additional bacteria contributing for faster biodegradation (Singh and Kamboj 2018).

6.2 The Anaerobic Microbial Inoculum

The anaerobic microbial inoculum (consortium) has been developed by enrichment of bacteria collected from biogas plants working in various hilly and low-temperature regions of India. The initial inoculum was taken from different biogas plants operational with cow dung, kitchen waste, and other agro-wastes in low-temperature areas of Himachal Pradesh, India (Singh and Kamboj 2018). A screening of effective bacteria working at temperature as low as 20 °C was carried out. The bacteria demonstrating better fermentation activities were gradually adapted to the lower temperature. The bacterial enrichment was performed in two phases, and the fermenters were fed with human waste in semi-continuous mode every day after diluting it with equal volume of water. The efficacy of fermentation by the bacterial consortium was determined through monitoring key variables such as pH, biogas production, and methane content and volatile fatty acids (VFAs). Subsequently, this consortium has further been adapted to work under varied temperature regimes, through a series of experiments. Also, it has been adapted to tolerate working doses of various toilet-cleaning agents. Currently, the mother culture is being maintained and multiplied at ~15 °C with agitation arrangement and being fed by cattle dung once a day (Fig. 2). Further, the technical know-how of generation of anaerobic microbial inoculum has been shared through licensing agreements with more than 75 industries located in different regions of India. The cumulative daily output of these industries is around 20,000 L inoculum which is sufficient to cater the need of 200 biodigesters per day (Singh and Kamboj 2018) (Fig. 3).

Fig. 2 Anaerobic microbial inoculum generation facility (14 m^3 capacity) in operation at Defence R&D establishment, Gwalior, India. *Source* Singh and Kamboj (2018)

Fig. 3 Temperature-controlled biodigester with toilet superstructure installed at snowbound location in Himalayan region of India. *Source* Singh and Kamboj (2018)

6.3 Biodigester for Indian Railways: A Success Story

Indian Railways (IR) with its vast network connects every city of the country. According to the IR statistics publications of year 2014–15 and 2015–16, it is the fourth largest railway network in the world, comprising 119,630 km (74,330 mi) of total track and 92,081 km (57,216 mi) of running track over a route of 66,687 km (41,437 mi) with 7216 stations at the end of 2015–16. In 2015–16, IR carried 8.107 billion passengers annually or more than 22 million passengers a day and 1.101 billion tons of freight annually. Flush toilets used in trains dump the human waste directly on the railway track thus making the railway stations, track, and surrounding environment unhygienic and unhealthy. Upon successful field evaluations at large scale in high-altitude low-temperature Indian Himalayan region, the biodigester technology was further customized to work on mobile platforms such as passenger coaches of IR to solve the problem of open excreta discharge on tracks.

For application in passenger coaches, various factors were taken into consideration like variability in number of users, availability of limited space for the fermentation tank, variation in ambient temperature with movement of train covering significant geographical length. To begin with, the laboratory investigations considering various concerns like working of anaerobic bacterial consortium under expected temperature variables, in the presence of detergents and phenyl and reduced treatment time (hydraulic retention time, HRT). Use of bacterial immobilization matrix was also explored. Various designs of biodigester were considered and evaluated. Finally, prototype biodigester was subjected to rigorous testing in the laboratory before actual trial in passenger coach. After protracted trial runs in various trains operating in different parts of the country and suitable design modifications in fermentation tank, the IR finally adopted the technology (Singh and Kamboj 2018). More than 75,000 biotoilets have so far been installed in various trains, and IR plans to have biotoilets in all its train coaches by 2019.

In the currently used version of biodigester for IR, the fermentation tank is made-up of stainless steel (SS316) having dimensions of 1150 mm (L) × 720 mm

Fig. 4 Biodigester fitted under an Indian train coach

(W) × 540 mm (H) for treating the human excreta originating from single toilet (Fig. 4). The fermentation tank weighs 110 kg and has total and working volumes of 400 and 300 L, respectively. The tank has six chambers separated by partition walls. Inner walls have immobilization matrix on both sides fixed with iron mesh. An effluent treatment chamber (chlorinator) has been provided on outer side of the tank. A sampling port is also provided in the chlorination chamber near the outlet for collecting effluent samples for testing of quality. The human excreta from the toilet enter the first chamber through P-trap and passes through various chambers wherein fermentation of the waste is completed. The effluent then enters the chlorinator and gets disinfected before being discharged through outlet on the rail track. Various parts of biodigesters are shown in Fig. 5 (Singh and Kamboj 2018).

7 Degradation Efficiency of Anaerobic Microbial Consortium

As it has been mentioned, the anaerobic digestion of organic waste is a complex bioconversion process. The performance of anaerobic biodigester is affected by several factors such as quantity and composition of feedstock, design of fermentation tank, and operating conditions. Temperature and pH are among the most critical operating conditions. Optimum temperature of mesophilic digester for biogas production has been reported to be 35 °C (Cioabla et al. 2012). A study reported that human waste (nightsoil) degradation at low temperature resulted in decreased biogas production and digester failure due to accumulation of volatile fatty acids (Singh et al. 1993). In addition, repeated exposures of temperature fluctuation on anaerobic digestion of nightsoil have been found to reduce the counts of hydrogenotrophic methanogens in inoculum (Singh et al. 1999). According to Samir et al. (2010), in the mesophilic range, the activity and growth rate of bacteria

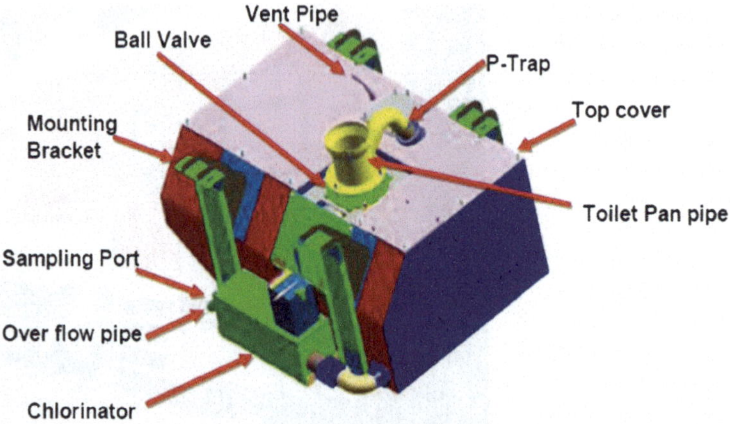

Fig. 5 Railway biodigester. *Source* Singh and Kamboj (2018)

decrease by 50% for each 10 °C drop. Moreover, the fall in biogas production is observed when the temperature decreases to 20 °C, and the production even stops at 10 °C (Samir et al. 2010). Increasing the temperature level up to 37 °C reduces the time required for the digestion process. Further increase in temperature decreases the rate of biogas generation (Cioabla et al. 2012). Among different groups of bacteria involved in the process, methanogens are the most fastidious organisms which are highly sensitive to the microenvironment including H ions, O_2, nutrients, metabolites, acetate, propionate, and other short-chain fatty acids (Dhaked et al. 2003). In fact, propionate has been found to be toxic for biomethanation of human waste particularly at low temperature with toxicity being concentration-dependent (Dhaked et al. 2003). The pH of the anaerobic digestion process is another important variable that tends to significantly influence the digestion process (Kumar 2010; Romano and Zhang 2011). During the hydrolytic stage, the acidogenic bacteria need a pH in the range of 5.5–7.0 (Nikolić 2006). On the contrary, during the final stage, methanogenic bacteria require a pH value ranging between 6.5 and 8.0. The major challenge in the initial stages is the reduction of pH in fermenter due to production of volatile fatty acids (Landine et al. 1983) which hinders the activity of methanogenic bacteria. The optimum pH range in an anaerobic digester has been found to be 6.8–7.2. However, generally the process can tolerate a range of 6.5 up to 8.0 (Cioabla et al. 2012). Anaerobic biodegradation of nightsoil is also influenced by sulfate and nitrate. In a study, presence of the sulfate found to promote the utilization of propionate, butyrate and valerate by sulfate-reducing bacteria and of acetate by acetoclastic methanogenic bacteria,whereas the nitrate inhibited the methanogenesis. Moreover, this effect increased directly with volatile solids concentration and inversely with temperature (Sai Ram et al. 1993). The performance of anaerobic microbial inoculum developed by DRDO is highly robust suiting to significantly a wide range of environmental variables.

8 Effluent Treatment

United States Environmental Protection Agency defines the effluent as 'wastewater—treated or untreated—that flows out of a treatment plant, sewer, or industrial outfall.' The untreated effluent coming from sewage treatment plants poses twin challenges. At one hand, it causes pollution to the environment, and at the other hand, the precious water resource is lost. If this wastewater is properly treated and recycled, it can address the major problems of environmental pollution and water scarcity. Today, a large number of Indian cities across River Ganga release sewage water with high-organic load, making it fifth most-polluted river of the world. This concern is further compounded by the fact that the River Ganga provides water for various purposes to about 40% of India's population across 11 states serving an estimated population of 500 million people exposing them to higher level of health risk (Wax 2007). A study conducted by Hamner et al. (2006) estimated that in Varanasi city alone, 200 million liters daily or more of untreated human sewage is discharged into the Ganges River. Further, river water monitoring over the past 12 years has demonstrated fecal coliform counts up to 108 MPN (most probable number) per 100 ml and biological oxygen demand levels averaging over 40 mg/l in the most-polluted part of the river in Varanasi (Hamner et al. (2006). It is well established that human fecal contamination of water transmits microorganisms that cause diarrhea, cholera, hepatitis A, and many other diseases. The effective treatment of sewage effluent can therefore significantly minimize the transmission of waterborne disease.

Mekonnen and Hoekstra (2016) estimated that two-thirds of the global population (4.0 billion people) live under conditions of severe water scarcity at least 1 month of the year out of which nearly half of the population live in India and China. Half a billion people in the world face severe water scarcity all year round (Mekonnen and Hoekstra 2016). In the wake of wide gap in availability and demand of water for domestic consumption compounded by polluting potential of untreated sewage, the effluent treatment of sewage water assumes greater significance as the large volume of sewage water offers immense scope for the recycling of water within the cities and reduces their dependence on bulk freshwater sources. Considerable efforts have been initiated by various governmental agencies and civil societies in this direction. The National Mission on Sustainable Habitat (NMSH), one of the eight National Missions under the National Action Plan on Climate Change of Government of India, has also envisaged addressing these challenges (CPHEEO 2014).

Tare and Bose (2014) have compiled information on various sewage treatment technologies that may be adopted in India. The treatment technologies have been classified into four categories based on comprehensive analysis of their performance and resource (energy, cost, land, etc.) requirement. The technologies belonging to category I are those showing good performance and require low energy and low resources but greater land. On the other hand, technologies with good performance, high energy requirement, high resource requirements and associated costs, and moderately low land requirement fall in this category. Other categories of technologies have moderate or marginal performance and relatively lower resource requirement (Tare and Bose 2014). Considering the uniqueness of the DRDO-biodigester technology, it may not be logical to fit it into any of these categories as this technology is characterized by high performance and relatively very low resource (energy, cost, and land) requirements.

In order to protect and improve the environment, Government of India enacted the Environment (Protection) Act in 1986 with the mandate of preventing environmental pollution in all its forms and to tackle specific environmental problems that are peculiar to different parts of the country. Under the ambit of the act, several rules/standards pertaining to water quality criteria for uses, recyclability, and quality of effluent discharge have been notified (Tables 3 and 4). Keeping pace with the advancement in waste treatment technologies, the stringency levels have also been tightened. More recently, a draft notification on standards for sewage treatment plants along with timeframe for implementation has been issued by Government of India under the Environment (Protection) Act. It has also been emphasized that none of the sewage treatment technologies currently used in India produce water of recyclable quality barring a few instance where the treated effluent can be reused for industrial purposes (Tare and Bose 2014). In this regard, the DRDO-biodigester technology when combined with other appropriate tertiary treatment strategy holds immense potential for producing water of recyclable quality, which needs serious consideration and exploration.

Table 3 Effluent discharge standard for sewage treatment plant

Parameters	Standards for new STPs (design after notification date)[a]
pH	6.5–9.0
Biochemical oxygen demand	10
Chemical oxygen demand	50
Total suspended solids	20
Ammoniacal nitrogen	5
Total nitrogen	10
Fecal coliform (MPN/100 mL)	<100

Note (i) All values in mg/mL expect for pH and fecal coliform. (ii) These standards will be applicable for discharge in water resources as well as for land disposal. The standards for fecal coliform may not be applied for use of treated sewage in industrial purposes

Source http://envfor.nic.in/sites/default/files/Draft%20notification%20of%20Sewage%20Treatment%20plan.PDF. Accessed 7 July 2017

[a]Achievements of standards for existing STPs within 0–5 years from date of notification

Table 4 General standards for discharge of environmental pollutants

Parameter	Standards			
	Inland surface water	Public Sewers	Land for irrigation	Marine coastal areas
Suspended solids (mg/l, max.)	100	600	200	(a) For process waste water100 (b) For cooling water effluent 10% above total suspended matter of influent
Particulate size of suspended solids	Shall pass 850 micron IS Sieve	–	–	(a) Floatable solids max. 3 mm (b) Settleable solids max. 850 μm
pH	5.5–9.0	5.5–9.0	5.5–9.0	5.5–9.0
Temperature	Shall not exceed 5 °C above the receiving water temperature	–	–	Shall not exceed 5 °C above the receiving water temperature
Oil and grease (mg/l max.)	10	20	10	20
Total residual chlorine (mg/l max.)	1.0	–	–	1.0
Ammoniacal nitrogen (as N) (mg/l max.)	50	50	–	50
Total Kjeldahl Nitrogen (as NH_3) (mg/l, max.)	100	–	–	100
Free ammonia (as NH_3) (mg/l, max.)	5.0	–	–	5.0
Biochemical oxygen demand (3 days at 27 °C) (mg/l max.)	30	350	100	100
Chemical oxygen demand (mg/l, max.)	250	–	–	250

Source CPCB, Government of India, http://cpcb.nic.in/GeneralStandards.pdf. Accessed July 7, 2017

9 Conclusions and Future Prospects

Disposal of human fecal waste in ecofriendly manner is one of the prime concerns before the environmentalists and policy makers. Concerted efforts of the researchers across the world have led to development of a large number of fecal waste management technologies suiting requirement of different geographical regions with varying degree of performance. Despite various challenges, the anaerobic digestion process has proven its appropriateness for management of human fecal waste. Nevertheless, among all currently available technologies, it appears more logical to choose a technology that can meet the twin challenges of environmental pollution and water scarcity worldwide. The DRDO-biodigester technology is a simple, ecofriendly, cost-effective, decentralized, and odor-free on-site human waste treatment technology that has been developed based on principles of anaerobic digestion. It has been successfully implemented for the onboard human waste treatment in passenger coaches of Indian Railways as well as for on-site waste treatment in rural and urban areas in India. Moreover, this technology with reed bed system has been evaluated as a cost-effective alternative to sewage treatment plants at certain places in India. If adequately propagated and implemented, the DRDO-biodigester technology can play a pivotal role in achieving the targets of United Nations Sustainable Development Goals related to sanitation. Nevertheless, with increased stringency levels in regulatory standards of sewage treatment and effluent quality, further research efforts are going on for making improvements in the technology with respect to improved fermentation efficiency, reduction in inoculum volume, recycling of water through more effective effluent treatment, and harvesting and use of biogas for energy generation.

References

Ben-Amor K, Heilig H, Smidt H, Vaughan EE, Abee T (2005) Genetic diversity of viable, injured, and dead fecal bacteria assessed by fluorescence-activated cell sorting and 16S rRNA gene analysis. Appl Environ Microbiol 71:4679–4689

Bojanova DP, Bordenstein SR (2016) Fecal transplants: what is being transferred? PLoS Biol 14 (7):e1002503

Caincross S, Feachem RG (1993) Environmental. Health engineering in the tropics: an introductory text, 2nd edn. Wiley, Chichester, UK

Choi E, Oa SW, Lee JJ (1997) Nightsoil treatment plant converted into sequencing batch reactor to improve removal of pollutants and nutrients. Water Sci Technol 35:233–240

Cioabla AE, Ionel I, Dumitrel G-A, Popescu F (2012) Comparative study on factors affecting anaerobic digestion of agricultural vegetal residues. Biotechnol Biofuels 5:39

CPCB (2016) National status of waste water generation and treatment, vol I. Central Pollution Control Board Bulletin, Updated Dec 2016

CPHEEO (2014) National mission on sustainable habitat: adaptation & mitigation measures in water supply & sanitation sub-sector. Central Public Health and Environmental Engineering Organization, Ministry of Urban Development, Govt of India. http://moud.gov.in/upload/uploadfiles/files/NMSH%20Advisory%20on%20Adaptation%20and%20Mitigation%20Measures.pdf. Accessed 7 July 2017

Daisy A, Kamaraj S (2011) The impact and treatment of night soil in anaerobic digester: a review. J Microbial Biochem Technol 3:043–050

Dhaked RK, Waghmare CS, Alam SI, Kamboj DV, Singh L (2003) Effect of propionate toxicity on methanogenesis of night soil at phychrophilic temperature. Bioresour Technol 87:299–303

Ding SY, Rincon MT, Lamed R, Martin JC, McCrae SI, Aurilia V et al (2001) Cellulosomal scaffoldin-like proteins from *Ruminococcus flavefaciens*. J Bacteriol 183:1945–1953

Ferguson DT (2014) Nightsoil and the 'great divergence': human waste, the urban economy, and economic productivity, 1500–1900. J Global History 9:379–402

Flint HJ, Scott KP, Duncan SH, Louis P, Forano E (2012) Microbial degradation of complex carbohydrates in the gut. Gut Microbes 3(4):289–306

GBD (2016) Mortality and causes of death collaborators. Global, regional, and national life expectancy, all-cause and cause-specific mortality for 249 causes of death, 1980–2015: a systematic analysis for the global burden of disease study 2015. Lancet 388:1459–1544

Hamner S, Tripathi A, Mishra RK, Bouskill N, Broadaway SC, Pyle BH, Ford TE (2006) The role of water use patterns and sewage pollution in incidence of water-borne/enteric diseases along the Ganges river in Varanasi, India. Int J Environ Health Res 16(2):113–132

Jackson D, Winkler M, Stenström TA (2015) Sanitation safety planning: manual for safe use and disposal of wastewater, greywater and excreta. World Health Organ. http://www.who.int/water_sanitation_health/publications/sspmanual/en/. Accessed 7 July 2017

Jindou S, Borovok I, Rincon MT, Flint HJ, Antonopoulos DA, Berg ME et al (2006) Conservation and divergence in cellulosome architecture between two strains of *Ruminococcus flavefaciens*. J Bacteriol 188:7971–7976

Kabel MA, Yeoman CJ, Han Y, Dodd D, Abbas CA, de Bont JA et al (2011) Biochemical characterization and relative expression levels of multiple carbohydrate esterases of the xylanolytic rumen bacterium *Prevotella ruminicola* 23 grown on an ester-enriched substrate. Appl Environ Microbiol 77:5671–5681

Kumar S (2010) Performance study for anaerobic digestion of municipal solid waste in a single phase reactor. Int J Environ Pollut 43:16

Landine RC, Brown GJ, Cocci AA, Virara H (1983) Anaerobic treatment of high strength, high solids potato waste. Agric Wastes 7(2):111–123

Macfarlane GT, Englyst HN (1986) Starch utilization by the human large intestinal microflora. J Appl Bacteriol 60:195–201

Mahowald MA, Rey FE, Seedorf H, Turnbaugh PJ, Fulton RS, Wollam A et al (2009) Characterizing a model human gut microbiota composed of members of its two dominant bacterial phyla. Proc Natl Acad Sci U S A 106:5859–5864

McInerney MJ, Bryant MP, Stafford DA (1980) Metabolic stages and energetics of microbial anaerobic digestion. In: Stafford DA (ed) First international symposium on anaerobic digestion, A. D. Scientific Press, Cardiff, Wales

Mekonnen MM, Hoekstra AY (2016) Four billion people facing severe water scarcity. Sci Adv 2 (2):e1500323

Mosey FE (1981) Anaerobic biological treatment of food industry waste waters. Water Pollut Control Fed 80:273

Nikolić V (2006) Producerea şi utilizarea biogazului pentru obţinerea de energie (Suport de curs). http://www.nikolicivasilie.ro/lucrari-stiintifice/Biogaz%20curs.pdf

Pretorius WA (1983) An overview of digestion processes. Water Sci Technol 15:1–6

Qi M, Wang P, O'Toole N, Barboza PS, Ungerfeld E, Leigh MB et al (2011) Snapshot of the eukaryotic gene expression in muskoxen rumen—a metatranscriptomic approach. PLoS ONE 6:20521

Rodda JC, Ubertini L (2004) The basis of civilization—water science? International Association of Hydrological Sciences, International Association of Hydrological Sciences Press, p 161
Romano RT, Zhang R (2011) Anaerobic digestion of onion residuals using a mesophilic anaerobic phased solids digester. Biomass Bioenerg 35(10):4174–4179
Rose C, Parker A, Jefferson B, Cartmell E (2015) The characterization of feces and urine: a review of the literature to inform advanced treatment technology. Crit Rev Environ Sci Technol 45:1827–1879
Sai Ram M, Singh L, Alam SI (1993) Effect of sulfate and nitrate on anaerobic degradation of night soil. Bioresour Technol 45:229–232
Sakuta T (1914) Shimogoe (Nightsoil), Tokyo, Yûrindô Shoten, p 8, cited in Kajo Tajima, The marketing of urban human waste in the Edo/Tokyo metropolitan area: 1600–1935. Ph.D. thesis, Tufts University, 2005 p 91
Salyers AA, Vercellotti JR, West SE, Wilkins TD (1977) Fermentation of mucin and plant polysaccharides by strains of Bacteroides from the human colon. Appl Environ Microbiol 33:319–322
Samir KK, Rao YS, Tian CZ, Buddhi PL, Tyagi RD, Kao CM (2010) Bioenergy and biofuel from biowastes and biomass. American Society of Civil Engineers, Reston, p 44
Schouw NL, Danteravanich S, Mosbaek H, Tjell JC (2002) Composition of human excreta—a case study from Southern Thailand. Sci Total Environ 286(1):155–166
Sender R, Fuchs S, Milo R (2016) Are we really vastly outnumbered? revisiting the ratio of bacterial to host cells in humans. Cell 164:337–340
Silvester KR, Bingham SA, Pollock JRA, Cummings JH, O'Neill IK (1997) Effect of meat and resistant starch on fecal excretion of apparent N nitroso compounds and ammonia from the human large bowel. Nutr Cancer 29(1):13–23
Singh L, Kamboj DV (2018) Biodigester: an innovative technology for on-board disposal of human waste in Indian Railways. Daya Publishing House, New Delhi, India
Singh L, Maurya MS, Sai Ram M, Alam SI (1993) Biogas production from night soil—effect of loading and temperature. Biores Technol 45:59–61
Singh L, Alam SI, Ramana KV (1999) Effect of fluctuating temperature regime on psychrophilic anaerobic digestion of night soil. Def Sci J 49:135–140
Sleat R, Mah R (2006) Hydrolytic bacteria in anaerobic digestion of biomass, p. 15
Snyder W, Cook M, Tipton I, Nasset E, Karhausen L, Howells G (1975) Reference man: anatomical, physiological and metabolic characteristics (Report of task group on reference man-international commission on radiological protection). Pergamon Press, New York, NY
Stephen AM, Cummings JH (1980) The microbial contribution to human faecal mass. J Med Microbiol 13:45–56
Tap J, Mondot S, Levenez F, Pelletier E, Caron C, Furet JP et al (2009) Towards the human intestinal microbiota phylogenetic core. Environ Microbiol 11:2574–2584
Tare V, Bose P (2014) Compendium of sewage treatment technologies. National River Conservation Directorate, Ministry of Environment & Forests, Government of India. http://nmcg.nic.in/writereaddata/fileupload/15_Technologies%20Involved.pdf. Accessed 7 July 2017
Tajima K (2005) The marketing of urban human waste in the Edo/Tokyo metropolitan area: 1600–1935. Ph.D. thesis, Tufts University, USA
Van Haandel AC, Lettinga G (1994) Anaerobic sewage treatment: a practical guide for regions with a hot climate. Wiley, Chichester, UK
Wax E (2007) A sacred river endangered by global warming. Washington Post. Accessed 25 June 2017
Wolgast M (1993) Rena vatten. Om tankar i kretslopp. Crenom HB, Uppsala (in Swedish)
Xue Y (2007) A "fertilizer revolution"? A critical response to Pomeranz's theory of "Geographic luck". Mod China 33(2):195–229

Author Biographies

Dr. M.K. Meghvansi, PhD is currently working as scientist at Biotechnology Division, Defence R & D Establishment, Gwalior, India. His research area includes biodegradation of organic waste/nightsoil, sewage treatment, and plant-pathogen-soil interaction. He has to his credit more than 50 publications of national and international repute. He also edited a volume under Springer *Soil Biology* series.

Prince Kumar, MSc is currently working as Senior Research Fellow under the supervision of Dr. D.V. Kamboj at Biotechnology Division, Defence R & D Establishment, Gwalior, India. His research area include biodegradation of nightsoil, microbial dynamics of anaerobic digester, and role of bacteriophages in sewage effluent treatment.

Dr. V. Vasudevan, PhD has ~10 years research experience in the field of anaerobic biodegradation of human waste and microbial community dynamics of anaerobic digester. He has many research publications to his credit.

Arvind Tomar, MSc is currently working as scientist at Biotechnology Division, Defence R & D Establishment, Gwalior, India. His area of work include anaerobic biodegradation under stress conditions.

Dr. D. V. Kamboj, PhD is currently working as scientist & head, Biotechnology Division, Defence R & D Establishment, Gwalior, India. His research area includes biodegradation of organic nightsoil, effluent treatment, and microbial dynamics of anaerobic digester. He has contributed to design and development of several variants of Biodigesters for different regions across the country. He has more than 55 research papers and 15 patents to his credit.

Dr. Lokendra Singh, PhD has superannuated in 2017 after serving as Director, Defence R & D Establishment, Gwalior, India. He led the research team working on degradation of human waste and developed the Biodigester Technology. He has a vast research experience in the field of anaerobic microbiology. He has more than 100 research publications, several patents and books to his credit.

Author Index

S. J. Varjani et al. (eds.), *Waste Bioremediation*, Energy, Environment, and Sustainability, https://doi.org/10.1007/978-981-10-7413-4

The manufacturer's authorised representative in the EU is Springer Nature Customer Service Centre GmbH, Europaplatz 3, 69115 Heidelberg, Germany. If you have any concerns regarding our products, please contact ProductSafety@springernature.com

Printed and bound by CPI Group (UK) Ltd, Croydon, CR0 4YY
25/09/2026
02222884-0003